The Management of
Sustainable Growth

Pergamon Titles of Related Interest

Coomer   THE QUEST FOR A SUSTAINABLE SOCIETY
Dolman   GLOBAL PLANNING AND RESOURCE MANAGEMENT
King/Cleveland/Streatfeild   BIORESOURCES FOR DEVELOPMENT

Related Journals*

ENVIRONMENT INTERNATIONAL
THE ENVIRONMENTAL PROFESSIONAL
EVALUATION AND PROGRAM PLANNING
FUTURICS
HABITAT INTERNATIONAL
UNDERGROUND SPACE
WATER SUPPLY AND MANAGEMENT
WORLD DEVELOPMENT

*Free specimen copies available upon request.

PERGAMON
POLICY
STUDIES

ON BUSINESS AND ECONOMICS

The Management of Sustainable Growth

Edited by
Harlan Cleveland

**Published in Cooperation with
The Woodlands Conference**

Pergamon Press

NEW YORK • OXFORD • TORONTO • SYDNEY • PARIS • FRANKFURT

Pergamon Press Offices:

U.S.A.   Pergamon Press Inc., Maxwell House, Fairview Park,
                     Elmsford, New York 10523, U.S.A.

U.K.   Pergamon Press Ltd., Headington Hill Hall,
                     Oxford OX3 0BW, England

CANADA   Pergamon Press Canada Ltd., Suite 104, 150 Consumers Road,
                     Willowdale, Ontario M2J 1P9, Canada

AUSTRALIA   Pergamon Press (Aust.) Pty. Ltd., P.O. Box 544,
                     Potts Point, NSW 2011, Australia

FRANCE   Pergamon Press SARL, 24 rue des Ecoles,
                     75240 Paris, Cedex 05, France

FEDERAL REPUBLIC Pergamon Press GmbH, Hammerweg 6, Postfach 1305,
OF GERMANY   6242 Kronberg/Taunus, Federal Republic of Germany

Library of Congress Cataloging in Publication Data

Woodlands Conference on Growth Policy, 3d, 1979.
    The management of sustainable growth.

    (Pergamon policy studies on business and economics)
    Selection of papers from the conference held
Oct. 28-31.
    1. Economic development–Congresses.   2. Industry
and state–Congresses. I. Cleveland, Harlan.
II. Title.   III. Series.
HD73.W66   1979   338.9   80-24162
ISBN 0-08-027171-5

Printed in the United States of America

CONTENTS

PART FOUR: THE LIMITS TO GOVERNMENT

FOREWORD
The Woodlands Conference on Growth Policy

In 1974 George Mitchell, Chairman and President of the Mitchell Energy & Development Corporation, resolved to set in motion a ten-year process to encourage the rethinking of growth policy by sponsoring five biennial conferences, and, with his wife, Cynthia, offered the Mitchell Prize "to those individuals demonstrating the highest degree of creativity in designing workable strategies to achieve sustainable societies."

The first such Conference, in 1975, was the product of the quite sudden concern, in the early 1970s, about the physical *limits to growth*. The question posed was: ". . . how might a modern society be organized to provide a good life for its citizens without requiring ever-increasing population growth, energy resource use, and physical output?"[1] In the autumn of 1977, the Second Woodlands Conference, still viewed as part of a continuing effort to define a "steady-state society," focused on *alternatives to growth*.

Meanwhile, academic researchers, business analysts, and government planners—in both industrial and developing countries—were asking hard questions not only about the consequences of indiscriminate economic growth, but also about a "no-growth" philosophy based on the prospective exhaustion of nonrenewable resources. Renewable "bioresources" and expandable "information resources," when added to new estimates of what nonrenewable resources might become available, seemed likely to make some continuing economic growth possible—and the fairer distribution of wealth and income made it clearly necessary.

The United States and, indeed, other industrial societies seemed already to be in transition toward a new concept of growth policy that would be neither indiscriminate material growth nor "no-growth" (both measured by GNP). Public attention, therefore, shifted toward the more complex and integrative questions of purpose and human needs—"Growth for what?" "Growth for whom''—and toward the apparent incapacity of social institutions to cope with the complexities of affluence, inflation, and fairness.

In planning the Third Woodlands Conference, responsibility for management was placed in the hands of the University of Houston. David Gottlieb, sociologist and dean of social sciences, became its chief administrator and Professor James C. Coomer of the Future Studies Program at the University's Clear Lake campus served as executive officer for the Woodlands Conference.

The original sponsors (Mitchell Energy & Development Corporation and the University of Houston) then brought into consultation and cosponsorship John and Magda McHale, whose Center for Integrative Studies had been moved to the University of Houston, and the Aspen Institute for Humanistic Studies, whose Program in International Affairs had been collaborating with the McHales in a series of studies on basic human needs, supply potential, and the dynamics of development.

The Third Conference, held at The Woodlands (near Houston) from October 28 to 31, 1979, was the culmination of an eighteen-month process that generated ten commissioned Mitchell Award papers, nine winning open-competition Mitchell Prize papers, and eight advance workshops. These included consultations in Europe, Japan, and Mexico; meetings in Houston of experts on bioresources, work and education, information and communications, the economics of the future and the future of economics; and a summer workshop on "The Limits to Government," held in Aspen, Colorado. In 1978 and 1979, this was certainly the nation's broadest, deepest, and most exciting nongovernmental inquiry on growth policy.

The present volume[2] pulls together the key commissioned (Mitchell Award) papers, with two keynote statements delivered at the Conference by William S. Sneath, Chairman of Union Carbide Corporation, and Maurice F. Strong, the creative Canadian businessman who sparked the Stockholm Conference on the Human Environment and served as the first Executive Director of the U.N. Environment Program. To round out the book's global perspective, the editor has also included, as Chapter 10, a paper on the political economy of international security prepared in early 1980 for the Presidential Commission on World Hunger by Thomas W. Wilson, Jr., who had also co-authored the Aspen Institute book on growth policy that served as a basis for this whole ambitious process of inquiry.[3]

It is not to be expected that a book authored by 16 strong-minded scholars and practitioners of public affairs would present a monolithic view of what's wrong with U.S. growth policy and what should be done about it. Yet there is striking agreement here on the direction of the transition we are already in and on the propositions that sustainable growth is physically attainable and morally imperative.

The hard part, we seem to be agreed, is managing ourselves. And the hardest part of the management of sustainable growth is finding ways to make complex decisions that do not require government to handle all aspects of governance for a necessarily pluralistic society in an increasingly interdependent world. Lao Tzu said it: "Ruling a big country is like cooking a small fish"—that is, too much handling will spoil it.

Harlan Cleveland
James C. Coomer

NOTES

1. Dennis L. Meadows, ed., *Alternatives to Growth—I: A Search for Sustainable Futures* (Cambridge, Mass.: Ballinger, 1977), p. xvii.

2. In a companion volume edited by James C. Coomer, the winning nine papers in the 1979 Mitchell Prize competition are published with a contextual introduction by the editor. That book is titled *Quest for a Sustainable Society*, and is also published by Pergamon Press.

3. Harlan Cleveland and Thomas W. Wilson, Jr., *Humangrowth: An Essay on Growth Values and the Quality of Life* (New York: Aspen Institute for Humanistic Studies, 1978).

PART ONE

RETHINKING
GROWTH POLICY

Chapter 1

WE CHANGED OUR MINDS IN THE 1970s[1]

Harlan Cleveland

Something very important happened in the 1970s. It will prove, I think, to have been more important than the decade's other landmarks—the U.S. constitutional crisis called Watergate, the multiplication of nuclear overkill, the rush of scientific discovery, the wars in Southeast Asia and the Middle East and Southern Africa, the emergence of the oil cartel, the opening of China, or the revolution of reaction in Iran. It happened just below the surface of "events," as defined by their media coverage. What happened was that we, the people of the United States, changed our minds about growth. And that may be good news.

As the decade began, we were preoccupied with multiple dangers—from modern technology in hostile hands, from the apparently widening gap between population and resources, and from a growth ethic that seemed to produce each year uglier cities, more impoverished land, more intractable human poverty, and a thickening cloud of pollutants. We were also living with some dubious illusions—that our national economy could be "fine-tuned," that the dollar was fundamentally strong, that the global drive for fairness could be safely palliated and otherwise ignored, that others might be dependent on us but we weren't interdependent with them, that we might have to bargain with the rest of the world but we didn't have to share, and that our institutions were capable of coping with the exponential growth of complexity.

As we now face the 1980s, some of the most dangerous trends have been sharply and surprisingly reversed—by thinking hard about them. And some of our fondest illusions have turned out to be "inoperative." These turnaround trends and forgettable fancies can be lumped together—we have done so in this

3

book—and viewed as a major shift in "growth policy," a shift not yet announced by established institutions but already sniffed out by survey researchers, social philosophers, future-minded economists, scientists curious about the social fallout of science, and reflective practitioners of business and public affairs.

"Growth policy" is, of course, connected with everything else—with security and money and poverty and cities and food and fairness and international order. Thinking about "growth policy" requires us to reconsider or reconfirm our personal and social values, and that makes "growth policy" an analytically promising way to think about the situation as a whole. Consider, then, eight happenings during the sobering 1970s that together point to new directions in "growth policy" in the 1980s and beyond.

Growth

A new growth ethic is clearly emerging. The shift in attitudes is worldwide, but is naturally most pronounced in the United States, which had been the most addicted to the old growth ethic.

The old ethic was clear enough: rapid material growth, powered by technological innovation, supported by an exuberant optimism, and measured and symbolized by gross national product (GNP)—an index of economic activity that is neutral as to purpose. In such an index, more good nourishment or more bad drugs both show up as pluses; smaller cars or better health (reduced medical billings) show up as minuses. Somehow, we got in the habit of thinking that pluses were signs of success and minuses were indexes of failure.

Not many years ago, a meeting such as the Third Woodlands Conference on Growth Policy would have been unthinkable—or at least unfundable. Growth was our secular religion. Growth was to be planned and practiced and pursued, not to be interrupted by protests and placards or slowed down by querulous queries.

The Woodlands Conferences were born of a backlash, a reaction to the environmental and social impacts of a growth guided mostly by "effective demand." We flirted then, briefly, with no-growth as an alternate philosophy. But that flirtation did not survive the recession of the mid-1970s, in which we tried no-growth, though not by design, and saw how little could be done about achieving more fairness, in the United States or in the world, without an increment of economic product to redistribute.

Now the American people, and a rising number of people in other countries, are groping in earnest for a new growth ethic. If indiscriminate growth and no-growth are both going out of fashion, to what political and ethical standard do we repair? If "more" isn't "better," what is?

Near the end of the decade of the 1970s, the Chairman of the U.S. Federal Reserve Board was still measuring purpose in quantitative terms. The only alternative to "more" is "less," Paul Volcker said in effect; we are all going to have to accept a lowered standard of living. But, on this issue, as in so many others, the people seemed to be ahead of the government. Every time the pollsters take our national pulse, they discover a deeper alienation from the idea that more is necessarily better. The people in their wisdom already sense that a qualitatively better life can be achieved with less energy and waste, and more stability, than the linear pursuit of quantitative growth seems to provide.

It is not too much to say that we the people are changing our minds about the very purposes of life and work. Daniel Yankelovich and Bernard Lefkowitz, in Chapter 2, document for the United States this sea-change in attitudes. William Lee Miller, in Chapter 3, tackles the same task, not with interviews and statistical analysis, but with an impressionist's brush. Robert Hamrin, in Chapter 4, considers how economics, which in the earlier era served society as oracle and pathfinder, can accommodate our shifting purposes and the resulting shifts in the content of growth. Eric Zausner, in Chapter 5, outlines the difficult choices that will add up to a national energy policy.

The centerpiece of "growth" has to be "work"—and our shifting attitudes toward the one are a reflection of our shifting definitions of the other. Willard Wirtz and James O'Toole, in Chapters 6 and 7, contrast the rapidly changing nature of work with our sluggishly changing education for work.

A majority of Americans, the survey researchers tell us in Chapter 2, now agree that "doing without something and living a more austere life would be a good thing." By a margin of more than four to one, Americans would prefer "teaching people to live with basic essentials" to "reaching a higher standard of living." By a similar margin, Americans would place greater stress on "learning to get our pleasure out of nonmaterial experiences" than on "satisfying our need for more goods and services."

The pollsters themselves warn that what people say to a polltaker is not a very accurate predictor of what they will themselves do in real life; after each popular election, we rediscover the grains of salt that should always be taken with sample polls and straw votes. Also, people are much more willing to counsel restraint in the behavior of others than to practice it themselves; the sluggish progress in energy conservation bears witness. (As manager of a university system, I was always struck by the amount of litter left behind by students after a demonstration in favor of a cleaner environment.)

But the survey research findings about attitudes toward growth are so lopsided and so consistent that Yankelovich, after reviewing all the data from Gallup, Roper, Harris and his own firm, was able to sum it up in this way: "A new synthesis is forming, one that meshes three elements: the pursuit of economic stability (even at the cost of reduced consumption) with more modest expectations and with the drive to establish maximum control over one's own destiny. The combination of these three elements creates a novel American outlook."

President Carter, in his energy speech of July 1979, was beginning to march to the same drummer: ". . . owning things and consuming things," he said on national television, "is no substitute for meaning."

Population and Resources

The race between world population and world resources doesn't look nearly so scary in 1979 as it did in 1970—or in 1972, when *The Limits to Growth* was published, or even in 1975, when the First Woodlands Conference was held.

What then appeared as an exponential explosion of population now seems more to resemble the familiar biological S-curve. Early in the decade, demographers could snatch headlines, sell books, and be taken seriously if they predicted the world's population would grow to 7.5 or even 8 billion by the year 2000. The current U.N. median projection barely reaches 6 billion by then. The forecasters had committed the original statistical sin, which is to mistake current trends for human destiny. They were mousetrapped by countervailing trends: development, chemicals, women's instincts, and hope. All proved hard to fit into a computer program.

I don't want to be misunderstood about this. Population growth is still the primary engine of world poverty. Two billion more mouths to feed, three-quarters of a billion jobs to create—these are still massive assignments to tackle in a short generation. But the task just might be manageable. By contrast, the assignment implied by the earlier projections seemed unmanageable—looked so discouraging, in fact, that we were exposed to all that popularized doom-saying about lifeboats and *triage*.

On the resources side of the equation, we have also gained in knowledge, insight, and wisdom during the 1970s. The panic about limits to growth in the early 1970s was a useful wakeup tonic. But it soon became evident that the "problematique," as the Club of Rome called it, was not a shortage of physical resources as such, but a shortage of imagination, an underdevelopment of appropriate institutions, and a failure of political will to control our human selves in using the biosphere's rich and versatile endowment.

Even the nonrenewable resources, the familiar fossil fuels and hard minerals, would present no real supply problem over the next generation *if* conserving attitudes and international cooperation are not in short supply. And beyond the nonrenewables lies a hugely underutilized biomass (one-fifth of it microorganisms, an inconceivably numerous army of workers now underemployed in making cheese and sauerkraut and fermenting beer and wine), plus a supply of solar radiation which is, for practical purposes, infinite. And these renewable resources are disproportionately available in the world's tropical regions, which are home for most of the absolutely poor.

Ecological thinking (the product of the life sciences) and the advertised future shortages of nonrenewable resources have brought bioresources to center stage. The new emphasis on bioresources will induce big changes in our thinking about the promise and purpose of new technologies. The bioresource may, by its nature, enhance equity by permitting more choice. By the same token, it may encourage decentralization, redevelop rural life, change our concept of waste, diffuse the personal responsibility for social outcomes, require of us a greater effort to understand the broader (which is to say, the ecological) framework for our individual actions, push us to think in longer ranges (that is, to take into account the interest of future generations, show us that mistakes need not be irreversible), and generally expand the "limits to growth."

The implications for management are also enormous. More stress on biore-sources accelerates the trend toward more horizontal systems, requires feedback information to be more widely available (with important implications for the ancient tension between secrecy and openness), places a premium on multi-ple–objective organizations and longer-term planning, requires a much broader analytical system, puts more explicit emphasis on values, and aims not at "equilibrium" but at the sustainability of a dynamic flow system.[2]

Beyond the physical and biological resources, *information* has, in the 1970s, come to be regarded as a resource, too. It is not marginal; Peter Drucker says "knowledge is now the main cost, the main investment, and the main product of the advanced economy and the livelihood of the largest group in the popu-lation." It is not depletable; John McHale taught us that information expands as it is used. It is not scarce; Lewis Branscomb says information "is in quan-titative surplus. To be sure, there are great gaps in human knowledge that have yet to be filled by research and study. But the yawning chasm is between what some have learned, yet others have not yet put to use."[3]

The central problem for economists, planners, managers, and political leaders has been scarcity and the allocation of scarce resources—things, services, jobs, skilltraining, money. But, if information is chronically in surplus, the infor-mation technologies—the revolution that embraces the microprocessor, the com-puter, and the satellite—have their own implications for economic, social, political, and international developments. These implications seem strikingly parallel to the implications of our greater reliance on the bioresource: more horizontal systems, widespread access to feedback, more organizational com-plexity, potential for more synoptic views, longer ranges, and more compre-hensive analytical systems.

The sudden new prestige of bioresources and information, the renewable and the expandable resources, does one more thing for us, and it may be the most important thing. Their management requires whole populations of thinking men and women and, therefore, focuses our attention on people as the boundless resource. In one of the discussions at the Woodlands in October 1979, Willard Wirtz put it this way: " 'Growth' includes whatever contributes to the common opportunity to make the highest and best use of the human experience."

We are, it seems, not about to "run out of" resources. But that is because we are expanding our ways of thinking about resources.

Economic Theory

We began, in the 1970's, to realize the need for a new, wider, more flexible, and more compassionate general theory of employment, interest, and money

than either of the contending theories—monetary management or demand management—seem able to provide. The demonstrable inadequacies of modern
economic theory leave plenty of room for fresh thinking about a new economics.[4]

We used to define economics as the science of cyclical social change. Inflation
was to be found at one end of the business cycle and recession at the other end.
But, for most of the 1970s, inflation and recession have been glued together at
one end of a system whose oscillations stump the experts, and which may not
even be cyclical.

Governments that reflate to create more jobs sometimes succeed in creating
more inflation and bringing more people into the job market to offset the jobs
they create, leaving the unemployment figures strangely static. Governments
that raise interest rates and try to restrict the money supply sometimes succeed
in inducing recession without controlling inflation. And governments that, like
the United States, try to pull both macro levers at once have to admit—as
President Carter once candidly did on national television—that contemporary
economic theory fails to explain the way the real world works.

"In the long run, we are all dead," said Lord Keynes in one of his most
familiar quotations. In the long run, it turned out, Keynes *was* dead. And the
confusion of signals from the political capitals of the industrial world suggests
that his successor is not yet out of graduate school.

The obsolescence of those useful Keynesian theories, which justified macroeconomic "fine-tuning" during the 40 years from the 1930s to the 1970s,
cannot be explained by a single cause. It is not just that during the 1960s we
tried to have our guns and our butter too, or that in the 1970s our politics of
equity came to mean single-issue factions pursuing their narrow economic interests at the expense of the general welfare (not for the first time in our history).

The profound socioeconomic transformation which Robert Hamrin describes
in Chapter 4, and which he says makes necessary a rethinking of economics,
is an alloy of many components—the global explosion of population, the mi-
croprocessing revolution, the sudden spread of communications, the energy
crunch, the potential of biological resources, the uncertain swings of climate,
the changing nature of work, the shifting requirements for education, the curiously stabilizing peril of nuclear weapons, the thickening web of international
interdependence, and the growing dangers of faraway poverty in a shrinking
world.

What traditional economics treated as externalities (factors outside one's analytical system, yet disturbingly relevant all the same) seem to have become the
core of the new system of thinking we now require. We can continue to call the
new system of thinking "economics" only if the profession quickly internalizes
the externalities, clasps to its bosom the criteria of judgment that have been kept
at arm's length because they were so messy, so hard to quantify, so loaded with
the politics of value and the values of politics: equity and fairness, employment
and education, energy and the environment, personal privacy and international

security, a planetary perspective in national and local decision making, and the interest of that ultimate silent majority, the future generations still unborn.

Until the 1970s, civilization was well served by the blooming of macroeconomic analysis. In the 1980s and beyond, economics—"the hardest of the soft sciences"—will have to come to terms with wider concepts of qualitative growth, social fairness, ecodevelopment, and what used to be called "political economy."

Interdependence

Even in the United States, the reality of interdependence has sunk in. During more than half of our two centuries of national life, Americans were citizens of an underpopulated nation spreading across a continent full of energy, arable land, and other natural resources. Comparative isolation was the natural condition because it did not have to be otherwise. Only a few decades of our short history have been spent in reluctant reconciliation with the idea that international relations, though uncomfortable, are unavoidable.

But, as E. B. White wrote long ago, ". . . there's no limit to how complicated things can get, on account of one thing always leading to another."[5] In the 1970s, world-circling technologies, and intercontinental weapons, world trade, oil shortages, and global inflation, and the interacting revolutions of information and communication conspired to educate even the most isolationist American. In Chapters 8, 9, and 10, aspects of "growth policy" are viewed in global perspective by Walter Orr Roberts and Lloyd Slater; Lincoln Gordon; and Thomas W. Wilson, Jr.

That international relations are a two-way street was hard, at first, for Americans to grasp. Everyone else seemed to want our aid, our technology, weapons, blue jeans, hard rock, and TV programs. It was not equally clear to us that we needed an influx of energy, raw materials, investment funds, brains, and international cooperation. The Arabs brought the matter to our attention with their oil embargo in 1973, and, by the end of the decade, "interdependence" had almost achieved the status of a cliché.

A remarkable number of Americans now understand that, in the early 1970s it got very cold in the Ukraine and a few months later Soviet wheat purchases had pushed up the price of a loaf of bread at the corner supermarket. We realize now that the disappearance of anchovies off the west coast of South America can so shrink the supply of protein that the United States has to deprive its soybean customers, including Japan, of their normal supplies. A revolution in Iran has a visible and immediate effect on the price of gasoline and heating oil. We are even beginning to understand that the burning of oil and coal in the industrial nations, plus the overcutting of forests in the preindustrial societies, leaves more and more carbon dioxide in the atmosphere and risks overheating the world of our grandchildren.

Cognition of interdependence has been helped along by satellite technologies. The greatest single instrument of education for interdependence just now is the exposure of whole populations to synoptic pictures of weather systems, taken from satellites and displayed in our living rooms every evening as part of the TV weather forecast. If you now ask an American about the weather he will no longer moisten his finger, or feel his joints, or even glance at the sky. He will tell you, in semiprofessional language, about the cold front coming in over Canada and scheduled to arrive in his own community at a predictable time. Everyone now has the tools to visualize how the weather develops and to understand, without being formally "taught," that events are interconnected, nations are interdependent, and the biosphere is more or less round.

Development and Human Needs

During the decade of the 1970s, the world's "North-South" relations—commerce and confrontation between the industrial nations and the developing nations—became the most dynamic engine in world politics. More and more they are driving those other big wheels which used to occupy center stage in U.S. foreign policy: our Atlantic and Pacific alliances and our relations with the Soviet Union.

The motive power for this new kind of politics is rapid change in the developing "South"—the familiar race between population and "modernization," and a newly explosive factor: the collision of modernization with traditional cultures and untraditional aspirations for economic growth with equity.

This triple collision was dramatized for us by the televised revolution in Iran: the mullahs in their long robes, spurred by a Paris-based leader communicating with them by audiocassette, teaming up with left-wing students in the streets of Teheran to set fire to tanks and automobiles—the hardware, and also the symbols, of modernization. But the three forces which collided to produce the turnover in Iran are present, in widely differing tactical forms, in the politics of a hundred societies in Asia, Africa, and Latin America.[6]

If the United States turns its back on the developing world, the fact that world poverty achieved center stage in the 1970s will make the 1980s and 1990s a dangerous passage for the only nation with a truly global reach. But Americans have a long record of adjusting pragmatically to newly-perceived realities. The dangers of poverty, discrimination, and frustration in the world's "South" may increasingly be perceived, in U.S. politics, as part of a widened concept of security—not just because of our need for oil imports or food markets, but because claims for "fairness" and "basic human needs" resonate in the American political psyche.

After all, science and technology *have* made possible—and, therefore, necessary soon or late—the meeting of human needs at a level nearly everyone would agree is a human right.[7] Moreover, we have seen in the 1970s a fusion of demands for the meeting of human needs with demands for political and civil rights, in an enlarged and invigorated international human rights movement. The way a government treats its own people, the issues of fairness in the "domestic politics" of nation-states, used to be considered privileged, protected by national sovereignty. But, in the surprising 1970s, a president of the United States could stand in the U.N. General Assembly and declare that "No member of the United Nations can claim that mistreatment of its citizens is solely its own business." He was talking, he made clear, not only about torture of political prisoners but about "unwarranted deprivation" of the poor. Nobody jeered.

The wobbly launching of the Carter human rights initiative (its impact on arms control negotiations, the president said later, was a surprise) should not obscure its historic quality. To fuse, as the president has fused, political and civil rights the state can assure by simply *not* oppressing its citizens (not arresting them without charge, not convicting them without trial, not persecuting them for dissent, not censoring them for unorthodoxy), *and* the economic and social ambitions requiring affirmative action to meet "basic human needs," provides a strong doctrinal basis for the world's first truly global revolution—all the previous revolutions (including the Christian, the Moslem, and the Marxian) having had to settle for primary influence in one or another world region.[8]

Planetary Politics

The 1970s saw the dawn of planetary politics. It started with the U.N.'s Stockholm Conference on the Human Environment in 1972, and continued in Bucharest on world population, Rome on world food, Mexico City on the status of women, Vancouver on the human habitat, Geneva on employment, Buenos Aires on world water resources, Nairobi on desertification and, in the summer of 1979, a conference in Vienna on science and technology for development.

These consciousness-raising sessions are not to be judged by the pale and wordy compromises they issued as "plans of action" just before the delegates left for home. They are devices for forcing national governments to pay attention to neglected subjects—sensitivity training on a global scale. And they work. Before Stockholm, no nation had a cabinet-level ministry mandated to protect the natural environment; now, eight years later, there are more than 70. Before Bucharest, many developing nations were denying that overpopulation was a problem, or claiming that excitement on the subject was a conspiracy to withhold help from the poor. Today, the population policy of every nation is an accepted topic of international discourse.

The recognition of global risks and the presence of global technologies creates a new kind of politics to which the adjective international, the traditional fear of "losing" sovereignty, and zero-sum, win-lose scenarios simply do not apply. When it comes to assessing environmental risk, controlling epidemic diseases, collecting data about the atmosphere, channeling appropriate investment from one country to another, or inventing appropriate technologies for development, the institutions are likely to be transnational; sovereignty is not "lost" but *exercised* by being pooled, as in the World Weather Watch; and the alternative scenarios tend to be, not win-lose, but either win-win or lose-lose. The resulting planetary politics is something new under the sun. In a world preoccupied with international conflict, that has to be good news.

Science and Technology

The worm has turned on science and technology. The idea used to be that, if we could think it up, we should surely manufacture and deploy it. But this "inner logic" of technological change is being shoved aside by the notion that the future directions and purposes of technology are matters for social determination—that is, for us the people to decide.

The watershed was the decision not to build a U.S. supersonic airliner, despite the French/British decision to go ahead with the Concorde. (Before the decade ended, the French and British had cooled off on the idea, too.) The environmental movement's shouts of "Hey, wait a minute!" began in the 1970s to drown out technology's inner logic; NEPA, the National Environmental Policy Act, came into effect on January 1, 1970. The deal with the Soviets not to make any more antiballistic missile systems, even though both sides had developed prototypes, was another straw in the same wind. (The nondeployment of chemical and bacteriological weapons was an even earlier example of technological self-control.) So is the dwindling enthusiasm for nuclear energy as a way to make electricity, and the tightening regulation of carcinogenic foods and drugs.

The new public attitude can be caricatured. The City Council of Cambridge, Massachusetts, caricatured it by asserting municipal jurisdiction over the pace and direction of genetic research at Harvard University. But a caricature not only exposes a prominent feature to ridicule; it calls attention to the feature's prominence.

Limits to Government

This new attitude parallels and contrasts with another shift of opinion in the 1970s. We the people have discovered the limits to government, the incapacity to cope of the sovereign nation-state, which had been regarded, for generations past, as the prime instrument for getting things done in the public interest. In the United States, the straws in this wind are clumsy initiatives—California's Proposition 13, the proposed constitutional convention on balancing the budget. But underneath them is another of those tidal waves of the 1970s—and the boat of national government is leaking in the storm.

We are going to need more public-policy decisions—to contain, channel, and control new and "improved" technologies—and, at the same time, we want to get along with less government. How are we going to "do less and do better"? William Sneath asks in Chapter 13. Murray Weidenbaum has some suggestions in Chapter 11, Ted Gordon considers "the revival of enterprise" in Chapter 12, and Maurice Strong proposes some nongovernment initiatives in Chapter 14.

How do we manage more governance with less government? Not, surely, by an adversary proceeding between government and the rest of society. Not by trying to maintain a fictional line between "public" and "private," a wall mostly honored in the breach. John Gardner suggested at the Third Woodlands Conference that we are institutionalizing the inside track for special interests inside the bureaucracy's walls.

Americans are going to have to face honestly an uncomfortable fact: a century of effort to draw a thick line between "public" and "private" is no longer serving us well. Some of the special pleadings by public-interest groups—for more antitrust enforcement and for tightening the conflict-of-interest statutes, to cite only two of many examples—are futile attempts to separate them further. The Woodlands Conference multilogue suggests another route: to infuse the leaders of the private sector with enough of a sense of public responsibility to qualify them as partners, with the government, in the wider processes of governance. As things stand, the missing "interest group" is a lobby for the *general* interest, a public constituency that enables our political leaders to get beyond brokerage and adhockery to vision and consensus.

People-in-general, perceptive as usual, have noticed that something is wrong. By overwhelming margins, current survey research records a rapid decline in the people's confidence that their government can do very much that is effective about what bothers them. In 1964, two-thirds of the American public had faith in the competence of government leaders—that is, they would agree with the poll-taker's statement: "They know what they're doing." By 1978, the ratio of Americans holding this view had dropped to two-fifths. Two decades ago, in the late 1950s, a 56 percent majority of the public agreed that "you can trust the government in Washington to do what is right most of the time." Now this index of public trust has dropped to 29 percent.

This state of affairs is often presented as tragic news in TV newscasts and newspaper editorials, and sometimes in the commentary of the survey researchers themselves. But I read these and similar polls differently: in thus telling it like it is, the American public is showing signs of robust mental health.

The fact is that our institutions are *not* coping. The government *doesn't* know what to do about the non-Keynesian fusion of inflation and recession. Many business leaders—do we have to look farther than steel, or small cars, or color television?—*do* seem to have lost that old spirit of American enterprise. Labor unions and other special-interest groups *can't* be assumed to put the general welfare ahead of their specialized welfares. If, in these circumstances, people were nevertheless persuaded that our leaders have the situation well in hand and told the pollsters that everything was fine, wouldn't that be a symptom of mental illness, of detachment from reality?

So . . . maybe . . . the perception that our institutions are not coping is itself the beginning of a new wisdom.

We are, in Daniel Yankelovich's phrase, "working through" to new work-ways, new life-styles, new ways of defining our most basic aims and ambitions. But we are already living in tomorrow's world today, using yesterday's ideas. As we "work through," there is an important place for the practical, venturesome experimentation that Maurice Strong urges on us in Chapter 14. But the tightest bottleneck in modern civilization, just now, is relevant theory. By far the most practical thing we can do about "growth policy" is to think hard about it.

The developments that made the 1970s a watershed decade—I have mentioned eight of them—are, above all, changes in our ways of thinking about the dangers and uncertainties mankind is heir to. It is the new thinking, not the old dangers and uncertainties, that constitutes the good news in this story. The objective "facts" of national and international life are no less threatening now than they were a decade ago. There are, indeed, more and bigger weapons, more obstacles to sound production and fair distribution, more uncertainties about energy, more baffled economists, more unavoidable interdependence, more Third World tur-bulence, more political prisoners, more absolute poor, more global environmental risks, more morally ambiguous scientific discoveries and technological achieve-ments, and more governmental incapacity in 1980 than in 1970. The difference is that we can now see how to think our way out of the mess we're in.

NOTES

1. Portions of this writing appeared, under the title "A Rich Vein of Good News," in *The Christian Science Monitor*, December 27, 1979, pp 12-14.

2. These social implications of the bioresource are more fully discussed in a book resulting from the International Conference on Bio Potentials for Development held in Houston in November 1978; that Conference also served as one of the eight workshops building up to the Third Woodlands Conference on Growth Policy in October 1979. Harlan Cleveland and Alexander King, eds., *Bioresources for Development* (New York: Pergamon Press, 1980), especially the editors' introductory essay, "The Renewable Way of Life."

3. Peter Drucker, *The Age of Discontinuity: Guidelines to our Changing Society*, (New York: Harper and Row, 1978), p. 264; John McHale, *The Changing Information Environment*, (London: Paul Elek, 1976); Lewis M. Branscomb, "Information: The Ultimate Frontier," *Science* 203 (1979): 4376. One of the preparatory Woodlands Workshops was on "Information as a Resource"; notes on that workshop are available from The Woodlands Conference, 2201 Timberloch Place, The Woodlands, TX 77380. See also Harlan Cleveland and I. H. Abdel Rahman, "Dynamism and Development," in *World Development*, April 1980.

4. These comments are adapted from my Preface to Robert Hamrin, *Managing Growth in the 1980s: Toward a New Economics*, (New York: Praeger, 1980).

5. E. B. White, *Quo Vadimus, Or the Case for the Bicycle* (New York: Harper & Brothers, 1938), p. 26.

6. A fuller development of this concept appears in a published lecture *The Triple Collision of Modernization* (Austin, Texas: The University of Texas, Lyndon B. Johnson School of Public Affairs, 1979).

7. John McHale and Magda Cordell McHale, *Basic Human Needs: A Framework for Action* (New Brunswick, N.J.: Transaction Books, 1978).

8. A full discussion of the new human rights doctrine and its antecedents will be found in Alice Henkin, ed., *Human Dignity*, a book published by the Aspen Institute for Humanistic Studies in 1979.

Chapter 2

THE PUBLIC DEBATE ABOUT GROWTH

Daniel Yankelovich and Bernard Lefkowitz

I
CHANGED EXPECTATIONS

At America's birth, Alexander Hamilton expressed his faith in the country's limitless future by writing: "There is in the genius of the population . . . a peculiar aptitude for mechanic improvements."[1] During the next 150 years, Americans, with unflagging energy, applied this "peculiar aptitude" to the construction of the world's most powerful industrial machine.

Self-evident Growth and U.S. Dominance

Shortly before he took office for his first term, President-elect Eisenhower urged the public to rededicate itself to maintaining America's economic preeminence. "What matters to the average citizen is—what can *he* do? A carpenter or a farmer or a bricklayer—what is *his* role?" President Eisenhower said. "Every individual has to understand and he's got to produce. He's got to work harder than ever. And he's got to understand why."[2]

Up to then, the *why* had seemed self-evident. Rapid economic growth—as measured by the familiar indicators of productivity and output, more jobs, products, services, and rising tide of economic progress—did, in fact, improve conditions for the majority. Progress yielded both material and social rewards. It offered the individual the promise of economic security. It provided tangible

16

symbols of success: the new car, the home in the suburbs, access to higher education for one's children, and a retirement virtually free of economic concern. Economic expansion also paid off in psychological dividends. As the country advanced, so too did the individual's social standing. He enjoyed the respect and esteem of his fellow workers, his neighbors, and his family.

Throughout the post-World War II period and up to the early 1970s, the expectation of growing affluence was not shaken by the occasional economic setback. If the economy faltered, it was assumed that new investments would regenerate the industrial machine; technological innovation—"the aptitude for mechanic improvements"—combined with financial acumen and skilled and dedicated manpower was seen as able to defeat any challenge mounted by foreign competitors and to compensate for scarcities of natural resources. New markets and new products would provide a job for all who wanted to work. Increases in productivity, vigorous competition, and mass consumption would restrain inflation. As the economy expanded, the real value of the dollar would steadily rise. The individual and society had joined in a congenial partnership. Both were moving in the same direction, toward economic progress and growth.

What was seen as America's unlimited capacity for growth provided a kind of psychological insurance against surprise and shock. The power of the economic engine, its velocity and bigness, was such that no economic or social claim could derail it. The administration of the City of New York might experience a twinge of apprehension when the annual welfare bill came in, but the revenue and taxes generated in the financial capital of the world were sufficient to support a generous humanitarian social program. The individual taxpayer might bridle at an increase in taxes to pay for foreign aid, but the next pay raise would cover it—and then some. Drought might reduce the grain yield one year, but the next year the sun would shine and the harvest would be bigger and better than ever. Americans were sustained and motivated by their confidence that adversity, whether it be unemployment or inflation or natural calamity, would be temporary and quickly forgotten. The country was moving on. From a psychological perspective, then, steady growth brought America within reach of that rarest of luxuries—the privilege of stability.

But few Americans who heard President Eisenhower speak in 1952 could anticipate how great the claims on growth would be in the next 25 years. When sociologist Amitai Etzioni speaks of this country's "core project"[3] he means its capacity to provide the individual with material security and stability. That capability, challenged seriously once before in this century by the Great Depression of the 1930s, was severely tested three times in the 1960s and 1970s.

The first claim against the primacy of the "core project" was the demand for a vastly expanded system of social supports and services. The assumption is often made that this claim was lodged on behalf of the poor and minorities. But, in fact, the public sector was being asked to invest massive resources in a social welfare program that served most of American society. In 1976, more than 100 billion was provided by the federal, state, and local governments under the rubric of social insurance. This included, among other programs, social security, unemployment insurance, public employees' retirement, veterans' disability, military retirement, and workmen's compensation. These funds do not go primarily to the poor. By comparison, only $12.7 billion was targeted specifically for the underprivileged. The federal contribution to social welfare had increased 600 percent over what Washington was spending a decade ago. In that period, federal assistance grew at a rate that was two-thirds faster than the rate of growth of the entire federal budget.[4]

The second claim was phrased in social and psychological terms but it, also, had significant economic implications. In essence, many Americans were revising the traditional definition of success. They expressed reservations about wasteful materialism, questioned the value of embarking on an intense, lifelong competition for wealth and privilege, and began to look toward nonmaterial standards of personal growth, such as self-awareness and interior development. But it is important to remember that, at the same time, Americans were not prepared to accept a lowered standard of living or a reduction in basic goods and services. While they were reexamining the meaning of psychological fulfillment, they also demanded economic choices. People might not feel compelled to buy a prestigious car to advertise their prosperity; instead they might spend their money on another—and more expensive—symbol of success, such as a second home in the country or at the seashore where one could lead a "simpler" life. The mink coat might lose favor as a status symbol, only to be replaced by a more refined symbol, a housewide stereo system which signified that its owner had both money and taste.

In the 1970s, another claim was lodged against the economic core product. Its thrust was to seek stability and security by pressing for a safer and healthier environment. But, like the other claims, it too was double-edged. People might have wanted factories to stop polluting the environment, but they didn't want the factories to go out of business. As the public expressed growing concern about the safety hazards posed by nuclear power plants, it was also vociferous in its criticisms of the high price of conventional sources of energy. The message from the public to national leadership was: We don't want the factory to pollute the environment, but neither do we want to lose the jobs provided by the factory and the products it manufactures, nor do we want prices to rise.

But these were not the only strains imposed on America's fundamental commitment to economic expansion. Two other formidable challenges developed almost simultaneously in the last ten years. One was the pervasive and deep-seated mistrust of the dominant institutions in society, particularly big business and government. Public criticism of business focuses essentially on health and safety issues—the public's fear that the large corporation will put its own short-term profit interests ahead of the health and safety of its customers, its employees, and the citizens of the towns and cities where it has its plants. The public's disenchantment with government stems more from concern about waste, inefficiency, and unresponsiveness than about corruption or tyranny.

From the early 1930s to the early 1970s, a majority of Americans had accepted the premise that an activist government was a sound response to rising expectations and the ups and downs of an unplanned private sector. In the 1970s, however, increasing numbers of Americans began to fear that the solution might prove more damaging than the "problems" it was supposed to solve. Americans grew increasingly skeptical about government effectiveness and good faith. They began to develop a "take back" psychology—a conviction that at least some of the powers assigned to government should now be taken away, and returned to the people.

Finally, the American vision of economic growth, grounded in the assumption that the United States is and will continue to be dominant in the world economic order, has been seriously shaken by the continuing impact of inflation, successive encounters with scarcity, and a shifting locus of international political and economic power. A sober reassessment of America's economic position is advanced by Philip Caldwell, president of the Ford Motor Company:

At the end of World War II, we were the undisputed leader in almost every field—economic, technological, productivity growth, military and political. Our products are not necessarily the best anymore, and we are no longer the only or most logical supplier of many products.

. . . we have three things going against us—higher labor costs in absolute terms, lower rates of productivity improvement, and higher inflation than all but a few of the leading industrial nations. For the first half of the postwar era, we had major advantages in two of these areas—productivity and inflation—and we made sizable gains in many foreign markets. We were able to compensate for our higher labor costs. Those advantages now have disappeared. Unless we regain at least rough parity in all three areas, or a major advantage in at least two, we will continue to lose ground.[5]

The cumulative effect of these claims poses serious problems for planners and policymakers. To satisfy all of them—the demand for greater social supports and services, the heightened concern about the quality of environment, the growing constraints on business and government—requires a major commitment of resources. Should the country continue to respond to those claims that might be subsumed under the heading of a "quality of life" society? Or should the United States seek, instead, to replenish its aging industrial stock and heavily reinvest in economic development to regain its economic leadership in the world community? Is the goal of quality, in fact, *dependent* on economic growth to pay the costs of social and environmental programs? Should we—can we—try to do both?

Etzioni estimates that a major investment in economic redevelopment would cost between $574 and $581 billion in an average year for the next ten years.[6]

He also calculates that, in the next decade, the cost of a high-priority "quality of life" commitment would average approximately $117 to $162 billion a year. In addition, the cost of basic needs, such as food, continuing expenses (payments on the national debt), and existing programs (support of veterans and farmers) range between $100 billion and $400 billion a year.[7] Etzioni concludes that, even under the most optimistic projections of gross national product growth, it would not be possible to advance both economic development and a "quality of life" society. Even a modest response to the claims made against the central economic product would result "in an acceptance of underdevelopment."

> Already being on a downward path, as most clearly reflected in productivity decline and in GNP declining growth rates, a decision not to grant redevelopment a high priority is, essentially, a decision for a slow-growth society. . . . Practically speaking, the choice is hence for a high-powered redevelopment drive and a rather thin quality-of-life program for the next decade—or, a quite effective quality-of-life program with growth underdevelopment.[8]

Etzioni maintains that, for both economic and social-psychic reasons, choose we must.

Other analysts, such as Herman Kahn of the Hudson Institute, believe that it may theoretically be possible for the economy to support both projects, but that would result in a clash between "traditionalists," who favor a high-consumption, steadily expanding economy, and proponents of the "new values," who support a slow-growth, high "quality of life" policy. Ultimately, Kahn says, the adherents of restrained growth will prevail. To support this conclusion, he cites various new trends which he says are reshaping American values and attitudes. These include an emphasis on public welfare and social justice, an animus toward technological innovation, a propensity toward regulation of the economy, and a personal self-involvement that emphasizes nonmaterial values. "Eventually we expect that the new values, attitudes and goals will become quite widespread—even if modified—and largely replace the goal of economic growth," he says. "We expect this process to be largely suppressed and then re-emerge during the next decade and to have gone quite far by the year 2000."[9]

Whatever their views on restraint and expansion, analysts tend to agree on one point—the importance of the social values, attitudes, and beliefs of the American people in shaping the character and extent of our economic growth in coming years. Many theories have been advanced about the changes in Americans' attitudes and beliefs and how they impinge on growth. But few facts have been presented either in opposition to or in support of these theories. The purpose of this chapter is to examine a wide array of attitude survey data from many sources to learn whether American attitudes toward economic growth have changed, and if so, what the nature of the change is and what it means for the future direction of the American economy and society.

Themes and a Caveat

Our review of a massive amount of public opinion survey data gathered over the past quarter century leads to four general observations. These are:

A Shift in Expectation. The public is aware of what it regards as a change for the worse in the country's economic circumstances. From an expectation of steady growth, ever–increasing abundance, continuing improvement in the individual's standard of living, stable prices, and jobs for all who want to work, the majority has shifted grounds to expect, instead, economic instability: recession, depression, continued inflation, joblessness, and shortages.

"Working Through." The public is still in the early stages of what psychologists call a "working through" process. People have yet to reconcile themselves to the conflict and disappointment created by the need to adapt to new, unwelcome conditions. They have yet to find new strategies for coping based on lowered economic expectations. Suppose, for example, someone has lived for years with the expectation that his retirement will be economically secure, and then learns with pained surprise that it will not be. If an economically secure retirement is as important to him as it is to most Americans, it will typically take a long time and much anguish to work through the shock of the new circumstance and adapt to it realistically.

In the American people today, there are few signs of panic, but many signs that the strong emotions and confusions associated with the "working through" process are surfacing. Consequently, the public is now in an unstable and unsettled state of mind, but also moving toward a resolution.

Care in Interpreting the Survey Data. This unstable state, elaborated later in this chapter, makes it difficult to interpret individual attitude survey findings with confidence. When Americans have completed the "working through" process, when the emotion has receded and the adaptation has been made, then peoples' responses to survey questions will more accurately reflect their behavior. But it is misleading at this early stage in the "working through" process to take what people say in response to any one question as a literal and faithful guide to what they are doing now or will do in the future. When, for example, Harris reports that 76 percent of the public agrees that we should be placing more stress on "learning to get our pleasure out of nonmaterial experiences" versus only 17 percent who opt instead for "satisfying our need for more goods and services," (see table 2.3) it would be a false leap to conclude that the American people have pushed materialism aside by almost a five to one margin. That is not Harris' interpretation; it is not ours; it would not be the interpretation of any reasonably sophisticated analyst. But, unfortunately, it is implied by the raw data itself because of our conventional model of using survey data. We are so accustomed to thinking of survey data in the form of pre-election surveys where a 42 percent preference for a political candidate is mentally translated into a 42 percent vote, that there always exists an implication that attitudes reported on surveys in answer to single questions are predictive of behavior.

We must, therefore, be careful to analyze available survey data as a whole so that we can take contradictions and inconsistencies into account. The public's response to various single questions seems to imply that Americans have already chosen the nonmaterial over the material. But an analysis of the survey data as a whole shows that this is not the case. In fact, Americans are only at the earliest stage of recognizing that painful choices may be required.

A New National Emphasis. With some reservations, we have arrived at our own tentative interpretation of where the public is likely to come out when it does complete the "working through" process. In an historic change of emphasis, the American public seems to be moving toward a provisional acceptance of a trade-off, where some growth is given up in favor of economic stability and nonmaterial values. Most Americans are disturbed by the prospect of economic instability, doubtful about excessive reliance on material values, and eager to hold onto the gains that they feel they have made in recent years. We may be moving toward a new national emphasis which downgrades growth and consumption, and elevates the importance of holding onto past gains, avoiding waste, minimizing risk, and finding greater satisfaction in activities that do not depend so heavily on acquisition.

In Section II of this chapter we will discuss the reasons for this different national emphasis, and the need for effective political leadership to guide Americans toward a resolution of ambivalence. Now let us examine some of the data that supports these conclusions.

Survey Findings

Our analysis of public opinion data suggests that, in the last ten years, a significant psychological shift has taken place—from an optimistic expectation of an open-ended and unlimited future, to a fear of instability and a new sense of limits. Where once people looked to the future with hopefulness, they now anticipate tomorrow with some sense of foreboding. In the 1950s and 1960s, they believed the future promised gain; at the end of the 1970s, they expect loss and seem more concerned with preserving what they have than in expanding their frontiers. Our concern here is with describing and documenting this transition in national outlook—a reluctant surrender of the core belief in a personal and collective future of unbounded material prosperity and the emergence of changed, and lowered, expectations.

Twenty years ago, when the United States went through a period of recession or a downturn in the economy, people generally retained their sense of optimism about the future. They were, for the most part, hopeful that things would get better for themselves and for the country as a whole. That this has changed is evident in the answers people give to questions about their personal future and their expectations about the American economy.

A number of important themes appear to influence the day-to-day existence of a majority of Americans. One is a heightened sense of foreboding and concern about the nation's economic future. Another is the individual's diminished confidence about his personal future which is reflected in an increasing emphasis on reducing expenditures and maintaining the present standard of living.

Survey findings from a wide variety of sources document the shift from an optimistic to a pessimistic majority.

• Gallup shows that, while only one out five Americans in the early 1970s (21 percent) believed "next year will be worse than this year," now a 55 percent majority holds this pessimistic outlook (table 2.1).

• Gallup's surveys also show that the number of Americans who believe "this year will be a year of economic difficulty" has increased from 48 percent in 1969 to a 69 percent majority now (table 2.1).

• Survey findings from Yankelovich, Skelly, and White tell us that the number of Americans who believe we are entering an era of enduring shortages (as compared to those who think that shortages are only temporary) increased from 40 percent in the mid-1970s to 62 percent in the late 1970s (table 2.3).

• In this same time period, Harris' findings show that the number of Americans expecting prices to rise more rapidly in the future has jumped from a 30 percent minority to a 52 percent majority (table 2.2).

• In the brief, two-year time span from 1976 to 1978, Michigan Survey Research Center finds that those who expect that their income will not keep pace with prices increased from 35 to 45 percent of the population, and those who expect the country to be plunged into periods of widespread unemployment or depression has shot up from a 32 percent minority of the public to a 53 percent majority (tables 2.2 and 2.1).

• In 1979, Yankelovich, Skelly, and White surveys find a full 70 percent of the public believing that a recession is likely in the next five years (table 2.1).

• Roper reports a 58 percent majority who believe that our natural resources are likely to deplete themselves rapidly in the near future (table 2.3).

Perhaps the single most persuasive documentation of the shift in the American mood comes from an ingenious scale developed by Hadley Cantril in the 1950s and used subsequently by various organizations, including Potomac Associates.

From the 1950s to the late 1960s Americans characteristically believed the present to be a better time for the country than the recent past, and anticipated that the future would inevitably improve over the present. In 1971, the pattern changed. Then Americans saw the past in a rosier light than the present, but anticipated that the future would once again brighten up for the country.

In 1978, for the first time, the pattern of the 1950s totally reversed itself. Now Americans believe that the past was a better time than the present, and they anticipate that the present, however bad, is likely to be better than the future. This is, indeed, an historic shift away from traditional American optimism to an uncharacteristically bleak outlook (table 2.4).

The Individual's Economic Future. In surveys conducted in the early 1970s, the public often distinguished between the nation's economic well-being and their own personal prospects for a satisfying life. While they believed that the country as a whole was in for hard times, they were personally optimistic. Inside their homes, within their families and communities, the future held bright promise. Now the pessimism they feel about the country shadows their own lives. So while, in 1971, only three out of ten (30 percent) voiced dissatisfaction with the future facing themselves and their families, today the number has increased to more than half (51 percent) (table 2.1). The Index of Consumer Sentiment, developed at the Institute for Social Research at the University of Michigan, measures attitudes about prosperity and economic security. When the index is at a high level, the consumer is delivering a strong vote of confidence in the economy. In the late 1960s, the Index stood consistently between 95 and 100. In 1974, in the middle of the recession, the Index was below 60. In 1978, although there was no recession, the Index rose only to 75 (table 2.5). Dr. George Katona, who devised the Index, has written:

During the great recession of 1974-1975 consumers were considerably more optimistic about future business conditions than they were about the past; while past trends were thought to have worsened substantially, the downturn was not expected to last. The end of the recession was thus indicated much earlier in surveys of consumer expectations than in economic conditions themselves. In terms of economic conditions in 1979 and 1980, it must be of some concern to policymakers that over the past year and a half just the reverse has occurred: expectations about the economy have worsened to a greater extent than have people's evaluation of past trends. . . .

What demarcates the 1970s from the fifties and sixties is that in the current era, public confidence in the government—and especially in the effectiveness of government economic policy—has practically vanished, and as it has, economic confidence has also crumbled.[10]

Dr. Katona points out that this erosion of confidence "was the case for all population groups . . . for higher– and lower–income respondents and thus for respondents differing widely in education and sophistication."

The individual's perception of ever-narrowing limits is most clearly expressed in the response to the two survey statements. One is: "Our current standard of living may be the highest we can hope for." In 1979, a 67 percent cross section of Americans agreed entirely or in part with that statement (table 2.1). The other statement is: "Americans should get used to the fact that our wealth is limited and most of us are not likely to become better off than we are now." In 1979, 62 percent of the American public agreed with that statement.

Such agreement is virtually uniform across all demographic groupings. There are no differences between men and women, black and white, or urban versus rural dwellers. Geographical differences exist but are quite modest: The South is the most pessimistic (67 percent agree with the statement), and the West the least (a 58 percent level of agreement). These differences are barely worth mentioning. There are similar shadings of differences between blue-collar and white-collar workers. Also, lower income groups tend to be somewhat more pessimistic than the highest income groups. But the spread rarely exceeds 10 percentage points. The largest differences relate to age and education, with older people being more pessimistic than younger ones. The least pessimistic group in the country are college students, only 47 percent of whom agree with the statement as compared to young people in the same 18-24 year old age bracket who are not attending college. Among this latter group, more than six out of ten (63 percent) subscribe to the idea that most of us are not likely to become better off than we are now—a profoundly new idea for Americans to hold in the post-World War II era (table 2.12).

It is important to emphasize that the new economic pessimism is based on direct personal experience. A majority of the Americans report:

• They are finding it harder to make ends meet in their personal budgets than in earlier years (69 percent).
• Inflation has reduced their family standard of living in the past few years (52 percent).
• Compared to the recent past, they have had to put off buying luxuries (70 percent) and even put off buying essentials (53 percent) (table 2.2).

The number of people who are caught in a tight financial squeeze has grown steadily over the past five years. More and more Americans report that they are having great difficulty in keeping up with their bills and in saving for the future. From 1974 to 1979, for example, the number of people who reported deep concern with being able to pay for the upkeep on their home or rent of their apartment more than doubled—from 23 to 48 percent (table 2.2).

The Struggle To Adapt. Earlier, we observed that Americans now find themselves in the early stages of a "working through" process. "Working through"
suggests struggle and lack of resolution. The public is still struggling with the
idea of limits and economic instability and is reluctant to accept it. Opinion has
not yet consolidated on either side of the growth issue. The public is still
weighing the arguments, pro and con, and still assessing the virtues and drawbacks of economic expansion. Americans are not sure what concessions or
sacrifices they will have to make if they enlist on one side or the other.

The "working through" concept also connotes the need for a certain amount
of time required to complete the various phases of the process. People need time
to digest the full implications of new ideas, particularly if these are disagreeable
and if there is reason to believe they may not be valid. They need time to
investigate, evaluate, and debate. They need political leadership that will define
the terms of the debate and propose real choices and national priorities for the
future.

The "working through" concept also connotes an outpouring of emotional
responses, inconclusiveness, and inconsistency. As the process unfolds, one
expects to find varied expressions of emotion such as anger, confusion, disbelief,
denial, barely suppressed panic, scapegoating, grasping at straws, depression,
overreaction, exaggeration, fatalism, instability of attitudes (saying one thing
one day, and another the next), lack of realism, inconsistency between attitudes
and behavior, and indulgence in Pollyanna-ish wishful thinking that everything
will turn out to be for the best, without requiring any real change. At the early
stages of the process, people often overreact or refuse to accept realities which,
deep down, they know to be true.

As part of an adjustment process, these are not pathological or abnormal
responses: they are the temporary—and perhaps inevitable—human concomi-
tants to unanticipated and threatening changes in the assumptions about how one
is going to live one's life. They are signs of the huge amount of effort it takes
for people to keep panic at bay when they first feel threatened and before they
have made a constructive adaptation to new circumstances.

The "working through" process also implies movement toward a goal. It is
a dynamic process, not a static one. It implies that when the sound and fury of
the struggle has abated and people have had time to digest the implications of
new realities, they will do so and find appropriate new strategies for dealing
with them. Resolution and consensus will then replace ambivalence, conflict,
and instability.

Energy provides one of the most dramatic but certainly not the only example
of a growth-related issue which has created public anxiety and confusion. The
public is actively engaged in the process of working through to a realization
that the energy shortage may, in fact, be real. At one level of consciousness,
people know full well there is an energy problem and that it is costing them
money at the gas station and in their home heating bills. But they don't know

how real it is and they don't know what to do about it. Because they doubt the veracity of the information provided by industry spokesman and government representatives, many suspect that there is no shortage, or that such shortages as do exist are manipulated to exploit them and enrich the energy companies at their expense—with the full compliance of the government. Their response is anger, uncertainty, confusion, and resistance to change. These forms of behavior characterize public reaction to many other growth issues, whether inflation or nuclear energy.

The reasons for the public's unwillingness to accept the reality of an energy shortage are concrete and understandable. The public has serious doubts that business or government is committed to a conserving strategy. There is no tangible or immediate system of rewards offered to the public for limiting consumption. (People try to reduce their use of electricity but their bills keep rising.) The information the public receives is often confusing and contradictory. (The Secretary of Energy says there is an oil shortage; the CIA says there is not.) Often there is no real alternative to a high rate of consumption. (It is difficult to persuade the public to limit the use of cars when, in most localities, there is no reliable areawide mass transit system.)

Despite these obstacles, the notion that the energy shortage may be real is making slow headway. Roper reports that, over a five-year period, the number of people who believe the gasoline and oil shortage is real and is likely to get worse in the future has increased from 21 to 31 percent, and the number of people who believe there never was any real shortage ("it was contrived for economic and political reasons") has decreased from 53 to 43 percent. Also, when the idea of an energy shortage is shifted from the immediate present to the future, the level of belief in its reality rises. Thus, we find a 63 percent majority who believe that there will be a serious shortage problem in energy supplies by the year 2000, 50 percent who feel that by the end of this century there will also be serious shortages of water supplies, and slightly smaller numbers who believe we confront serious problems of shortage in food supplies (table 2.3). Clearly, we are still far from universal acceptance of the need for embarking on the "moral equivalent of war" which President Carter believed was necessary to combat the present energy shortage.

Perhaps the clearest hint as to why the public continues to resist the idea that the energy shortage is real comes from a recent finding in a Yankelovich, Skelly, and White survey. An overwhelming 81 percent majority of the public subscribes to the view: "It is difficult to believe that a country as large, rich and industrial as ours will be unable to solve its energy problems" (table 2.9). At one level of consciousness, Americans simply cannot believe that we are unable to deal effectively with this problem. Yet coexisting side by side with this act of faith is an equally widespread expression of fear. An impressive 72 percent of the public concur with the view: "We are fast coming to a turning point in our history where 'the land of plenty' is becoming 'the land of want' " (table 2.3). In short, the public is caught between two feelings: the emotion that "somehow it cannot be'' and the emotion, ''my God, it may be worse than we think.'' Because Americans find it difficult to believe the United States is powerless to do anything about the oil shortage, they suspect conspiracy, greed, gross lack of leadership, or a combination thereof. On the other hand, the dawning awareness that their worst fears—that the land of plenty may be changing before their eyes to a land of want—may be true, breeds an underlying fear approaching panic.

It is difficult to judge how much of this reaction is warranted by the facts, and how much is an emotional overreaction bred by an understandable reluctance to change a way of life built around the presupposition of plenty. Almost surely, both elements are present. To some extent, the public is overreacting and overly suspicious. At the same time, a strong argument can be made that the public attitude of disbelief is not wholly ill-founded.

As the public grows increasingly aware that the energy crisis is real and that expectations for the future may have to be lowered, how are these new attitudes being translated into behavior? The answer to this question yields a vivid picture of the public's present state of mind. On the one hand, the majority of Americans maintain that they are acting in a less wasteful manner than in the past. More than 60 percent say they are keeping their automobile longer, are setting their thermostats at a lower temperature in their homes, and are making a conscious effort to save possessions they would have ordinarily discarded in the past (table 2.3). An even higher proportion—70 percent—report being more selective and careful when they buy food. More than half report being more cautious about expenditures for clothes.

Yet, at the same time, both Roper Reports and Yankelovich, Skelly, and White trend data depict a sharp inconsistency between attitudes and behavior. Over the past five years, awareness of shortages and of limits has grown, while conserving behavior actually has decreased—according to the testimony of respondents themselves. Roper, for example, shows this pattern:

- In 1974, 52 percent of respondents reported using their cars less often than in the past; several years later this number had diminished to 33 percent.
- In 1974, 71 percent reported that they were driving at lower speeds than in previous years. Yet, several years later, only 56 percent reported driving at lower speeds (table 2.2).

Roper does show a gradual increase over a several-year period in the number of people reporting that they are being more careful in their expenditures for food (from 59 to 69 percent), and an almost equal increase in the number who say they are being more careful of their expenditures for clothing (from 52 to 58 percent). But, at the same time, he shows a decline from 60 to 56 percent in the number of people who say they are cutting down on the luxuries of life (table 2.2).

In the four-year period between 1974 and 1978, Yankelovich, Skelly, and White show an even more dramatic disparity between attitude and behavior. For example, the number of people who say they are driving more slowly on the highway than in previous years has declined from 72 to 60 percent. More startling is the finding that a 61 percent majority in 1974 reported taking fewer nonessential car trips than in earlier years and, yet, by 1978, this number has been slashed to a 36 percent minority. In the five-year period between 1974 and 1979 Yankelovich, Skelly, and White also show a net decline in the number of people reporting that they:

- keep a car longer before getting a new one—from 69 to 61 percent;
- keep thermostats at lower temperatures than in earlier years—67 to 63 percent;
- are more careful with clothing which they keep longer or patch up before discarding—from 60 to 48 percent;
- save things such as empty jars, containers, and wrappings that they would normally throw away—55 to 43 percent;
- repair rather than replace broken furniture—52 to 44 percent;
- use paper towels and dishes more sparingly—48 to 36 percent (table 2.3).

Perhaps peoples' overall attitude can best be summed up by this provocative finding: In 1973, just before shortages developed, a 55 percent majority of people argued "it is permissible to buy the things I want when I want them." In the next few years, as awareness of shortages and economic instability grew, paradoxically, the sense of entitlement about buying "what I want when I want it" did not decline but actually increased—from 55 to 61 percent (table 2.2).

The reasons for these striking inconsistencies between attitudes and behavior are varied and complex. They are partly traceable to circumstances. The 1973-1974 period was that of the Arab oil embargo and stagflation. This was the first time in the postwar era that Americans became aware that shortage and serious economic instability might be possible. In this same period, millions of Americans began to adjust their behavior to a new psychology of limits and avoidance of waste. But then, as if by magic, the long lines disappeared from the gas stations and television news broadcasts began reporting stories about a glut on the world oil market. Everything suddenly seemed to return to normal with no shortages—except that prices were higher and the oil companies reported higher profits. This experience made Americans believe that they had been tricked, and they quickly fell back to their old habits—almost.

We say almost because, while the reduced level of consumption that prevailed during the gas shortage and the recession has not been maintained, neither have Americans reverted to the unrestrained consumption of the preshortage period. The public opinion data does show that the impact of the shortage experience fades somewhat as time passes. But the public has not entirely forgotten what it went through when the gas supply dried up. Sixty percent still drive more slowly; 61 percent still keep their cars longer; 36 percent still think twice before using their cars for nonessential purposes (or at least claim to do so) (table 2.3). It is not unreasonable to expect that the cumulative effect of successive and occasionally simultaneous scarcities of gasoline, heating fuel, meat, coffee, bread, and other staples will reinforce the conserving pattern established during the embargo of 1973-1974.

Apart from this episode, we must also keep in mind the fact that the 1970s have been dominated by changing social values moving toward an increasing emphasis on self-fulfillment, self-actualization, and self-expression. Even though the new *economic* signals suggest some degree of instability, American *culture* has long been dominated by a growing psychology of entitlement and a greater emphasis on the freedom of the individual. Throughout the 1970s, Americans have come to question the need to make sacrifices that other generations of Americans had found to be necessary on practical grounds and desirable on moral grounds. One result of the new psychology of entitlement has been to catch Americans in a strange cross current of pressures between cultural trends that tell the individual "you have a right to a greater freedom of choice" and economic trends that signal the individual American that "the good times may be coming to an end." Caught between these cross pressures, Americans are not clear what they should believe. Until the confusion is dissipated, we can expect the pattern of inconsistency and contradiction between attitudes and behavior to persist.

The Role of Technology. As Americans struggle to sort out these confusing cross currents, several growth-related issues have moved to the center of the public agenda. One relates to what the role of technology should be. In general, public attitudes toward technology are favorable. For example, a whopping 78 percent of the public report in a recent Yankelovich, Skelly, and White study that they favor greater emphasis on technology to solve major health problems. A 1979 Roper Report shows that 60 percent of the public believe life to be better today than 60 years ago because of advanced technology, whatever its drawbacks. Yankelovich, Skelly, and White show an impressive 77 percent of the public who believe that solar energy will play a major role in solving the energy problem of the future, followed by a 53 percent majority who have faith in technologies revolving around better home insulation. Forty-five percent endorse technologies that will make automobiles more efficient. And much lower levels of the public have confidence in technologies associated with geothermal energy (15 percent), shale oil (13 percent), liquefied natural gas (10 percent) (table 2.9).

Overall, Yankelovich, Skelly, and White show that a narrow majority of the public (52 percent) endorse the view that "technology will find a way of solving the problem of shortages and natural resources" (table 2.9). Since the problem of shortages is of recent origin, trend data on the ability of technology to solve the shortage problem does not exist. Yet, by inference, it appears likely that this bare majority vote of confidence in a technological solution to the shortage problem represents a sharp falling off from the near universal and unqualified confidence in technology of the post-World War II period.

The slippage, moreover, comes from the best educated and youngest segments of the population—a finding of great importance. In virtually all of the other survey findings, reported demographic differences among various subgroups in the population were not of major significance. But, when it comes to confidence in technology to solve our natural resources problems, the country is sharply divided. The 52 percent level of support masks a deep split in faith in technology. Generally speaking, confidence in technology to solve our energy/natural resources problem diminishes with education and is stronger among the old than among the young. Among those 50 years and older, confidence in technology exists at the 62 percent level. Among those 34 years and younger, confidence falls to the 45 percent level. The split along education lines is even sharper. Among young people (18 to 24 years old) only 29 percent of those attending college are confident that technology will find a way to solve our shortages problem compared to almost twice this number (48 percent) among young people not attending college. The range of confidence in technology is, therefore, exceedingly wide—wider than in any other finding thus far reviewed. It ranges from a 69 percent level of confidence among the older, less well educated lower income segments of the population to a paltry 29 percent level of confidence among college students (table 2.12). Interpretations of why confidence in technology should vary so sharply by age and education should be advanced with caution. Disillusionment with technology among educated youth may reflect either a more sophisticated level of knowledge or new cultural values associated with the counterculture or the so-called "new class" described by Herman Kahn, Irving Kristol, and others.

It is impossible to know which factor is the more important one. We suspect that both contribute. College youth are more aware of the unwanted side effects of technology on the environment than older, less well educated segments of the public. At the same time, a certain antitechnology bias has come to be associated with the environmental movement, and this bias is widespread among educated youth. But whether due to insight or prejudice or both combined, skepticism about technology is likely to spread. Almost invariably, the young and well educated anticipate attitudes that spread to the larger society.

The Role of Government and Government Spending. The profile of public attitudes toward government is also a current focus of the "working through" process. One of the most fundamental changes in American life over the past several decades has been the decline of confidence in government. The decline has been swift, sharp, and all-encompassing.

- In 1964, a 69 percent majority of the American public had faith in the competence of government officials ("They know what they are doing"); by 1976, the number of Americans holding this view dropped to 44 percent; by 1978 it had dropped further to 40 percent.

- At the end of the 1950s, a 56 percent majority of the public expressed the view that "you can trust the government in Washington to do what is right most of the time." Two decades later, their level of trust had been cut virtually in half to 29 percent.

- In the mid–1960s, by a two to one margin (64 to 28 percent), Americans believed that the government was run for all the people rather than for the benefit of a few big interests. By 1978, this pattern had completely reversed itself by a comparable two to one margin (65 to 23 percent). Americans came to believe that government is run for the benefit of a few big interests rather than for all the people.

- In 1960, a 71 percent majority of the public rejected the statement "public officials don't care much what people like me think." By the end of the 1970s, only a minority of the public found itself able to reject this statement.

- The number of Americans believing that the Federal Government is "getting too powerful for the good of the country" has increased over approximately a quarter of a century from a 42 percent minority to a 68 percent majority (table 2.10).

The public's case against the government is multifaceted, but no facet of it is more prominent and distressing to the public than the belief that "the government wastes a lot of money we pay in taxes." This belief has increased from 42 percent in 1958 to 77 percent in 1978. Between 1976 and 1979, Harris shows the number of Americans favoring a major cutback in Federal Government spending increasing from 62 to 69 percent (table 2.8).

That the government wastes money is indisputable. But the public has grasped the waste issue with an enthusiasm that strongly suggests a large dose of wishful thinking. A two–thirds majority (69 percent) believes it is possible to cut taxes without decreasing services, if the government becomes more efficient and less wasteful. An NBC Poll found in 1979 that a 71 percent majority agrees that "the Federal budget could be balanced just by reducing waste and inefficiency" (table 2.8).

Americans believe fervently in the waste-in-government theory because reducing waste is seen as a method for reducing taxes without reducing favored government programs. According to a 1979 Harris survey, a majority would oppose any major cutback in Federal Government spending if it meant cutting back on help for the elderly, handicapped, and poor (78 percent); for health (75 percent); for education (73 percent); for defense (62 percent); for environmental protection (57 percent); and for help for the unemployed (54 percent). A Yankelovich, Skelly, and White survey shows that an overwhelming majority of the public believe it is the government's responsibility rather than anyone else's to "see that senior citizens get the health care they need (84 percent) and that products are safe (79 percent)." Yankelovich, Skelly, and White also show that large segments of the public believe that the government is spending too little on fighting crime (59 percent); health care (51 percent); and education (45 percent). The only place where people believe the government may be spending too much is on foreign aid (72 percent) and welfare and care for the poor (51 percent) (tables 2.7 and 2.8).

Here, once again, we see a familiar pattern reasserting itself. The public focuses on a real and legitimate issue—that of government waste. There is no doubt about the reality of this issue. But, almost surely, people are grasping at straws when they say that solely by cutting back on waste we can balance the Federal budget, reduce taxes, and maintain and increase the level of government services. We see in this public attitude toward government waste valid grounds for complaint, a political issue of considerable force, and also a not insignificant amount of wishful thinking.

The Moral Issue. The most difficult—and perhaps most important—element of the "working through" process is what weight to give to the view that shortages and limits may be morally "good for you" because they discourage waste, encourage efficiency, and lead people to live simpler, more frugal, less materialistic lives.

- The majority of Americans (55 percent) state that "doing without something and living a more austere life would be a good thing."

- By 79 to 17 percent, Americans think more stress should be placed on "teaching people to live with basic essentials" than on "reaching a higher standard of living."

- By a similar margin, (76 to 17 percent), Americans would place greater stress on learning to get pleasure out of nonmaterial experiences than on satisfying "our need for more goods and services."

- By 63 to 29 percent, Americans would stress learning to appreciate more human, less materialistic values versus finding ways to create more jobs for producing goods.

- By 59 to 26 percent, Americans would stress controlling inflation by buying less of the products in short supply rather than by attempting to produce more goods.

- Sixty-six percent of the public endorsed the view: "I'm not that unhappy about the possibility of shortages because I know it will encourage me to use everything efficiently and not wastefully[11] (table 2.3).

As in the responses to energy, technology, and government waste, we perceive a familiar pattern. Americans speak enthusiastically about the moral benefits of the simple, nonmaterialistic life, but they have yet to fully incorporate these benefits either in their day to day behavior or in their practical thinking and planning for the future.

II
TOWARD RESOLUTION

We have described the ambivalence and uncertainty Americans are experiencing as they try individually and collectively to confront the issue of growth. A certain amount of anxiety may be inevitable as people attempt to puzzle out their response to a problem as complex, poorly understood, and all-encompassing as growth. As Etzioni points out, a sense of divided and uncertain purpose in the economic domain creates stress in almost every sphere of life.

> What the resulting strains agitate for is not a neat monolithic pattern—which never existed anyhow, not even at the height of the industrial project—but for priorization: so young persons can know more clearly what is expected of them . . . so the community and its leaders know what to extol . . . and so authorities know what standards to uphold.[12]

He argues, then, that "from an economic and social-psychic viewpoint, the present fairly high level of ambivalence and lack of clear priority needs to give way over the next few years to either a decade of rededication to the industrial, mass-consumption society, or a clearer commitment to a slow–growth, quality-of-life society. *In the long run, high ambivalence is too stressful for societies to endure*" (emphasis added).[13]

Yet, the probabilities are that a high level of ambivalence is likely to persist for a long period before the public begins to order its national priorities. Given the present instability—the new economic pessimism, mistrust of business, anger against government incompetence and wastefulness, confusion about the sources of economic distress, the conflict between the cultural emphasis on entitlements and the limitations imposed by current economic conditions—it is possible that the feelings of disillusionment, frustration, and betrayal will harden, that the divisions between the public and the economic and political leadership groups will widen, and that the perception of inequity in the allocation of scarce resources and unfairness in bearing the burden of inflation will destroy all sense of a common social and economic purpose.

The Key Concern: Inflation

A look back at the last six years forces the observer to raise grave questions about the future. Before America felt the impact of scarcities and before it was shaken by recession, its first priority was expansion of the economic product. During and immediately after these dislocations, the public seemed ready to accept a reduction in consumption and tried to apply certain conserving principles to everyday life. But the urgency of uniting to defeat a common enemy diminished as the memory of hard times receded. The pattern today alternates between breathless consumption, the buy-before-the-price-goes-up syndrome, and an awareness of limits to scarce resources and to the pleasures of materialism.

The temptation is to go along with the currently fashionable pessimism which foresees sharper conflicts and deeper divisions in the next two decades. We don't believe that is the only or inevitable course open to Western industrial countries. As we suggested earlier in this chapter, the public appears to be moving, slowly and unevenly to be sure, toward accepting a hard bargain—a trade-off of high consumption in exchange for a period of relative economic stability and the preservation of the material gains they have achieved in recent years, if this is possible. While the evidence is not conclusive, there are indications that the public may reduce the consumption of certain nonessential goods and services and practice moderate restraint in the acquisition and use of scarce materials.

In return, however, the public demands social and economic policies that will cushion the individual against severe shock and instability. When Americans are asked in national polls to describe those elements they believe are most disruptive to a stable life, they include a scarcity of essential resources; mounting personal debt; unjustified corporate profits; the high cost of medical treatment; hazardous working conditions; deterioration of the environment; excessive and wasteful government spending; and the dangers posed by certain forms of technology, particularly nuclear energy.

But the one concern that always tops the list is inflation. Summarizing the public's economic concerns in 1978, Yankelovich, Skelly, and White reported: "Inflation is now a given in our society—we don't ask whether there will be inflation, we ask how much. . . ." After a decade of rising inflation rates, most Americans don't believe they can win the battle against inflation. The most they hope for is to come out even. The defensive psychological attitude that now permeates American society is summed up vividly by a 32-year-old black electrical engineer in California. Together, he and his wife earn more than $45,000 a year. That income puts them in the top ten percent of American families.

"It's not possible to get ahead of inflation," he told *The New York Times*. "Fifty thousand today just won't be $50,000 tomorrow. It's a battle just to stay where you are. Things only get higher."[14] The economic record of the 1970s supports his analysis. In the last decade, average income increased at a much slower rate than it did in the 1950s and 1960s, and even these gains were eroded by inflation. When inflation is factored in, the 1977 income level was $423 below what it was in 1973, before the recession hit.

In 1978, Harris offered the public two choices related to growth. Nine percent of the respondents said that they would prefer an expanding economy that offered the possibility of a higher individual income and standard of living, but which would be accompanied by rising inflation and periodic recessions. But *81 percent* said they would prefer an economy with little growth which would allow only a slow but steady increase in the standard of living and income, but which would also mean reduced inflation and few recessions.

The public's anxiety over inflation is so high that no new economic program can succeed without addressing it. If America is, indeed, moving toward a resolution of the present ambivalence, moving toward a consensus which emphasizes moderate growth and restraints on consumption, it must also succeed in restraining inflation. This presents planners with a difficult problem. Inflation was at its lowest levels during the period of most rapid economic growth. In the 1970s, when productivity and economic expansion declined, inflation soared. If the past is any guide, a shift in national emphasis toward moderate growth and reduced consumption may not result in easing inflation.

Nevertheless, the public believes (or hopes) it will. In a number of recent national surveys, the public supports measures that will reduce consumption for the sake of reducing inflation, including: restrictions that would make it harder for people to get mortgages (51 percent); mandatory wage and price controls (73 percent); imposing a ceiling on housing prices to prevent annual increases in cost (75 percent); and restricting the use of credit cards (58 percent).

The precise mixture of growth and restraint which will reduce the impact of inflation has yet to be determined. This and many other issues will not be resolved until the country moves further along the "working through" process and decides what its priorities are. The question is, then, how far along is the public toward resolution of the conflicts which stand in the way of a national consensus on growth. What preconditions have to be met before a new emphasis takes shape? Under what circumstances will the public respond to an appeal for restraint? We must attempt to answer these questions before we can discuss the specific terms of the new consensus which may emerge after the thinking-through process is completed.

Preconditions for Resolution

We believe that there are three overall preconditions for forming a new consensus. Together, they provide the basis for reaching agreement on very different national priorities.

The first, and perhaps most important, is that the country has to *feel* the necessity for making hard choices. These choices cannot be abstract or theoretical. People must believe that their decisions will have a direct, immediate, and significant effect on their lives. Practically and emotionally, the public has to recognize that it cannot postpone or deflect these choices. The continuing experience of scarcities and, particularly, the effects of inflation have brought the public almost to the point where it is prepared to choose between conflicting themes in American society.

The second precondition is effective leadership to guide the public toward resolution. The way leaders manage events and conditions often shapes the public response. It is incumbent upon the national leadership to explain, in compelling terms, why the choices are necessary and to chart the new directions. Policymakers and others in leadership positions must encourage a comprehensive discussion by constituents of the various alternatives. If sacrifices are called for, the leadership will have to present solid evidence that the burden will fall equally on all groups in society.

The last precondition is that the public be given the opportunity to confront and think through the real choices. In the current climate of mistrust, people often sense that they have lost control over their lives. They feel more like subjects than citizens. On most issues, the public is content to turn over the responsibility for policy decisions to its leaders. But some decisions are so important, so central to the security and stability of the average citizen, that the public must be given the time to challenge and probe and question the proposal that is being advanced. The question of growth is so complex, technical, and all-inclusive that leadership will be tempted to parcel out different aspects of the problem to technicians, experts, and special interest groups. The temptation should be resisted. If the public feels it is being excluded from the decision making process, it will grow ever more frustrated, angry, mistrustful, and cynical. Under these conditions, the opportunity for compromise, civility, and positive politics will disappear; and, rather than achieving resolution, the country will remain mired in confusion and disorder.

We believe that resolution is possible if average Americans feel they are part of the decision making continuum, if they feel that their point of view is heard and responded to, if they feel that their fundamental values are respected. A unilateral decision by leadership is not going to stick. If they are being asked to play, citizens must have a chance to make the rules. They will not accept someone else's judgment that they, the public, have to sacrifice. They have to reach this conclusion for themselves.

In the recent past, we have seen situations where these preconditions were met and the public responded by carefully considering the choices open to it, supported the policies proposed by thoughtful and responsible leadership, and accepted the need to make certain sacrifices in the interest of the general welfare. We have also seen other situations where these preconditions were absent and the public was unable to respond in a coherent manner. What follows are two brief illustrations of how the presence (or absence) of these preconditions affects the public's reaction to growth-related issues.

Scarcity: The Social Limits to Growth

In his book, *The Twenty-Ninth Day*, Lester Brown has written:

> While coming to terms with nature we must also come to terms with ourselves, frankly recognizing our own limitations as we attempt to stabilize our relationship with nature. While it is tempting to couch discussion of growth prospects in physical terms because they are measurable, the more severe constraints are invariably human ones—the limits on the human capacity to change, the slowness and occasional irrationality of political processes, and the glacial pace at which institutions adapt.[15]

When these human restraints are discussed, the oil shortage is usually offered as a case in point. Less often cited is the coffee shortage, but it provides a telling illustration of how inadequate explanation, conflicting judgments, general suspicion, and perceived inequities influenced the public's reaction negatively.

In 1975, a severe frost damaged or destroyed half of the coffee crop in Brazil. Coffee production dropped from 22 million bags in 1975-1976 to 6.4 million bags in 1977. Coincidentally, a civil war in Angola, drought and earthquakes in Central America, and political strife in Uganda all diminished the world's supply of coffee.

When the shortage developed and prices doubled, various public representatives urged a consumer boycott. In New York, for example, Elinor Guggenheimer, the Commissioner of Consumer Affairs, asked the public to reduce its coffee consumption by half. Despite these appeals, there was no widespread participation in a boycott, and no significant adjustment in life–style to meet the shortage. One would have thought that a combination of high prices and shortages would have significantly reduced the demand for a product that is considerably less essential to the public welfare than gasoline or heating fuel.

One reason the public did not limit its consumption of coffee was that it had serious doubts that a real shortage existed. And these doubts were reinforced by the comments of various analysts and observers. A *New York Times* columnist concluded that the shortage was being used to manipulate prices. He contended that Brazil, facing billion dollar trade deficits, had decided to pass the costs on to the American consumer.[16] In Congress, Rep. Frederick Richmond, a member of the House Agriculture Committee, said, "This is a crisis dreamed up by the coffee exporting nations to gouge the American consumer."[17] On its editorial page, the *New York Times* said the belief in a conspiracy theory was "an unfortunate fringe effect of the oil price increase . . . it has given devil theories a good name."[18] The *New York Times* opposed a boycott, but it did not discuss the broader question of the need to limit consumption of scarce resources.

There are other explanations for the public's unwillingness to respond. The source of the shortage was remote. All the public knew about it was what it read in the newspapers or saw on television. When a consumer walked into a supermarket, he saw cans of coffee on the shelves. The price was much higher but, if you were willing to pay for it, it was available. The shortage of coffee on the world markets never became a visceral reality for the public. And that reality was made even more intangible by the conflicting policy statements and analyses. Finally, there was the feeling that the burden of reducing coffee consumption did not fall equally on the entire public. It fell primarily on people who had a hard time paying three dollars a pound. Those who could afford it could have as much as they desired. Why, the public asked, should people of modest means be asked to conserve—and nobody else? In the end the American public did not kick the coffee habit: consumption remained high and eventually prices went down, though not to the level which existed before the shortages.

In Marin County, California, and nearby San Francisco, the public did learn to slake its thrist for a commodity more basic than coffee—water. In 1977, after a two-year drought in this area of some 500 square miles populated by more than 12 million, the reservoirs contained less than one-quarter of their normal amount of water. In 1976, Marin County imposed a restrictive rationing program designed to reduce consumption by 25 percent. Soon after, the city of San Francisco instituted a water rationing program. During the drought period, Marin residents reduced their consumption by 65 percent: from a normal use of 30 to 33 million gallons a day, consumption declined to 8.5 million gallons. What is even more impressive is that 15 months after the drought ended, consumption levels remained 25 percent below what they were before rationing was introduced.[19]

In San Francisco, residents were required to reduce consumption by 25 percent. The residents responded by decreasing water consumption by 35 percent. Currently, although there is no water shortage, consumption is 20 percent less than what it was before the drought.

The drought in northern California seemed to generate a conserving attitude throughout the state. In May 1977, a statewide poll conducted by Field Enterprises[20] reported that 93 percent of the residents in the state were voluntarily cutting their water consumption, and that 58 percent of all residents favored statewide rationing. The San Francisco *Chronicle* editorialized that "the spirit of sacrifice, exemplified in Marin County, is apparently prevailing throughout the state."[21]

What evoked this response? Why did Californians adhere to the mandatory rationing plan, and why did they voluntarily continue to limit their water consumption? Unlike other shortages of oil, wheat, and coffee, the drought affected almost all residents equally. No one stood to gain by the drought. In addition, there was no doubt or suspicion that this was an authentic shortage; the evidence was there in cloudless skies and brown lawns. But the most important reason may be that a just and effective policy was presented to the public after a period of serious and thoughtful discussion.

When the effects of the drought began to be felt, Marin County proposed a rationing plan that would have banned all nonessential water use, such as sprinkling lawns or washing trailers, and raised water rates substantially. The plan was sharply criticized by consumers who objected to being told how to use water. Moreover, consumers recognized that the plan would be difficult to enforce because it required constant police patrols to identify offenders. The plan was revised, so that individuals and families could use their water allotment however they pleased. People could decide whether they wanted to use the water for their washing machines or for a bath. Dorothy Hughes, Director of the Marin County Mental Health Center, later noted, "People didn't feel powerless. They were given choices. They could determine how they were going to live their lives within limits. People felt after a while that it was almost a relief to pitch in and help."

Before the new plan was implemented, J. Dietrich Stroeh, the Marin County Water Director, allowed a two-month grace period. In that time, he conducted numerous meetings, met with state and federal officials to request aid (which he later received), negotiated with affected business and interest groups, and assigned specific responsibilities to his staff which were disclosed to the public. "Call it leadership, ego tripping, or whatever, somebody's got to get moving," he said later. "The people here cared about their community. In fact, the life of the community was at stake. Once I was able to communicate what the stakes were, people responded. They responded tremendously."[22]

The conservation program worked because a number of elements came together: civic pride and concern, creative leadership and an obvious self-interest for all of the residents. People knew this was a real shortage which had not been contrived for the gain of any individual or group. It continues to work—people are still conserving—because residents want to guard against future shortages. What may be more significant for the future development of a conserving ethic in the United States is that, in San Francisco and Marin County, people learned that they could get along with less, that their lives would remain essentially intact, and that what they were doing would advance their own welfare and the general good.

What the water drinkers were offered—and the coffee drinkers were not—was a reasonable alternative to high consumption, an alternative that was sensible, workable, and fair. The presence of an alternative is a basic precondition for developing a consensus in support of a new approach to growth. In this regard, Economist Burkhard Strumpel writes:

> In order to effect behavioral changes away from expenditures on material goods there must be a foundation for an alternative life-style; smaller cars, public transportation, durable consumer goods, effective health care and income security must be within reach of the average consumer. There must be also financial incentives for the change

such as price reductions. And finally there must be changes in belief systems and ideologies, a new conception of the merit and role of alternative styles of consumption. . . .

It seems that publicly emphasizing or dramatizing the common challenge, making provisions for alternative consumption goods and patterns, and creating or preserving a climate of equity and solidarity, would enable government and other institutions to foster effective changes in consumer tastes and demands.[23]

In other words, if the public is asked to make a fundamental shift in outlook and behavior, it must be offered a real alternative—a choice.

Alternative Models

Assuming that the preconditions for change are present, in what direction is the public likely to go? Because the public has just begun to conceptualize the choices, it would be premature to say that the direction is certain. But, at this early stage, various ideological forms can be dimly discerned. It is beyond our purview to describe them in depth, but they can be summarized as follows:

- *A period of austerity and sacrifice during which the country committed its material resources to the rebuilding of the economic infrastructure.* This is the recapitalization priority described by Etzioni. For many reasons, it seems doubtful that the public is prepared to undergo an extended period of self-denial to reconstruct the high-consumption, heavily materialistic environment of the 1950s, even though the improvements in material well-being made in that period are widely recognized.

- *An emphasis on government intervention and social supports.* The public will not eliminate programs that it feels entitled to, such as social security, unemployment insurance, veterans' benefits, and medicare. At the same time, people are in a "take back" mood, as exemplified by the Proposition 13 movement. Public support for foreign aid and assistance to the poor, which has never been high, has dropped even more sharply in recent years. There are other publicly supported programs in education, assistance for the elderly, job training, and health, about which the public is uncertain. These programs will probably not be eliminated; but the level of support for them may not increase substantially. The liberal activist consensus, which provided the basis for government intervention through the 1960s, has splintered. It is unlikely, then, that the public would back a wider role for government in social welfare.

- *A society built around the principles of "voluntary simplicity."* There is a small group of Americans who have been influenced by some of the countercultural innovations of the 1960s. While they tend to resist conventional political labels, many are active in consumerism and the environmental movement. They are deeply committed to a simplified life-style which rejects materialistic values and seeks to limit possessions to a bare minimum. While some aspects of their philosophy appeal to the general public, such as their concern for physical well-being, it is unlikely that the majority of the public will voluntarily embrace an outlook which rejects many of the conveniences and comforts offered by the most technologically advanced country in the world.

● *A more conserving society with a greater balance than now exists between consumption and nonmaterialistic values.* We believe that the public opinion data presented thus far, and in Section III below, indicate growing support for this form of resolution, although no firm consensus has developed yet. The movement toward a more conserving society is accelerated by three significant psychosocial developments: the heightened emphasis on economic security, lowered economic expectations, and new values. All three may be converging to create a new fusion of the search for economic stability with the increasingly powerful impulse to claim control over one's own destiny.

The New Values

Starting in the late 1960s and gathering momentum in the 1970s, Americans began to change their philosophy of life—their sense of what is important and what isn't. The essence of this change is that, in the past, people were motivated mainly by earning more money, adding to possessions, gaining economic security, and providing material comforts for the family. Increasingly, throughout the 1970s, new self-fulfillment motives have gained in importance. Moreover, the traditional material incentives and the new self-fulfillment ones have moved in somewhat separate directions. People still want material rewards but they no longer feel that it is necessary to give so much of themselves to achieve them. Also, they often see their self-fulfillment as something different from material success, and of equal or greater value. So they want self-fulfillment *in addition* to financial security. To oversimplify a little, people want more but are willing to give less for it. They desire more personal freedom, more time off, more self-expression, more flexibility, more of a say of how things are done, more variety, more opportunity. What they want is more of everything. In essence, they want to assert ever-greater control over how they live their lives. Money and status still count, but they are not as powerful or universal as they used to be. Increasingly, Americans want something different out of life than in the past.

Many of the new values have their roots in the social movements that began in the 1960s, such as the student movement and the women's movement. Most of the student-generated ideas of the 1960s, centering on political radicalism, the drug culture, and the counterculture, have been rejected by the public. But at least three ideas have been carried over to the present and have been appropriated by a majority of Americans.

- First is the growing conviction that what is regarded as a "nose-to-the-grindstone" way of life, with its hard work, its unquestioning loyalty to employers, and its suppression of desires that conflict with obligations to others, is too high a price to pay for material success. Besides, many people have come to believe that it may not be necessary to make such sacrifices.

- Second is the feeling that we have devoted too much of our time and attention in this country to the task of how to make a living and not enough to the question of how to live; hence, the current preoccupation with finding a "lifestyle" that precisely expresses each person's unique individuality.

- Last is the belief that what counts most in life is that "I keep growing" as an individual, that "I have the opportunity to fulfill my potentials," and that "I have a moral obligation to myself to do so." This is a startling new conception of moral obligation.

These attitudes and values represent a large shift in the traditional American ethos. Most people appear unsure of these new beliefs and are torn between the old philosophy and the new one. But despite their uneasiness and anxiety, the public psychology continues to change.

The new values are now merging with the fear of economic instability. A new synthesis is forming, one that meshes three elements: the pursuit of economic stability (even at the cost of reduced consumption) with more modest material expectations and with the drive to establish maximum control over one's own life and destiny. The combination of these three elements creates a novel American outlook.

The core fantasy today is not the pot at the end of the rainbow or the dream of sudden wealth; if the price of a lifetime of shoving, struggling, and competing—even if it results in your becoming the most powerful man in the sales division or the wealthiest man in the neighborhood—is a nervous breakdown at 35 or a heart attack at 40, it's not worth it. The drive to reach the top has been replaced by the need to keep one's life on a relatively even keel. Success is defined in relative terms: achieving a relative degree of commitment to work and career.

The wealth-at-any-cost fever has given way to the more palliative "stay well and be happy." In other words, have a *nice* day. This theme of a relatively decent, happy, nice existence is the new connection between social and economic classes. When the shared goal is a life that has diversity and excitement but which does not career wildly between failure and success, between sudden catastrophe and brief spurts of fulfillment, then the different groups in the middle socioeconomic range seem to have much in common. They all want job security; they all want an income whose real dollar value increases, however modestly; they want to preserve those material gains they have worked for and which represent an improvement over what their parents and grandparents had, and they want to keep those social supports which they have come to regard as entitlements. (At the same time, they are reluctant to keep paying for the support of groups who do not share their values; for example, they don't want to reward anybody for *not* working.) They are choosing security, stability, and control over risk, adventure, and expansiveness.

One final example will help to illustrate how this new mood and ethic expresses itself. Earlier, we discussed the ambivalent attitudes Americans now hold toward technology. When, however, the uses of technology shift toward more qualitative goals, much of the ambivalence disappears. Once Americans looked to technology to speed up everyday life, to improve efficiency, and to streamline the communications process. Now people emphasize other uses for technology. Rather than applying technology to accelerate travel, 88 percent of the respondents in a 1977 Harris poll say they would prefer that scientific and technical research be used to improve existing modes of travel. Instead of improving and accelerating mass communications, 77 percent of the public would prefer to develop a communications system "that allows people to spend more time getting to know each other better as human beings." In the past, the majority of Americans believed that the bigger and more efficient a system was—whether the system involved communications or travel or information—the better it was. Now they are not so sure. By a majority of 66 percent to 22 percent, Americans say they would rather "break up big things and get back to more humanized living through technology" than to develop bigger and more efficient ways of doing things. Efficiency as a goal in itself has grown problematic.

These themes—living on a human scale, devoting time to interpersonal relationships, improving existing facilities rather than breaking new ground—are repeated on many levels of the growth debate. They underlie much of the current discussion about the role of government and business in the lives of individual citizens, the emphasis on community and family, the desire to restore close personal relationships that may have been threatened by a preoccupation with career and social mobility, and the interest in living more simply.

As a practical matter, we do not believe that America is about to enter into a grand debate on the significance of materialism, the application of the new values, or the desirable rate of growth in the last decades of the twentieth century. What is more likely is a continuing discussion of many different proposals, programs, and strategies presented at all levels of government and business, from town hall to Congress, from the executive boardroom to the company cafeteria. Whether these hundreds or perhaps thousands of proposals will collectively express the kind of dynamic consensus America enjoyed in the postwar period will depend on how consonant they are with the public's emerging value priorities and with the public's ability to confront painful choices—and make them.

Opinion survey data show that, at the present time, the public remains somewhat unrealistic about all that it wants. But the data also suggest that, at a deeper level of consciousness, Americans have begun to reconcile themselves to the need to accept greater, and different, limitations than in the past. The stage is set, therefore, for the next phases of "working through" as the nation moves toward some resolution of what it has to give up and what it insists upon retaining.

By way of conclusions, therefore, we summarize a brief profile of the priorities of the American public.

1. The public wants to retain the gains it won in the affluent years. It seems unwilling to sacrifice these in an effort to regenerate the economic machine or to expand vastly the existing system of social supports.

2. The public has little taste for self-denial and austerity for the purpose of adding new consumer products to the market or increasing efficiency. What sacrifices it is prepared to make will be directed toward maintaining economic stability and a reliable and fair supply of essentials (e.g., gasoline), even if the amount is reduced.

3. Americans want to strike a new balance between hard work and leisure. They are prepared to slow down their pursuit of luxuries if that gives them the freedom to explore the possibilities of self-fulfillment and allows them a measure of economic stability.

4. Americans appear to want to halt the expansion of certain government services if this will reduce their taxes and protect their privacy. They do want, however, to retain those services that they consider entitlements, such as social security and pension benefits. And they would like to be protected against the high costs of catastrophic illness.

5. The public has clearcut priorities about what forms of consumption it is prepared to give up and what it considers essential to its well-being and freedom. People don't want to give up their family cars, their central heating, their own homes, or their washing machines. They are prepared, however, to make modest cutbacks in the use of energy, to keep their old model cars longer, to waste less, to reduce consumption of meat and clothing, and to reduce their use of items that can't be recycled.

6. A new antiwaste morality is gaining momentum. The public is unlikely to support any government program that it regards as potentially wasteful, and

there is support for regulations that curb wasteful practices by the public itself. This attitude extends to the consumption of products that are in relatively short supply. So the public would go along with such measures as closing gas stations on Sunday to limit consumption and restricting the use of credit cards. It should be pointed out that, in the public's mind, the rising rate of inflation is fused with wastefulness. The public would support higher taxes on foreign-made "luxury" goods, such as color television sets, in the hope that it would reduce inflation and discourage the purchase of nonessential products.

There is no guarantee that these kinds of trade-offs—the bargains the public is willing to make—will bring people what they want. Americans are understandably reluctant to give up some things that they regard as essential to a comfortable life. They are just beginning to suspect that such choices may lie in the offing, which is an unnerving prospect. The result is a profound ambivalence and conflict. No one can state for certain how this conflict will be resolved, but we believe that Americans will not wish to turn the clock back to the great period of dynamic growth in the two decades following World War II. For all of its attractions and accomplishments—it was in many ways a golden age for America—there is a deep-seated conviction in the public that we overdid the materialist thrust of a consumer society. America now yearns for a more balanced life-style in which the needs of the spirit (including the demand for morally meaningful sacrifice) will be in better balance with materialist aspirations. Whatever the future may be, for social, psychological, as well as practical economic reasons, it will not recapture the past.

III.
ATTITUDES TOWARD GROWTH—A SURVEY OF SURVEYS

The tables that follow present data from the major U.S. national polling organizations, in quantitative and comparative detail. Sources of data included the Gallup Organization, The Roper Organization, Louis Harris and Associates, the Survey Research Center of the University of Michigan, the NBC News Poll, the CBS/*New York Times* Poll, American Council of Life Insurance (MAP—Monitoring Attitudes of the Public), National Opinion Research Center of the University of Chicago, and Yankelovich, Skelly, and White.

Our interpretation of these data has been summarized in the first two sections of this chapter. Readers are welcome—indeed, encouraged—to formulate their own.

Table 2.1. Perceptions of the Economic Future

Public Attitude	Source	Period of Time	Percent/Change
1. Feel *no real* confidence in the future prosperity of the country.	Yankelovich, Skelly, & White	Feb. 1975-April 1979	14%-23%
2. Believe that the condition of the American economy during the next five years will:	Yankelovich, Skelly, & White	1979	
Improve			32%
Worsen			44%
Remain the same			34%
3. Say they are *dissatisfied* with the future facing them and their family.	Gallup Surveys	1971-78	30%-51%
4. Think that this year will be not as good as last year for the country.	Roper Report 79-1	Oct. 1971-Dec. 1978	12%-31%
5. Think that this year will be a year of economic *difficulty*.	Gallup Surveys	1969-79	48%-69%
6. Think that next year will be *worse* than this year.	Gallup Surveys	1972-79	21%-55%
7. Believe the country *will be* in a recession a year from now.	Harris Surveys	April 1976-April 1979	46%-61%
8. There is likely to be a recession in the country in the next five years. (*Agree*)	Yankelovich, Skelly, & White	1979	70%
9. There is likely to be a depression in the country in the next five years. (*Agree*)	Yankelovich, Skelly, & White	1979	50%
10. Expect the economy during the next ten years to grow more *slowly* than it did during the 1950s and 1960s—*or not at all*.	Scarcities Study (Yankelovich)	1975-77	66%

Table 2.2. Inflation

Public Attitude	Source	Period of Time	Percent/Change
PERCEPTIONS OF INFLATION			
1. Expect prices in a year to be rising *more rapidly* than now.	Harris Surveys	Nov. 1974-Jan. 1979	30%-52%
2. Expect income will go up *less* than prices.	Survey of Consumer Sentiment (Survey Research Center, U. of Michigan)	Nov. 1976-Nov. 1978	35%-45%
3. Name inflation as one of the most serious problems facing the country that particularly worries them (Open-ended question).	Yankelovich, Skelly, & White	March 1976-April 1979	46%-61%
4. Two choices:	Harris Surveys	Oct. 1978	
a. Expanding economy that would give a chance for a higher income and higher standard of living, but would be accompanied by rising inflation and periodic recessions.			9%
b. An economy with little growth, which would allow only a slow but steady increase in standard of living and income, but would also mean little inflation and few recessions.			81%
REACTIONS TO INFLATION			
5. Inflation has reduced my family's standard of living in the last few years (*Agree*)	NBC Polls	Dec. 1978	52%
6. Finding it *harder* to make ends meet in personal budget compared to a year ago.	Harris Surveys	Sept. 1978-April 1979	56%-69%
7. Compared to a year ago have *had to*:	Harris Surveys	Oct. 1978	
a. Put off buying luxuries			70%
b. Put off buying essentials			53%
c. Buy things on credit			30%
d. Borrow money to pay bills			22%

Table 2.2 (cont.) Inflation

Public Attitude	Source	Period of Time	Percent/Change
8. Items will cut back on first if prices continue to rise during next year:	NBC Polls	Dec. 1978	
a. Vacations			46%
b. Savings			21%
c. Clothing			11%
d. Auto expenses			8%
e. Food			5%
9. Worry a *lot* about:	Yankelovich, Skelly,	May 1974-	
a. Saving for the future	& White	April 1979	41%-50%
b. Being able to pay for the upkeep/rent of my house			23%-48%
c. Not being able to keep up with the bills			31%-42%
10. What planning to do in the next few months to cope with the high cost of living:	Roper Surveys	March 1977- Sept. 1978	
a. Being more careful in expenditures for food			59%-69%
b. Being more careful in expenditures for clothing			52%-58%
c. Cutting down on the luxuries of life, nonessentials			60%-56%
d. Putting off buying things for the house			39%-45%
e. Cutting out or dipping into savings			20%-27%
f. Using the car less			23%-26%
g. Supplementing income by working longer hours or getting a second job			18%-23%
h. Putting off household repairs			16%-21%
11. Reported changes in purchasing patterns and life–styles during the past year:	Roper Surveys	May 1974- May 1976	
a. Cut down electricity use			66%-62%
b. Drive at lower speeds			71%-56%
c. Economize on food			51%-51%
d. Economize on clothing			36%-44%
e. Spend more time at home			39%-38%
f. Economize on recreation and entertainment			34%-37%

Table 2.2 (cont.) Inflation

Public Attitude	Source	Period of Time	Percent/Change
g. Cut down on eating out in restaurants			35%-47%
h. Use or repair things you normally would replace			28%-37%
i. Use the car less often			52%-33%
j. None of the above.			6%-9%
12. It's permissible to buy things that I want when I want them. (*Agree*)	Yankelovich, Skelly, & White	1973-77	55%-61%
13. Owing money is a fact of life I must learn to live with. (*Agree*)	Yankelovich, Skelly, & White	1973-77	52%-54%
14. Are deeply/quite concerned about money they owe.	Harris Surveys	Oct. 1978	44%
15. Steps families have taken in order to deal with inflation:	Yankelovich, Skelly, & White	April 1979	
a. Drew upon savings to make ends meet			46%
b. Delayed replacing appliances because they cost too much			39%
c. Cut down on your contributions to charities because you had to			37%
d. Wife has gone to work as a result of cost of living			32%
e. Bought a major appliance or car now because it will cost more later			23%
f. Postponed buying a house or cooperative because you couldn't find one you could afford			22%
g. Head of household has taken a second job in order to increase the family income			21%
h. Took a job you didn't like because it paid more			15%
i. Went into debt in order to pay for college			12%
j. Sold a house because of the profit you could make and put the money into a more expensive house			6%

Table 2.3. Shortages

Public Attitude	Source	Period of Time	Percent/Change,
PERCEPTIONS OF SHORT-AGES			
1. Believe we are entering an age of shortages (vs. short-ages are only temporary until new production techniques are developed that will get us back to normal).	Yankelovich, Skelly, & White	1974-78	40%-62%
2. We are fast coming to a turn-ing point in our history. The "land of plenty" is becom-ing the "land of want." (Agree)	Yankelovich, Skelly, & White	1979	72%
3. It is immoral for this country to consume so much of the world's resources. (Agree)	Yankelovich, Skelly, & White	1979	57%
4. Believe a *rapid depletion of our natural resources* is likely to happen in the coming years (chosen from a list of things likely to happen).	Roper Report 78-4	March 1978	58%
5. Believe the following *will* be serious problems in the year 2000:	Roper Report 79-1	1979	
a. Shortage of energy sup-plies			63%
b. Shortage of water sup-plies			50%
c. Shortage of food			46%
d. Overpopulation			44%
6. Alternative views of the gas-oline and oil shortage:	Roper Report 78-10	May 1974-Nov. 1978	
a. There is a very real short-age and the problem will get worse during the next 5 to 10 years.			21%-31%
b. There is a very real oil shortage but it will be solved in the next year or two.			12%-10%
c. There was a short-term problem, but it has been largely solved and there is no real problem any longer.			8%-9%

Table 2.3 (cont.) Shortages

Public Attitude	Source	Period of Time	Percent/Change
d. There never was any real oil shortage—it was contrived for economic and political reasons.			53%-43%
ATTITUDES TOWARD AUSTERITY			
7. Americans should get used to the fact that our wealth is limited and most of us are not likely to become better off than we are now. (Agree)	Yankelovich, Skelly, & White	1979	62%
8. I'm not that unhappy about the possibilities of shortages, because I know it will encourage me to use everything efficiently and not wastefully. (Agree)	Yankelovich, Skelly, & White	1979	66%
9. Believe that doing without some things and living a more austere life would be a *good* thing.	Roper Surveys	Dec. 1973-July 1976	56%-55%
10. See benefits in cutting back on the way we live.	Scarcities Study (Yankelovich)	1975-77	50%
11. Would be *very/somewhat satisfied* if standard of living were to remain at present level for next 5 to 10 years.	Scarcities Study (Yankelovich)	1975-77	42%
12. Think more stress should be placed on:	Harris Surveys	May 1977	
a. Teaching people to live more with basic essentials			79%
vs. Reaching a higher standard of living			vs. 17%
b. Learning to get our pleasure out of nonmaterial experiences			76%
vs Satisfying our need for more goods and services			vs. 17%
c. Learning to appreciate human values more than material values			63%
vs. Finding ways to create more jobs for producing goods			vs. 29%

Table 2.3 (cont.) Shortages

Public Attitude	Source	Period of Time	Percent/Change
d. Controlling inflation by buying less of the products in short supply			59%
vs.			vs.
Producing more goods.			26%
13. Believe Americans will cut back on their consumption of things over the next 25 years.	Harris Surveys	Sept. 1978	52%
LIFESTYLE ADAPTATIONS TO SHORTAGES			
14. Believe *compulsory* (vs. voluntary) measures need to get Americans to reduce amount of gasoline and fuel oil they use.	Harris Surveys	February 1979	58%
15. Now doing or seriously considering doing:	Yankelovich, Skelly, & White	1974-79	
a. Keeping a car longer before getting a new one			69%-61%
b. Keeping thermostat at a lower temperature			67%-63%
c. Being more careful with clothes so they'll last longer or doing more patching up			60%-48%
d. Saving things such as empty jars, containers, wrapping, etc., that you would normally throw away			55%-43%
e. Repairing rather than replacing broken furniture			52%-44%
f. Using paper towels and tissues more sparingly			48%-36%
g. Reusing kitchen wraps, such as aluminum foil, plastic wrap, etc.			43%-36%
h. Using fewer electric appliances or using them less often			40%-36%
i. Having garage sales rather than throwing things out			28%-33%
j. Using telephones more sparingly			27%-22%
k. Stocking up on food products that are likely to be in short supply			27%-20%

Table 2.3 (cont.) Shortages

Public Attitude	Source	Period of Time	Percent/Change
l. Saving gold and silver coins			16%-15%
m. Using paper plates, cups, etc. more sparingly			23%
n. Buying a car which consumes less gasoline			41%
o. Installing better insulation in home to conserve heat			35%
p. Installing wood or coal–burning stove			13%
q. Using a fireplace heat circulator			12%
r. Installing solar heating			10%
16. Lately I find myself saving things that I ordinarily would have discarded in the past on the chance that I might have a need for them in the future. (Agree)	Yankelovich, Skelly, & White	1979	66%
17. Changes in life style made because of the energy problem:	Yankelovich, Skelly, & White	1974-78	
a. Trying to use less electricity			71%-73%
b. Maintaining a lower temperature at home			70%-73%
c. Driving more slowly on highways			72%-60%
d. Trying to use less gas at home			48%-45%
e. Taking fewer nonessential car trips			61%-36%
f. Economizing on food, clothing			34%
g. Buying a smaller car			20%-29%
h. Installing new insulation			28%
i. Delaying a purchase of a new car			19%-24%
j. Devoting more time to activities at home			24%-18%
k. Using mass transit more often			13%-15%
l. Cancelling plans for major auto trips			32%-13%
m. Using a car pool more often			18%-13%

Table 2.3 (cont.) Shortages

Public Attitude	Source	Period of Time	Percent/Change
n. Delaying major pur-chases			13%
o. Trading in or selling a gas-guzzling car			7%-12%
p. Converting home to a dif-ferent form of heating			7%
q. Moving to a warmer cli-mate			6%
r. No changes			4%-6%
18. Conservation behavior to cut down on use of electricity:	Roper Surveys	Jan. 1974-Feb. 1976	
a. Been more careful about turning off lights			71%-72%
b. Kept house temperature lower than usual			74%-66%
c. Shut off some rooms of the house to avoid heating them			36%-37%
d. Used major appliances less frequently			26%-29%
e. Started using lower watt bulbs in some lamps			20%-28%
f. Weather stripped doors and windows			19%-26%
g. Installed new storm win-dows or doors			10%-17%
h. Used fireplace more			13%-13%
i. None of these			8%- 8%

Table 2.4. National Ladder Ratings: The Future of the Country

"Here is a picture of a ladder. Let's suppose the top of the ladder represents the best possible situation for our country; and the bottom, the worst situation for our country. Please show me on which step of the ladder you think the United States is *at the present time*?"
"On which step would you say the U.S. was *about five years ago*?"
"Just as your best guess, if things go pretty much as you now expect, where do you think the U.S. will be on the ladder, let us say, *about five years from now*?"

	Past	Present	Future
1959	6.5	6.7	2.4
1964	6.1	6.5	7.7
1971	6.2	5.4	6.2
1972	5.6	5.5	6.2
1974	6.3	4.8	5.8
1976	6.0	5.5	6.1
1978	5.8	5.4	5.3
White	5.9	5.4	5.2
Black	5.1	5.5	6.0

SOURCE: *Public Opinion* (November/December 1978), p. 38.

Table 2.5. Index of Consumer Sentiment (Survey Research Center, University of Michigan)

ALL FAMILIES							
DATE		*INDEX****	*INDEX COMPONENTS***				
			1*	2*	3*	4*	5*
FEB-MAR	1973	81.4	106	118	100	72	136
MAY	1973	76.0	103	111	83	69	154
AUG-SEPT	1973	72.1	93	106	78	67	127
OCT-NOV	1973	75.7	105	116	79	68	124
FEBRUARY	1974	60.9	88	100	45	46	121
MAY	1974	72.0	99	111	75	65	117
AUG-SEPT	1974	64.5	85	102	63	60	108
OCTOBER	1974	58.4	83	99	45	53	96
FEBRUARY	1975	58.0	81	103	45	47	100
MAY	1975	72.9	87	118	91	65	111
AUG-SEPT	1975	75.8	93	109	97	75	120
OCT-NOV	1975	75.4	98	119	94	64	117
FEBRUARY	1976	84.5	103	124	118	79	128
MAY	1976	82.2	100	118	120	76	124
AUG-SEPT	1976	88.8	105	119	127	92	139
NOV-DEC	1976	86.0	96	115	117	105	129
FEBRUARY	1977	87.5	99	121	115	97	141
MAY	1977	89.1	108	115	118	92	149
AUG-SEPT	1977	87.6	107	116	115	86	149
NOV-DEC	1977	83.1	107	107	104	84	139
FEBRUARY	1978	84.3	106	109	108	89	139
MAY	1978	82.9	105	111	99	79	147
AUGUST	1978	78.4	102	112	82	70	145
NOVEMBER	1978	75.0	106	105	74	67	135

PERSONAL FINANCIAL ATTITUDES
1* BETTER OR WORSE OFF THAN A YEAR AGO
2* EXPECT TO BE BETTER OR WORSE OFF IN A YEAR

EXPECTED BUSINESS CONDITIONS
3* NEXT 12 MONTHS
4* NEXT 5 YEARS

MARKET CONDITIONS
5* BUYING CONDITIONS FOR LARGE HOUSEHOLD GOODS

** PERCENT SAYING "GOOD TIMES" (OR "BETTER"), MINUS PERCENT SAYING "BAD TIMES" (OR "WORSE"), PLUS 100

*** 100 = FEBRUARY, 1966

Table 2.6. Freedom and Individual Autonomy

Public Attitude	Source	Period of Time	Percent/Change
1. Feel people like themselves would/would not be ready to go along with the following measures to curb inflation:	Yankelovich, Skelly, & White	April 1979	[*Would - Would not*]
a. Add a tax to foreign made goods such as colored television sets, which would make them as expensive as American made models.			78% - 15%
b. Close gasoline stations on Sundays to limit the consumption of gasoline.			77% - 20%
c. Put a ceiling on housing prices which would mean that their value would not increase each year.			75% - 21%
d. Institute mandatory wage-price controls despite the arguments against them.			73% - 20%
e. Place a restriction on people's use of credit cards.			58% - 36%
f. Raise the prices on all oil–related products— fuel, gasoline, etc.—to discourage use.			55% - 39%
g. Make it harder for people to get mortgages.			51% - 39%
h. Increase taxes so that people will not have much money to spend.			25% - 70%
2. Are not willing to see public funds used to build and improve mass transit if it would mean less money available to improve and construct highways.	Yankelovich, Skelly, & White	1976-1978	34% - 47%
3. Defined as a luxury:	General Mills	1975-77	
a. Having a new car	American Family		72%
b. Paying someone to do housework	Report 1974-75 (Yankelovich		59%
c. Taking a vacation each year	Skelly, & White)		55%

Table 2.6 (cont.) Freedom and Individual Autonomy

Public Attitude	Source	Period of Time	Percent/Change
d. Going out to eat in a nice restaurant			45%
e. Having a second car			33%
f. Going on weekend trips with family			31%
g. Going to a hairdresser regularly			29%
h. Belonging to clubs			26%
i. Having a color television			24%
j. Drinking liquor			23%
k. Playing golf/tennis			21%
l. Smoking cigarettes			20%
m. Going to the movies			20%
n. Giving money to charity			18%
o. Buying a winter coat			15%
p. Having meat with meals			14%
q. Subscribing to magazines			11%
r. Having company for dinner			11%
s. Having a baby-sitter			10%
t. Taking a Sunday drive			8%
u. Having a telephone			7%
4. Things people find hard to give up: (chosen from a list)	Scarcities Study (Yankelovich)	1975-77	
a. The family car			94%
b. Washing machine			91%
c. Central heating			88%
d. Own home			84%
e. Personal physician			63%
f. Television set			62%
g. Clothes dryer			61%
h. Second car			58%
i. Having a large family			54%
j. Air conditioning			50%
k. Meat once a day			48%
l. Annual vacation			42%
m. Dishwasher			28%
5. Sacrifices people would be *willing* to make to cut down on amounts they consume:	Scarcities Study (Yankelovich)	1975-77	
a. Eliminate annual model changes in automobiles			90%
b. Reduce amount of bags, tissues, cups, napkins and other disposables			87%

Table 2.6 (cont.) Freedom and Individual Autonomy

Public Attitude	Source	Period of Time	Percent/Change
c. Do away with changing clothing fashions every year			85%
d. Eat more vegetables and less meat			79%
e. Cut down sharply on plastic bags and packaging			78%
f. Stop feeding all-beef products to pets			77%
g. Drive cars to 100,000 miles before junking them			74%
h. Use wood and natural fibers for packaging			66%
i. Wear old clothes until they wear out			58%
j. Prohibit the building of large houses with extra rooms that are seldom used			56%
k. Do away with "second homes" for weekends and vacations			51%
l. Make it much cheaper to live in multiple unit apartments than single homes			44%

Table 2.7. The Psychology of Entitlement

Public Attitude	Source	Period of Time	Percent/Change
1. What people think they have a right to expect from society:	Scarcities Study (Yankelovich)	1975-77	
a. A good school for their children			65%
b. The best medical care, regardless of cost			48%
c. The right to take part in decisions that affect their jobs			47%
d. A secure retirement			33%
e. A right to work guaranteed by the government			21%
f. A minimum income			17%
g. An interesting job			13%
2. It's up to the government (vs. the individual/family) to see to it that all Americans get adequate health care.	General Mills American Family Report 1978-79 (Yankelovich, Skelly, & White)		41%
3. The government is spending too little on:		Oct. 1978	
a. Fighting crime			59%
b. Health care			51%
c. Education			45%
d. Transportation			33%
e. Defense			31%
f. Aid to the cities			26%
g. Welfare & care for the poor			25%
4. Would oppose a major cutback in Federal Government spending if it meant cutting back in spending on:	Harris Survey	Feb. 1979	
a. Help for the elderly, handicapped, and poor			78%
b. Health			75%
c. Education			73%
d. Defense			62%
e. Environmental protection			57%
f. Help for the unemployed			54%

Table 2.7 (cont.) The Psychology of Entitlement

Public Attitude	Source	Period of Time	Percent/Change
5. Believe it is the government's responsibility (vs. individual's/parents'/business/school's) to:	General Mills American Family Report 1978-79 (Yankelovich, Skelly, & White)	1978	
a. See that senior citizens get the health care they need			84%
b. See that products are safe			79%
c. See that children of minority groups get proper health care			40%
d. See that workers on the job have proper health care			21%
e. See that pregnant women get proper prenatal care			16%

Table 2.8. Waste

Public Attitude	Source	Period of Time	Percent/Change
GOVERNMENT WASTE			
1. Believe the government wastes a *lot* of the money we pay in taxes.	Michigan General Election Studies	1958-78	42%-77%
	CBS/N.Y. Times Survey	June 1978	80%
2. Main reason for the increase in taxes over the last few years:	Yankelovich, Skelly, & White	Oct. 1978	
a. People want more from government			11%
b. Government waste			44%
c. Inflation			42%
3. Believe it *is* possible to cut taxes without decreasing service if the government becomes more efficient.	Yankelovich, Skelly, & White	Oct. 1978	69%
4. Federal budget could be balanced just by reducing waste & inefficiency (*Agree*)	NBC Polls	Feb. 1979	71%
5. *Favor* a major cutback in federal government spending	Harris Surveys	1976-79	62%-69%
6. The government is spending *too much* on:	Yankelovich, Skelly, & White	Oct, 1978	
a. Foreign aid			72%
b. Welfare and care for the poor			51%
c. Aid to the cities			29%
d. Defense			27%
e. Transportation			24%
f. Education			22%
g. Health care			17%
h. Fighting crime			9%
7. Believe it's possible to balance the federal budget in the next few years	NBC Polls	Feb. 1979	42%

Table 2.8 (cont.) Waste

Public Attitude	Source	Period of Time	Percent/Change
CONSUMER/INDUSTRIAL WASTE			
8. I have become *less* wasteful in the past year.	Scarcities Study (Yankelovich)	1975-77	62%
9. Materials causing the most serious consumer solid waste problems	Yankelovich, Skelly & White	1977-78	
a. Aluminum cans, containers			65%-62%
b. Glass bottles			53%-54%
c. "Tin" (steel) cans			47%-46%
d. Aerosol containers			50%-41%
e. Plastic bottles			47%-41%
f. Plastic packaging			40%-39%
g. Large consumer durables, such as automobiles, appliances			43%-38%
h. Wastepaper			39%-38%
i. Paper plates paper napkins			28%
j. Disposable diapers			25%
k. Organic household garbage			28%-23%
10. Materials causing the most serious industrial solid waste problems:	Yankelovich, Skelly & White	1977-78	
a. Toxic industrial waste (mercury, cadmium, lead, etc.)			71%-68%
b. Radioactive wastes			52%-45%
c. Industrial wastes such as "slag" mine tailings			53%-43%
d. Industrial packaging materials			30%-29%

Table 2.9. Technology

Public Attitude	Source	Period of Time	Percent
PRO-TECHNOLOGY			
1. Believe life is better today than 50 years ago because of advanced technology (despite certain drawbacks of technology).	Roper Report 79-1	1979	60%
2. It is difficult to believe that a country as large, rich, and industrious as ours will be unable to solve its energy problems. (*Agree*)	Yankelovich, Skelly & White	1979	81%
3. Technology will find a way of solving the problems of shortages and natural resources. (*Agree*)	Yankelovich, Skelly & White	1978	52%
4. Favor more emphasis on technology. It's the only way we will solve our major health problems. (vs. technology has been the cause of many of our health problems)	General Mills American Family Report 1978-79 (Yankelovich, Skelly, & White)	1979	78%
5. Technologies that will play a major role in solving the energy problem (chosen from a list):	Yankelovich, Skelly, & White	1978	
a. Solar energy			77%
b. Better home insulation			53%
c. More efficient automobiles			45%
d. More nuclear plants			43%
e. New developments in oil production			33%
f. Greater use of coal			33%
g. Electric automobiles			16%
h. Geothermal energy			15%
i. Shale oil			13%
j. Alcohol			12%
k. Breeder reactors			11%
l. Liquefied natural gas			10%
ANTI-TECHNOLOGY			
6. Oppose all the emphasis on technology. It is responsible for many of the health problems we have today. (vs. item 4 above)	General Mills American Family Report 1978-79 (Yankelovich, Skelly & White)	1979	22%

Table 2.9 (cont.) Technology

Public Attitudes	Source	Period of Time	Percent
7. Would place more emphasis on:	Harris Surveys	May 1977	
a. Improving those modes of travel we already have *OR* Developing ways to get places faster.			82% vs. 11%
b. Spending more time getting to know each other better as human beings on a person to person basis *OR* Improving and speeding up our ability to communicate with each other through better technology.			78% vs. 15%
c. Breaking up big things and getting back to more humanized living *OR* Developing bigger and more efficient ways of doing things.			66% vs. 22%

Table 2.10. Confidence in Institutions

Public Attitude	Source	Period of Time	Percent/Change
ATTITUDES TOWARD GOVERNMENT			
1. Have a great deal of confidence in:	Harris Surveys		
a. Executive branch of Federal government		1966-79	41%-17%
b. State government		1973-78	24%-15%
c. Local government		1973-78	28%-19%
d. Congress		1966-79	42%-18%
2. Always/most of the time feel they can trust the government in Washington to do what is right.	Michigan General Election Studies	1958-78	56%-29%
3. Believe most people running the government are smart people who know what they're doing (vs. don't seem to know what they're doing).	Michigan General Election Studies	1964-78	69%-40%
4. People like themselves have no say in what the government does. (Agree)	Michigan General Election Studies	1960-78	27%-45%
5. Public officials don't care much what people like them think. (Disagree)	Michigan General Election Studies	1960-78	71%-44%
6. The people running the country don't really care what happens to me. (Agree)	Harris Surveys	1966-77	26%-60%
7. The government is pretty much run by a few big interests (vs. run for the benefit of all).	Michigan General Election Studies	1964-78	28%-65%
8. Most people with power try to take advantage of people such as myself. (Agree)	Harris Surveys	1966-77	33%-60%
9. The federal government is getting too powerful for the good of the country and the individual person. (Agree)	Michigan General Election Studies	1964-76	42%-68%
10. Feel a great deal of confidence in major companies.	Harris Surveys	1966-79	55%-18%
11. Business strikes a fair balance between profits and the public interest. (Agree)	Yankelovich, Skelly, & White	1968-77	70%-15%

Table 2.10 (cont.) Confidence in Institutions

Public Attitude	Source	Period of Time	Percent/Change
12. Business should do more to prevent inflation.	MAP American Council of Life Insurance	1968-75	72%-86%
ATTITUDES TOWARD OTHER INSTITUTIONS			
13. Feel a great deal of confidence in:	Harris Surveys	1966-79	
a. Medicine			73%-30%
b. Higher education			61%-33%
c. Organized religion			41%-20%
d. The military			62%-29%
e. The press			29%-28%
f. Organized labor			22%-10%

Table 2.11. Social Values

Public Attitude	Source	Period of Time	Percent/Change
PERSONAL VALUES			
1. Plays a *very important* role in my life:	Yankelovich, Skelly, & White	March 1978	
a. Family			92%
b. Understanding myself			91%
c. Work			78%
d. Money			35%
2. *Not* strongly committed to as personal goals:	Yankelovich, Skelly, & White	1974-78	
a. Job and Career			
To earn as much as I can			56%-57%
To be outstanding in my field of work			48%-58%
b. Marriage and Children			
To have a successful, happy marriage			19%-22%
To have successful children			35%-37%
c. Status in the Community			
To be the kind of person others respect			40%-45%
Do something that is beneficial to mankind			61%-65%
3. Would *welcome*:	Yankelovich, Skelly, & White	March 1978	
a. More emphasis on traditional family ties			84%
b. More emphasis on self–expression			69%
c. Less emphasis on money			55%

Public Attitude	Source	College Youth 1973	Adult Family Members 1974-75	Total Parents 1976-77
4. Very important personal values:	Yankelovich, Skelly, & White			
		%	%	%
a. Family		68	92	81
b. Education		76	76	71
c. Self-fulfillment		87	75	67
d. Hard work		43	77	62
e. Religion		28	62	52
f. Patriotism		19	59	43
g. Having children		31	60	43
h. Money		20	—	31
i. Financial security		—	74	—
j. Doing things for others		—	61	—
k. Not being in debt		—	60	—
l. Saving money		—	55	—

Table 2.11 (cont.) Social Values

Public Attitude	Source	Period of Time	Percent/Change
5. Feel strong need:	Yankelovich, Skelly,	1978	
a. To have something meaningful to work toward	& White		59%
b. To have a very clear idea of my objectives and goals in life			53%
6. Next to health, money is the most important thing in life. (*Disagree*)	NORC General Social Surveys	1973-76	66%-66%
WORK			
7. Attitudes toward working:	Yankelovich, Skelly,	1978	
a. I want to enjoy my work, earn good money, etc., but it is more important that my job not interfere with my family life.	& White		39%
b. I want to be intellectually and emotionally satisfied in my work. This is, to some dgree, more important than money.			27%
c. I really want to succeed, not necessarily just in terms of money, but in sense of rising to the top, becoming a leader.			10%
d. I really want to make a contribution to society in my job. I want my work to really help people.			10%
e. The most important thing to me is to earn more money. My greatest satisfaction in my work would be to earn a lot.			8%
f. I really don't care much about my job. The best and most important things in life are outside my work.			3%
8. Agree with:	Yankelovich, Skelly,	1974-78	
a. I am very much involved in my work and I don't feel the need for a lot of leisure time activities.	& White		13%-14%

Table 2.11 (cont.) Social Values

Public Attitude	Source	Period of Time	Percent/Change
b. When I finish my day's work, I am usually so tired that there's very little that I really feel I like doing.			19%-21%
c. I enjoy my work, but it isn't the main source of the satisfaction I get from life. I depend more on my leisure time activities for personal satisfaction.			60%-65%
9. Choices:	Yankelovich, Skelly, & White	1978	
a. It's important to me that I enjoy my job.			91%
b. I don't really expect to enjoy my job.			7%
a. If I didn't need the money, I wouldn't work.			37%
b. I would work even if I didn't need the money.			59%
10. Report as more important today than a year ago:	Yankelovich, Skelly, & White	1978	
a. Being with pleasant people whom you like.			57%
b. Finding what you do interesting most of the time.			49%
c. Work in which you feel you're always learning something new.			43%
d. Being recognized and appreciated by the people you work for.			41%
e. Facing new challenges in your work all the time.			36%
f. Knowing that what you do is of benefit to society.			30%
g. Knowing that you'll be missed if you're not there.			29%
h. Knowing you're making a contribution to the company/organization you work for.			27%

Table 2.11 (cont.) Social Values

Public Attitude	Source	Period of Time	Percent/Change
11. Reasons for working:	Yankelovich, Skelly, & White	1974-78	(Employed Women)
a. Primarily to derive income			24%-27%
b. Primarily as a source of enjoyment			44%-35%
c. Primarily to be independent			32%-38%
12. Work seen as:	Yankelovich, Skelly, & White	1974-78	(Employed Women)
a. A career			31%-34%
b. A job			69%-66%
13. Believe an employee who loses his job because he refuses to move to another city is in the right.	Yankelovich, Skelly, & White	1978	63%
14. Believe an employee who loses his job because he refuses a promotion is in the right.	Yankelovich, Skelly, & White	1978	65%
15. Tradeoffs:	Yankelovich, Skelly, & White	1978	
a. Take a job I wouldn't like and get $2,000 raise, OR			18%
b. Keep a job I like a lot			78%
a. Take a job I wouldn't like and get $5,000 raise, OR			32%
b. Keep a job I like a lot			62%
a. Give up some of my wages and work a few less hours/week, OR			17%
b. Keep things the way they are			80%
a. Relocate, get promotion and raise, OR			31%
b. Live where I want, give up promotion and raise			57%
FAMILY AND SEX ROLES			
16. Believe and want children to believe:	General Mills American Family Report 1976-77 (Yankelovich, Skelly, & White)	1977	(Total Parents)
a. It's not important to win, it's how the game is played			71%
b. The only way to get ahead is hard work			65%

Table 2.11 (cont.) Social Values

Public Attitude	Source	Period of Time	Percent/Change
c. Duty before pleasure			58%
d. Any prejudice is morally wrong			51%
e. There is life after death			51%
f. Happiness is possible without money			50%
g. Having sex outside of marriage is morally wrong			47%
h. Everybody should save money even if it means doing without things right now			42%
i. People are basically honest			37%
j. My country right or wrong			34%
k. People in authority know best			13%
17. No longer believes a man with a family has a responsibility to choose the job that pays the most rather than one that is more satisfying but pays less.	Yankelovich, Skelly, & White	March 1978	58%
18. Definition of a real man: Being a good provider	Yankelovich, Skelly, & White	1968-78 (Rank Order)	86%-67% #1 #3
HEALTH			
19. Doing more than a year ago:	General Mills American Family Report 1978-79 (Yankelovich, Skelly, & White)	1978-79	(Total adult family members)
a. Taking better care of your health			34%
b. Doing physical exercise			24%
c. Eating nutritiously			25%
d. Watching calories			26%
20. What most prefer to do during their leisure time:	General Mills American Family Report 1978-79 (Yankelovich, Skelly, & White)	1978-79	
a. Sit around and relax			51%
b. Get some physical exercise			24%
c. Go out to a movie, bar, or restaurant			25%
21. Feel Americans are more concerned about their health & preventing problems than a few years ago	General Mills American Family Report 1978-79 (Yankelovich, Skelly, & White)	1978-79	70%

Table 2.12 Confidence in Technology

	All this talk about running out of natural resources is ridiculous; technology will find a way of solving our shortage problems.		Americans should get used to the fact that our wealth is limited and most of us are not likely to become better off than we are now.		We are fast coming to a turning point in our history; the "land of plenty" is becoming the "land of want."		I'm not that unhappy about the possibility of shortages, because I know it will encourage me to use everything efficiently—not wastefully.	
	AGREE	DISAGREE	AGREE	DISAGREE	AGREE	DISAGREE	AGREE	DISAGREE
TOTAL	52	48	62	38	72	28	66	34
SEX								
Total Women	51	49	62	38	77	23	65	35
Total Men	53	47	62	38	67	33	67	33
AGE								
Under 21 years	44	56	57	43	66	34	67	33
21-24 years	46	54	60	40	79	21	69	31
25-34 years	44	56	53	47	71	29	62	38
35-49 years	55	45	58	42	74	26	60	40
50-64 years	63	37	73	27	67	33	60	40
65 & over years	61	39	80	20	80	20	83	17
YOUTH 18-24 YEARS OLD								
Attending college	29	71	47	53	70	30	65	35
Not attending college	48	52	63	37	72	28	66	34
RACE								
White	51	49	62	38	72	28	64	36
Black	65	35	65	35	79	21	79	21
SOCIO-ECONOMIC LEVEL[1]								
Prosperous/Upper middle	47	53	60	40	75	25	66	34
Middle	49	51	62	38	71	29	63	37
Lower	69	31	68	32	75	25	75	25

[1]An index derived from income and education

Table 2.12 (cont.)

INCOME								
Under $5M	66	34	70	30	75	25	78	22
$5M-$7,499	66	34	69	31	76	24	88	12
$7,500-$9,999	56	44	71	29	77	23	72	28
$10M to $15M	52	48	63	37	71	29	72	28
$15M to $20M	48	52	61	39	70	31	55	45
$20M to $25M	45	55	57	43	74	26	60	40
$25M to $30M	47	53	63	37	74	26	61	39
$30M to $40M	42	58	51	49	72	28	56	44
$40M to $50M	49	51	67	33	76	24	56	44
$50M & over	41	59	58	42	92	8	60	40
EDUCATION								
In school now:								
High School	48	52	55	45	58	42	66	34
College	28	72	49	51	68	33	61	39
Not in school:								
Not High School graduate	65	35	73	27	74	26	77	23
High School graduate	53	47	63	37	74	26	63	37
Attending college/graduate	45	55	56	44	72	28	60	40
MARITAL STATUS								
Married	51	49	62	38	73	27	63	37
Single (never married)	46	54	56	44	68	32	65	35
Divorced/Widowed/Separated	64	36	72	28	78	22	78	22

Table 2.12 (cont.) Confidence in Technology

	All this talk about running out of natural resources is ridiculous; technology will find a way of solving our shortage problems.		Americans should get used to the fact that our wealth is limited and most of us are not likely to become better off than we are now.		We are fast coming to a turning point in our history: the "land of plenty" is becoming the "land of want."		I'm not that unhappy about the possibility of shortages, because I know it will encourage me to use everything efficiently—not wastefully.	
AGES OF CHILDREN								
Under 6 years	46	54	56	44	69	31	60	40
6-12 years	50	50	59	41	70	30	59	41
13-17 years	56	44	62	38	77	23	62	38
OCCUPATION OF HOUSE-HOLD HEAD								
Professional/Manager	44	56	55	45	66	34	58	42
White collar	44	56	56	44	69	31	59	41
Blue collar	55	45	62	38	74	26	66	34
OCCUPATION OF WOMEN								
Housewife	59	41	68	32	76	24	66	34
Employed	47	53	59	41	79	21	64	36
RELIGION								
Protestant	54	46	64	36	75	25	67	33
Catholic	53	47	60	40	67	33	65	35
Jewish	51	49	54	46	65	35	54	46
None	43	57	63	37	73	27	59	41
POLITICAL PARTY								
Liberal Democrat	51	49	62	38	69	31	66	34
Conservative Democrat	53	47	65	35	75	25	67	33

Table 2.12 (cont.)

Liberal Republican	51	49	56	44	72	28	64	36
Conservative Republican	50	50	61	39	71	29	62	38
Independent	38	62	64	36	78	28	69	31
None	59	41	60	40	79	21	71	28
GEOGRAPHIC REGION								
Northeast	50	50	63	37	66	34	62	38
North Central	52	48	59	41	71	29	67	33
South	58	42	67	33	78	22	69	31
West	44	56	58	42	71	29	62	38
TYPE OF AREA								
Central city	51	49	60	40	72	28	66	34
Suburban area	51	49	61	39	68	32	63	37
Rural	55	45	66	34	78	22	70	30

NOTES

1. The quote from Alexander Hamilton's *Report on Manufactures* appears in the Second Annual Report of the National Commission on Productivity, March 1973, p. 26.

2. The quote by President Eisenhower appears in *The Glory and the Dream* by William Manchester (New York: Bantam Books, 1973), p. 653.

3. These and the following comments by Etzioni are from his paper, "Choose We Must," presented at the Franklin Lecture Series at Georgia State University on February 20, 1979.

4. The material on the growth of government assistance and social support programs is drawn from the Welfare Policy Project of the Institute of Policy Sciences at Duke University, the volume entitled, *Toward Income Opportunity: Current Thinking on Welfare Reform*, 1978, p. 1.

5. Philip Caldwell's comments appear in *Newsweek*, April 16, 1979, p. 16.

6. Etzioni, "Choose We Must," p. 15.

7. Ibid, p. 16.

8. Ibid, p. 17.

9. Herman Kahn's discussion of "new values" is from his report to the Joint Economic Committee of the U.S. Congress, *Current Medium and Long-Term Economic Prospects*, U.S. Government Printing Office, 1976, pp. 23-25.

10. Katona's quote is from his article, "Consumer Expectations as a Guide to the Economy," *Public Opinion*, February 1979, p. 17.

11. See Table 2.3.

12. Etzioni, "Choose We Must," p. 20.

13. Ibid, p. 21.

14. *The New York Times*, "Living With Inflation: Who Wins, Who Loses," April 23, 1979, p. 1.

15. Lester R. Brown, *The Twenty Ninth Day* (New York: W. W. Norton, 1978), p. 247.

16. From William Safire's column in *The New York Times*, January 13, 1977, p. 35.

17. Rep. Richmond is quoted in *Time* Magazine, January 17, 1977, p. 47.

18. *The New York Times*, January 7, 1977, p. 22.

19. The data on water consumption levels in Marin County are from an interview with J. Dietrich Stroeh, manager of the Marin Municipal Water District, March 20, 1979.

20. The results of the Field Enterprises poll were reported by the San Francisco *Chronicle*, May 1, 1977, p. 1.

21. Ibid., p. 1.

22. Stroeh interview.

23. From *Economic Means for Human Needs: Social Indicators of Well–Being and Discontent*, edited by Burkhard Strumpel (Ann Arbor, Michigan: Institute for Social Research, University of Michigan, 1976), pp. 278-79.

Chapter 3

THE PARADOX OF PLENTY: ECONOMIC GROWTH AND MORAL SHRINKAGE

William Lee Miller

C. Vann Woodward remarked as an aside in his 1977 Jefferson Lecture that many Europeans think Americans have not only *pursued* happiness; they have overtaken it. As with happiness, so with the other intrinsic goods of life: it sometimes seems that the economically advanced nations—with the United States in the lead—have run on beyond them. They have misplaced the ends of living that their elaborate apparatus for living was intended to achieve. "The bad news," says the modern pilot, "is that we are lost; the good news is that we are making very good time."

I want in this chapter to ventilate this awareness that the further reaches of economic growth do not necessarily correlate with what life is all about. Indiscriminately pursued, indeed, "economic growth" may eventually come to stand in contradiction to the truly human goods. Flannery O'Conner liked the quotation from a dwarf in the brothers Grimm: "I prefer something human to any amount of gold."

This chapter deals with the state of "values" in the economic "transition" in which we now find ourselves. I examined that state of values by gathering up my own ideas, on the convenient theory that what I have to say, although not necessarily representative, is certainly not original. It is drawn out of the smoggy atmosphere of contemporary life in the most economically advanced of countries.

The overriding theme is that, at some point, the tail begins to wag the dog.
I infer that increasing numbers of people perceive, dimly and grumpily, that the
"means" begin to obscure or determine the "ends." The "means" are not only
the material product of the economy but also the social apparatus for producing
it and the mental outlook that accompanies it. This last is the key to everything.

I do not mean to indict economic growth as such. One can understand the
impatience felt by many citizens of the Third World, and of poor people anywhere, with the sort of thing I will be saying about growth, abundance, an
advanced technology, and a modern outlook. It's all very well for you, the poor
may say, to deplore the limits and defects and hazards and ill effects of growth;
but that's a luxury for those of you who have already attained the riches of a
modern economy. The rest of the world can't afford it. Modern economic growth,
they may rightly insist, has brought a human happiness, and diminished human
misery, on a vast world scale; it has brought the basic human needs like food
and shelter, and the further need of interesting life, within reach of millions.
For all the reservations this chapter will state, given a choice human beings in
the main will head for the material wonders of the world of modernity.

And so one must insist on difference between the rich segments of humanity
and the human beings for whom the basic human needs have not been met. A
good meal for a hungry or undernourished person is different from the trillionth
Big Mac sold by fast food franchises in a rich country. A hungry man does not
have much of a problem of choice among economic goods. He knows what he
wants, and needs no national advertising campaign to tell him. During the famine
in Ireland, one would not have required billboards and commercials to sell
potatoes, if there had been any. Moreover, what a starving man *wants* is also
what he *needs*; the "good"—the food—that he seeks is also something required
by, and good for, him.

Economics and ethics share the word "good" and the now ubiquitous post-
Watergate word "value." In the case of food for the starving man, or water for
the person dying of thirst, or shelter or clothing or any other of the basic human
needs, where these are lacking, ethical good and an economic good coincide.
The goods the person wants, and surely would express an enormously powerful
"demand" for in the market (if the supply were there and he had the money
to make his "demand" "effective") is also good *for* him, and it is good that
he have it. Once we pass beyond the level of basic human needs, however, and
the commodities proliferate, ethical value and economic value diverge. But
habits of mind carried over from the condition of elemental need continue on
into the condition of plenty. Therein lies an important source of the current
spiritual confusion in the rich countries, and perhaps especially in the United
States.

Material well-being as a condition, the market as an institution, quantitative
measurement as a device, economics as a science, and modern technology as
an apparatus, are all, properly, subordinate and instrumental goods. But they
have a powerful tendency to obscure or distort the true ends of human life.

(They come as a cluster, and reinforce this tendency in each other.) Good and valuable in their subordinate and limited role, they become something ungood when overextended, unguided, out-of-place.

We can picture an imaginary turning point (there is, in fact, no such single point, but a long "incremental" process, as the economists would say) at which the distortion takes over: when the marvels of material achievement crowd out other goods; when the market begins to reach out into inappropriate territory and to turn *everything* into commodities; when economics begins to become not the social equivalent of plumbing but the metaphysics of the age; when the marvelous servant, modern technology, turns around and begins to give orders to its putative human master. Then the public at large begins to be aware of the faults that hitherto had been seriously noticed only by a handful of literary and religious people, social critics, and cranks. I suggest that is the stage we are approaching now in the rich sections of rich countries—the stage of a widening public awareness of the hazards and limits of that cluster of modern powers. The *New Yorker* reported one of those comments by an "unconscious genius" that unintentionally go to the heart of a large issue with a chance phrase, like the officer in Vietnam who said that we have to destroy this town in order to save it. A technician at Three Mile Island, asked whether a worker might not be able to fix a stuck valve inside the radioactive plant, replied: "In theory he can but in practice he can't." It is a sentence with a very wide application to the modern world's complicated effort to attain the good life.

The final reason the modern world gets out of shape has to do not with an economic or technological determinism but with belief. That is a point to which I will return briefly at the end of this chapter, although it should be evident throughout. The culprit in realm of belief is the reductionism that leaves these powerful forces unconstrained by moral criteria. The goal should be neither "Growth" per se, nor "No Growth" either: as E. F. Schumacher wrote, that is one emptiness substituted for another.[1] The point rather is the discovery of standards—of the moral and human "limits," if that is to be our word, on an economic-technological dynamic that has thrown off such limits and dominates the modern world because it dominates our minds. (Of course, human standards—human goods—should function not only as "limits" but also goals, claims, and creative possibilities.)

I mentioned above a cluster of phenomena, reinforcing each other, that, though good, are also properly subordinate and/or instrumental; it is the failure to subject these to a criterion that is the root of the problem here described. The first of these is the fact of prosperity itself.

PEOPLE OVERRUN BY PLENTY

Material well-being is itself, obviously, a fundamental good, not to be loftily dismissed by those who don't have to worry about it. Moreover, the values—or the traits and habits—that bring about economic well-being are also genuine goods. Most important, prosperity makes possible other goods. Here, in violation of the Harlan Cleveland rule of global perspective, I want to focus on the United States.

A modern purposive, activist, individualist, world-mastering ethic arose in Europe, in whatever subtle relationship to the Protestant Reformation the most recent dissertations on Max Weber have proven it to have. It was reformed and changed by the unencumbered scientific attention to the practical world brought forward in the Enlightenment. It was reinforced by the Industrial Revolution in England. We economically productive Americans were, in large part, the product of movements that began in Europe. Nevertheless, the United States came to be, after many years of English leadership, the nation that exemplified the characteristics of a society built around modern economic growth. And we have helped the recovery of those characteristics in Europe and their development in Japan.

The defining values of the American nation have been closely tied to the fact and the idea of an expanding economy: the work ethic has been tied to it; the idea of progress has been tied to it; our future-facing optimism has been tied to it; our success ethic has been tied to it. The legend of our mastery of our destiny—collective and individual—has been tied to it. David Potter absorbed the famous hypothesis that the frontier shaped American character into the larger idea of *abundance* shaping American character: *People of Plenty*.[2] The availability of free land, and the attraction to Americans of the West, of virgin land, of frontiers old and new, is just one aspect of the larger phenomenon of the land of economic abundance—of economic growth. The analysis made a quarter of a century ago by Potter was, on the whole, sanguine: riches were good for us—and we were good for riches. Economic abundance had shaped our national "character" (that is, our national makeup); and our national "character" had helped bring about that abundance. Our plentiful economy was seen to be in large part the result of important "American" values—work, enterprise, practicality, ingenuity. Potter wrote:

> Critics of this country have often dismissed the American standard of living as an example of luck *in excelsis*. They repeat the saying that there's a special providence which watches over drunkards, fools, and the United States, and they interpret the prosperity of this country as a simple reflex of nature's profusion. As they see it, the Americans are a people who have wandered unwittingly into a vast cornucopia whose plenty is accepted with moronic complacency.[3]

But that picture, said Potter, is false. The natural resources of the continent were available to the native American before the colonizers came. For a hundred years, England, with a very meager supply of nature's bounty, was the world's leading economic power. Among the regions of the United States, New England, nature's stepchild, had for a long time a standard of living as high as, or higher than, other regions of the country. In other parts of the world, other peoples have eked out a meager existence amid natural resources which were technologically meaningless to them and yet were comparable to the resources of North America. Potter quotes Isaiah Bowman:

> Man himself is a part of his own environment; his skill and knowledge are assets to him as definitely as that which nature provides in the raw. We cannot determine the capacity of a land from its physical aspects alone, its soil, its water supply, its temperature means and extremes, its forests or the presence or absence of fisheries and the like. Greenland was one kind of a country to the Vikings; it was another kind of country to the Eskimos; it would be still another kind of a country to us.[4]

Expert opinion, according to Potter, regards the qualities of a people as a vital factor in economic growth. After presenting the views of two such experts he wrote that "they seem not far from accepting a more refined equivalent of the old, popular assertion that determination, ingenuity, and hard work made America what it is."[5]

Perhaps most important, economic abundance makes possible other important social values. Abundance, Potter argued, made possible the American combination of equality and liberty; it made it possible to think of these two values—opposed to each other in much of European history—as combined in the idea of equality of opportunity. This was Potter's version of the familiar point we all learned in the 1950s: the positive virtues of economic growth—of an expanding Gross National Product—of the larger pie that allowed us to by-pass the hard problems that other countries, faced with scarcity, had to deal with. "Our practice, indeed," wrote Potter, expressing an idea that was very widely held back then, "has been to overleap problems—to by-pass them—rather than to solve them."[6] Potter carried the idea further: abundance makes democracy work. Because the economy was expanding, enough of the promissory notes of American democracy could be fulfilled to make that ideal convincing. Democracy and abundance were seen to be twin values, and together they are our American message to the world. Democracy and abundance, each reinforcing the other, not democracy alone, is the "mission," the message, that America has for the world. "We supposed that our revelation was 'democracy revolutionizing the world' but in reality it was 'abundance revolutionizing the world.' "[7] Potter quoted with approval Franklin Roosevelt's famous statement that, if he could show people of Russia one book about American society, that book would be the Sears Roebuck catalogue.

Now, 25 years later, we may still affirm much that Potter wrote, but not in the same confident tone of voice. And some of it we don't affirm at all. We have come to see the limitation and even paradox of the people of plenty. One part of it is simply material: plenty can be not only a blessing but also, eventually, something a society is stuck with. It closes doors as well as opening them; it makes problems even as it solves them.

Potter unintentionally provides a most electric and current reminder of this point. Extolling the achievements of American abundance, he cited favorably the immense increase in the supply of *energy* that this nation used. Between 1820 and 1930, he wrote, American increased *forty-fold* the supply of energy which it could command per capita. "This was," he wrote, "in effect a forty-fold increase in capacity to convert and utilize resources." He wrote all of this in a glowing tone. "It was achieved," he wrote, "not merely by accepting nature's proffered gifts but by the application of science, by the elaboration of a complex economic organization, by planning, and by toil."[8] We did it. Yes, we did. Now in gasless 1979 the sentences have an irony which Potter could not have predicted. By that enormous increase in the use of energy per capita we have created for ourselves dependence on some desert nations of the Middle East which can dry the tanks of our Buicks, stop our air conditioners, and cause mayhem at the corner gas station.

Our dependence on supplies of energy to fuel our cars and elevators and television sets and washing machines and airplanes and air conditioning units, preoccupying us now, pulls at the edges of the social order. It is all too graphic an indication that there's something lost as well as something gained with your forty-fold increases. It is, moreover, a major cause of inflation, which creates its own creeping social unhappiness; the form of inflation we have may, indeed, be exactly the disease of a people of plenty. And looming over are weapons that could only have been developed originally by a people of plenty, which weapons can destroy whole civilizations.

Reading the material about Three Mile Island, one felt that the scenario had been written by one of Jane Fonda's favorite script writers. At the time of that event a nuclear physicist said: "We have made a Faustian compact with society." Back in happier days of plenty, we did not realize that our soul might one day be required of us for the power that we are now given. But the modern picture of an economic growth that is rimmed with missiles, explosives, inflation, stagnation, the China syndrome, and pesticides that can make a silent spring conjures a Mephistophelian figure in the background who was never there in the days of a simpler progress.

We now realize that, though the Club of Rome may have been in error in picturing the end of economic growth with a big bang, the steady-state in economics will not be an easy or a likely achievement. Economic growth in the advanced countries may proceed with a lot of smaller bangs and a continuous whimper. Even the more conventional wonders that the abundant economy brings us are rather a mixed bag. People want to eat: we give them Big Macs, Jumbo Wimples, and banana royale sundaes. They want to recover from eating: we given them Pepto Bismol and Alka Seltzer. They want not to eat: we give them the Scarsdale diet, the Atkins diet, Metrecal. They want to be warm: we give them at first woolen underwear but later central heating. They want to be cool: we give them air conditioning. We ride (or once we did) from an air conditioned house in an air conditioned car to an air conditioned building, striding for one or two sweaty moments perhaps in the actual outdoor temperature. On my one visit to Las Vegas, I had the sense that neither night nor day nor the temperature of the sun nor the rain nor any other natural thing deterred the slot machines from the swift completion of their daily/nightly rounds. In a country lacking in consumer goods, the Sears Roebuck catalogue is an unanswerable argument; but Las Vegas isn't, or shouldn't be. And abundance means not only the Sears catalogue but also the one from Nieman Marcus, offering as Christmas presents one of the fifty states for him and one for her, a Dakota or a Carolina each.

In his book, *The Social Limits to Growth*,[9] Fred Hirsch describes the way economic growth can make some things less valuable. A middle class person says of a tropical resort: "I know it's no good now that I can afford to go there." Some kinds of "goods" do not increase in number, and may decrease in value, as the economy grows. His general illustration is this: A person in a crowd improves his view by standing on tiptoe. But if *everybody* stands on tiptoe, there is no gain. Moreover, I may add, the arches begin to ache.

The most important lesson about the defects of plenty, though, has to do with what we stand for. Potter's message, that abudance *shapes* our character, may be read more warily than he did. It decides who our heroes are; what our preoccupations are; what the goals of our life are. The picture, for just one example, of the missionaries of American abundance, with their Sears catalogues instead of Bibles, was rather attractive when Roosevelt stated it in the 1940s, and when Potter repeated it in the early 1950s: that homely profusion of everyman's consumer goods was a symbol and a vindication of American democracy. When, however, the decades that followed brought the "Americanization," the "Cocacolonization," of much of the world, the picture became a good deal less attractive. The question arose whether abundance was just a vindication of democratic ideals, or also a substitute for them, and even a threat to them. An article about American soldiers fighting in Vietnam—called "Why They Fight"—included the following paragraphs, not surprising perhaps but not very uplifting either:

> The overriding feature in the soldier's perception of America is the creature comforts that American life can offer. . . . The soldiers described the United States by saying it had high-paying jobs, automobiles, consumer goods and leisure activities. No other description of America came close to being mentioned as often as the high—and apparently uniquely American—material standard of living.
> Even among front-line combat soldiers, one sees an extraordinary amount of valuable paraphernalia. Transistor radios are practically "de rigueur." Cameras and other photographic accessories are widely evident and used. Even the traditional letter-writing home is becoming displaced by tape recordings. It seems more than coincidental that

American soldiers commonly refer to the United States as "The Land of the Big PX."[10]

"The Land of the Big PX" seems a rather inadequate rendering of the ideals of the Declaration of Independence and the Gettysburg Address.

WHAT COUNTS AND WHAT DOESN'T

As with "plenty," so with the free market and economic science. An elaboration of the familiar criticism of the concept of the Gross National Product will serve to symbolize the point. We are all aware that the GNP gathers up a great deal of economic activity that would be judged by almost any standard to be undesirable. If there should be an epidemic of good sense on the part of the American public, and we all stopped smoking cigarettes, walked when we could and drove our cars no faster than 55 miles an hour, stopped drinking, overeating, eating junk food, and eating in expense account restaurants and, therefore, had a great outbreak of good health, there would almost certainly be a decline in the GNP, because of the enormous expenditures it includes for medical care, for the tobacco industry, for the alcohol industry, for the ambulances, tow trucks, and police cars that cope with the stream of highway accidents. If the prodigal son stops his expenditures on riotous living and sobers up and goes home to his father and lives a quiet life, the one-time payment for the fatted calf isn't going to make up for the losses in the Riotous Living Industry.

If peace should break out; if, as all three national leaders, following Isaiah, recommended at the signing of the Israeli-Egyptian Peace Accords, the swords should be beaten into ploughshares; if the lion should lie down with the lamb; if SALT VII should bring the expenditures for stragetic arms down to zero and end the building of all missiles, bombs, bombers, anti-missiles, anti-bombers, and anti-bombs, the consequences would be an immense oversupply of ploughshares: little sleep, as Woody Allen said, for the lamb; and a major shock to the economies of the major powers.

If there should be a sudden outbreak of good taste and no one watched prime time television situation comedies or Monday night football, or any other of the network offerings for our evening's edification, including their hymns to dentures, sonnets on deodorants, and dramas to dog food, and the Nielsen ratings dropped to zero, the economy would have a very hard time making up the huge loss to the television and advertising industries themselves, not to mention the sales of the dentures, deodorants, and dog foods.

If there should be a sweeping conversion to righteousness, and crime dropped to zero, the effective demand for the services of policemen, of jailers, and even conceivably lawyers (although, given their agility at finding new territory, we couldn't count on that) would decline and presumably the GNP with it.

If suddenly we had peace, righteousness, good taste, law observance, health, good sense, moderation, self-restraint, consideration of others, and if the cardinal virtues were to come to dominate the whole society, there would be a big economic loss by the current measures of economic good. If we all lived sensible, quiet, healthy lives, reading inexpensive editions of the great classic literature, going to bed earlier and rising earlier and being therefore healthier and wiser, the nation as a whole as it now records its own condition would not be "wealthier." We would appear on the books to be poorer, because there would have been fewer and cheaper transactions.

What a national income account like the GNP measures, of course, is transactions: the moment when something is bought and sold, a "good" or a "service." The measures of agregate economic activity make no distinction among the kinds of transactions that they measure. A whole series of transactions called into being by catastrophe, malice, wickedness, and greed count as additions to the Gross National Product; some economists have had to invent a category of "goods" called "regrettable necessities" for one particularly embarrassing group of them.

I cite these familiar deficiencies of the GNP in order to lead to a larger point: it is not only the GNP that is faulty and vulnerable in this way, but the world of thought, and of life, of which it is a part. If one used other, less well known, national income concepts, one would not evade the defect I have dramatized, because one would still be engaged in counting *transactions*, without regard to the moral worth, or lack of it, of that which is bought and sold. And that defect applies not only to the measures of national income and national product but to economics as such—both to the science, dismal or otherwise, carried on by professionals, and to the outlook of the nonprofessional consumer and producer in the street. And most important of all: it applies to the *market*, the institution economics studies. What economics is about, and in a way what the market is about, is *counting*, and what are counted are transactions.

Do they deal instead with "utility" or "consumer satisfaction" or "choice"? Only in a narrowly circumscribed and eventually misleading way—misleading, that is, if they are not subordinated to and corrected by a supervening ethic. I believe it has not been so subordinated and corrected in the consumer economy of the twentieth century United States (perhaps in the other advanced economies as well). The implicit metaphysic or ethic of economic science fits all too well with a prevalent hedonism and egotism, which, in turn, are justified by a more prevalent pseudo-democratic or pseudo–libertarian relativism (let each person decide for himself and thus no one else will tell him what's good for him). The American reader will recognize the omnipresence of this idea, perhaps even in himself. It is a nest of misleading assumptions and applications.

The market, and (in their official capacity) those who study it scientifically, cannot discriminate among the goods and ills of life. A fool satisfied *is* "better" than Socrates dissatisfied, and a satisfied pig than a man dissatisfied, because "consumer satisfaction" is the measure. Whether the poor of South Side Chicago receive basic human needs or the rich of Grosse Pointe receive luxuries is all the same not only to national income accounting, but also to the market and to economic science; otherwise, one would engage in the unscientific practice of "interpersonal comparisons."

There is the matter just of *counting* itself: as with the GNP, so with our modern world. It is easily misled by the apparent precision of numbers. Social scientists sometimes say—and I am not always sure they are joking—"If you can't count, it doesn't count." Now all the important realities in the higher realms of human life—everything, by definition, *qualitative*—either cannot be counted or are mistreated if they are counted: how *many* ecstacies? But a civilization of the numbers tends either to ignore the domain of the qualitative or to subject it to its own rule.

Moreover, there is sometimes a corollary of the sentence quoted above: "If you *can* count it, it *does* count" almost without regard to what "it" is: therefore, the Nielsen ratings, a thousand polls, a million charts and graphs and tables. We count: we substitute the surface precision of numbers for the subtlety of truth. We begin with a swarm of qualifications, of course: the counting is only as valuable as the conception of what is to be counted. But once we start counting, and set up machines to carry on the counting for us, the initial qualifications slip out of our consciousness. The calculators click their digital certainties. A lot of the modern world is a story of very fancy work on the wrong track. Of course, the counters admit there is something limiting their counting and something beyond it. But once you are habituated to the counting, and unpracticed in that beyondness, the latter becomes pretty vaporous. Hard data, hard noses, hard news: modern "hardness" takes over.

Computer people use the phrase: "garbage in, garbage out." In other words, if you start with junk, no fancy process will keep the outcome from being also junk. That phrase applies much more widely than to the functioning of computers.

While I was writing this, I heard about a speech by John Connally. Mr. Connally declared that what these times ahead require is leadership that lives for the objectives: 'produce, produce, produce.' If one marshmallow cookie is good, a hundred is better, a million better still. The other side of produce, produce, produce, is consume, consume, consume. Here is Russell Baker in *The New York Times*, satirizing his deficiencies by that criterion:

> Got up late and consumed three newspapers, four cups of coffee, two cups of tea, one egg, two strips of bacon, small quantity of cheese and black forest ham and mustard . . . two aspirins, three innings of a televised baseball game, and an indeterminate quantity of natural gas and electric power, several gallons of water, one thirtieth of a month's rental space. . . . While simultaneously consuming yesterday's newspaper, I consume an article about one Martin Davidson, a veritable Ajax of consumption, a man who wants to consume "nothing but the best." And does! He has recently

consumed a Cadillac limousine, for example. What have I consumed? Toothpaste, aspirin tablets. So, a little Dover sole. What is that to consuming a Cadillac? Davidson is even now just finishing up consuming $150,000 worth of furniture for the space he is consuming in a Manhattan loft. This newspaper story becomes a rebuke to my own trifling powers of consumption, which, I now see, lack drive and motivation.[14]

Produce, produce, produce; consume, consume, consume. "The civilization of always more" lacks discrimination about "Goods": their true nature, their true value, and who gets them. But we count them well.

MARKETS AND MORALITY

I had an instructive experience at a recent seminar for business executives in the Colorado Rockies—an experience that suggests the true heresy of our time. One of the readings discussed at this seminar was the Communist Manifesto, and these capitalists had very little trouble with that. They dealt with the old flaming revolutionary document quite calmly, and even, in a way, sympathetically. Certainly, they didn't mind the idea that *economic* realities are fundamental, and the rest of life and culture a superstructure built upon them. In fact, that idea was not far from a view some had expressed themselves. A Soviet citizen, presumably a Communist, an official of the United Nations rumored to be the head of the KGB thereabouts, was a member of the seminar; he and a number of the American business executives not only had friendly relations but often had a common point of view on the subjects we discussed. Neither Communist words nor a live Communist aroused much passionate disagreement.

On the last day of the seminar, on the other hand, we read a piece of Schumacher's *Small is Beautiful*[11] which dealt with what he called Buddhist economics, and that touched a nerve. This short section found in the Buddhist concept of "right livelihood" a contrast to many modern presuppositions. A modern economic point of view, said Schumacher, regards labor as a necessary evil both for the employer and for the worker—to be reduced as much as possible. For the employer it's a cost; for the workman it's a "disutility," a sacrifice that one must make in order to obtain wages to pay for one's leisure and comfort. Schumacher wrote: "The ideal from the point of view of the employer is to have output without employees, and the ideal from the point of view of the employee is to have income without employment."[12] Over against this, Schumacher sets what he calls a Buddhist point of view: Work has the functions of using and developing a person's faculties, and joining with other persons in a common task, as well as that of bringing forth goods and services of the community needs. From this point of view, the organization of work so that it is meaningless, boring, or nerve-racking is thoroughly objectionable. The magnification of "output" is not the objective, certainly not the sole objective. As against those who use as the criterion simply the total quantity of goods produced,

"The Buddhist sees the essence of civilization not in a multification of wants but in the purification of human character." To make total production the sole test, says Schumacher, is to stand truth on its head—to consider goods more important than people and consumption more important than creative activity. "It means shifting the emphasis from the worker to the product of work, that is, from the human to the subhuman, a surrender to the forces of evil."[13]

Well, unlike the Communist Manifesto, that reading brought a sharp and angry response from the American business executives. Some of the reaction was a dismissive "You're another," "Go back to where you came from" "So's your old man" kind. "Isn't Burma a police state, a totalitarian regime?" "Aren't the leaders of Buddhist countries trying hard to imitate the West's industrialism?" "Would you rather live there or here?"

Schumacher, who was a Roman Catholic, was presenting an ideal. In the book, though not in our seminar reading, he said the ideal he presents could be derived as well from Christian, Islamic, or other religions of the East, as from Buddhism. The reason that his criticism caused anger, resentment, and resistance when the Communist Manifesto did not was that he did not use production and quantity as the measure, and rejected the idea that economics is the foundation of civilization. That's the real heresy.

C. E. Lindblom says in *Politics and Markets*[14] that the heroes of his book are Adam Smith and Karl Marx. They are also heroes of the modern world, taken as a whole, although usually they are not presented as a pair. Symbolically, they represent the modern article of faith, that economics is fundamental. (I don't know whether that is exactly fair to Smith, although it seems to be to Marx.) There are those of us who respect Smith and Marx but find neither of them to be a "hero" in this sense; they overdo economics. So does the modern world, whether in Marxist Institutes or in the University of Chicago Department of Economics.

Economic science is, of course, a remarkable human achievement. Smith, whose major work was issued, with symbolic neatness, in the year of the American Declaration of Independence, came to have a *practical* effect that few other intellectual workers can match. The science of which he is a father combines high intellectual endeavor with practical effect in a way that is quite unusual. That combination appeared again in our recent past in the work of John Maynard Keynes. One hopes for such a genius now to deal with stagflation.

As with economics as a science, so also with the free market and the price mechanism as social inventions: they are splendid in their place. They "solve" a central problem that every society must: how shall scarce resources be allocated among competing claims? They answer that question by the signals sent back and forth through the social order: this much value foregone, indicated by this price, to obtain this value. They deal with that central problem, with respect to commodities, better than alternatives—better by the criteria of efficiency and "consumer satisfaction." The market may be a better solution also with respect to a higher criterion, human freedom—freedom in the consumers' choices and the producers' choices—and perhaps also in some correlation between the free market and the institutions of civic and political freedom.

But both the market as an institution and economics as a science make assumptions, and encourage interpretations of the world that are of restricted application. The fault is not in their essence but in their being spread across the entire civilization without effective subordination to criteria superior to those built into their assumptions.

The story is told about the economists and the moralists who got together when the old Federal Council of Churches, with Carnegie money, set in motion its series of volumes on ethics and economic life. The economists all said that the moral problem was easy—the golden rule and all that sort of thing; the *real* problems are economic. The moralists, of course, said the opposite: the *real* problems are the moral ones; after those are solved then the merely "technical" work of their application to the economy can be turned over to the economists. They were both wrong, but the economists were a little more wrong. Economics ought to be subordinate, and the market used only where appropriate. The market is and presumably will remain *one* of the central ways to coordinate human behavior. But it ought to be *only* one, used where it works best with the items of our life that fit the competitive market (commodities) and not where it doesn't.

In order for the market to work as an institution and for economics to work as a science, "value," in any moral or other "non-economic" sense, must be set aside. Pushkin must be as good as poetry. The price of the rich man's cigar must be measured neutrally against the price of milk for the hungry child. The cost of a resource that is inexhaustible, like sunlight, is not distinguished from the cost of the fossil fuels which will one day be exhausted. The indiscriminability of resources and of commodities is essential to the idea of the market and to the science of economics. The setting aside of moral (noneconomic) judgments is essential to the intellectual and institutional achievement. Here economists admit, or assert, the limitation of their territory. Characteristically, they say something like this paragraph from Paul Samuelson's *Economics*:

> Basic questions concerning right and wrong goals to be pursued cannot be settled by science as such. They belong in the realm of ethics and "value judgments." The citizenry must ultimately decide such issues. What the expert can do is point out the feasible alternatives and the true costs that may be involved in the different decisions.[15]

The apparent modesty in statements like this—characteristic of the social sciences generally—is readily apparent. That realm of "ethics and 'value judgments' " to which the economist appears to defer often becomes to his students, and perhaps to him, pretty vaporous. The really solid stuff, with plenty of "value judgment" actually bootlegged into it, is in the hands of the economist.

Another economist, John M. Clark, with a sense of the real limitations of his discipline, noticed his colleagues' belief that there is no way to avoid sheer arbitrary prejudice or "opinion" except by "scientific measurement."

> In some classes the student may find ethical judgments stigmatized as "prejudices." Thus one who judges a certain patent medicine harmful on the basis of medical evidence would be expressing a "prejudice"; while a purchase of the same remedy brought about merely by a persuasive tone of the radio advertiser's voice is not only an economic "fact," as distinct from a "prejudice," but a fact having due weight as evidence of "welfare," medical evidence to the contrary notwithstanding.[16]

On *Wall Street Week* a market analyst, who doesn't smoke herself, calmly tells the investors what tremendous money-makers the cigarette companies are.

Economics has an implied picture of human beings. "Economic man" cannot be dead because economics needs him as a presupposition: the consumer and producer whose behavior is "rational" (i.e., materially self–interested). And economics has also as a presupposition a methodological individualism: separate individual consumers and producers, each maximizing his gain. The science arose in a seventeenth century setting in conversation with individualistic social philosophies; it grew in the nineteenth century in a close relationship to utilitarianism; it had a renewed twentieth century life linked to an analytical and skeptical British empiricism. It has, in other words, its philosophical associations.

Schumacher quotes that most prestigious of all modern economists, John Maynard Keynes, as saying that the time should come before too long (although not yet) when economists are reduced to the status of dentists.[17] Although not yet! The implication is that just a little more needs to be done before that time. But it will never come unless there is a change in belief. It isn't as though one needed to get up to a certain point of economic development before one could then have the luxury of reducing economists to a subordinate place. There is no end to the logic of economic development once you wholeheartedly endorse it, and it brings its own philosophical implications along with it.

One of these is an obscuring of the fact of power. The way human beings have, from the start, scrambled to protect their own livelihood from the unrestrained fluctuations of the market suggests strongly that something important is neglected in the classical economic picture of society.

As the free market has fit into the actual life of human beings, it has not applied in the way described in texts; and for a good reason: though we endorse the efficient allocation that the free market can accomplish in general, we do not endorse its application to ourselves with respect to our own livelihood, our role as producers, our career, our life, our children's shoes, food, and shelter; or to our own status and money. None of us wants the occupation around which our own lives are built to be abolished by a movement in the market. Groups, therefore, have found ways to protect themselves against the free market as it applies to themselves as producers: the minimum wage, collective bargaining, the claim of a "living wage" by labor unions; the giant corporation, with limited liability, a "personality" of its own (protecting the shareholders), and some control over its price for businesses (monopoly, oligopoly, and "price leadership," condemned in theory, are sought in practice). Tariffs protect the home industry (everyone was, in theory, for international free trade but, in fact, nations find ways to protect their own). The multinational corporation, operating in different national economies, and the polyglot corporation, mixing a hundred products, protect themselves by diversity. Farmers weak in economic power used their strength in political power to find a government instrument to protect themselves: government price supports. The corner drugstore protects itself by fair price laws against "unfair" competition.

All of us try to find ways to keep our own livelihoods going. More than that, we want our careers and incomes to keep rising. *And* we do not do that just by making a better mouse trap. We do it by combining with other mouse trap makers to be sure the price of mouse traps is such as to keep us in business even if ours should prove to be second best or third best. In other words, all of us want to do something "uneconomic" with respect to our own livelihood, the support of our children, our own economic position, because that, quite literally, hits us where we live. We want some *power* over that. Some of us obtain a lot of it. And the problem of power is tied, of course, to the perennial egotism of humankind. Reinhold Niebuhr once said that special privileges make men dishonest. That is true in every sort of economy. One may ask about any society: how does it deal with this perennial fact?

Ideally and theoretically the individualistic market-directed engine of abundance makes use of that self-interest, by-passes its worst effects, and controls it. It makes use of it, as the dynamic that moves the invisible hand to everybody's benefit; it by-passes its worst effects, as Potter said, by expansion; it controls it by "setting ambition to check ambition"[18] (the phrase of Madison applied to diverse centers of political power; it is even more important to have multiple and countervailing centers of economic power, and the separation of economic from political power).

But, as it has developed, the society of abundance has, alas, not quite solved the problem after all. In two ways it exacerbates it: It not only uses but also *encourages* and *rationalizes* self interest; and it creates huge new power centers. And these two points combine: people do more rigorously self-interested acts in behalf of huge organizations than they would in their own behalf. Men who wouldn't dump their garbage over the fence into their neighbor's lawn dump their corporation's garbage in the river and the air, fouling whole counties out of their duty to the stockholders of the big organization.

Our modern, advanced economies generate an ethos that rationalizes self–interest and resists moral obligations that might restrain it. Adam Smith wrote in an era in which the moral "sentiments" could be assumed to exist, and to correct and contradict the imperatives of the market. But the sentiments fade.

Douglas Jay, the English economist, presented in his book *The Socialist Case* a picture of the logic of the free market, as it would be when freed of moral restraint.[19] A man is dying of thirst in a desert. The merchant coming by with his canteens of water could require of the dying man everything he had, and everything he would even have in his whole life, in return for a drink of water. That's what the traffic would bear. That would maximize his profit. That price would indicate to other suppliers that there surely is a strong demand for water out there in the desert.

Now, you may say that is an unfair example because it's a monopoly situation, of which the market theory disapproves: One should imagine *two* merchants with canteens, so the dying man can get them to bid against each other competitively. But if we multiply merchants—suppliers—how much more should we multiply persons dying of thirst—''consumers''—bidding for the water. Two of them? Ten? A hundred? A thousand clamoring desperately around the two merchants with their water oligopoly? It is a picture not altogether unlike some we have seen, and do now see, in the real world. In that situation, in the real world—a desperate person lacking a basic human need, whose very life is threatened by its lack—do not other motives than the economic ones come into play? Compassion, charity, fellow human feeling? Yes, sometimes they do. There are *other* motives than economic ones. When the merchant, out of the goodness of his heart, gives the dying man a drink of water—a handout—he is doing something unbusinesslike. Now, it is evident that giving water, when one has it, to the man dying of thirst, is a plain duty. It is a moral obligation. It is an act of justice, not of charity. It has its roots in our common humanity, and in the most obvious requirements of our shared life on this earth. But if the economic logic has grown beyond its boundaries, we lose sight of this elementary point.

We may, then, create a private interest society, a society of vindicated selfishness. I read an item in the paper that reminded me of the debate about private fall-out shelters in the early 1960s. This had to do, as everything this year, with gasoline. It seems that many private citizens have built their own private gasoline tanks in which to hoard gas, and that one of these is the former Secretary of the Navy, J. William Middendorf. He has installed a 4,000 gallon gasoline tank in the front yard of his home in McLean, Virginia—storage for enough gas to supply the average car for seven years. As defense of this, the former public servant said that he is in a situation where he has to get places; that he is in constant demand from a political, social, and business point of view.

I thought it might have been Art Buchwald, but it was a sober State Department economist named Edward L. Marsen who remarked about oil, presumably with a straight face, that one of the problems is that many people making price decisions in OPEC were trained at places like the Harvard Business School, and as a result are able to come up with splendid rationales for why $20-a-barrel is good for us.

TECHNOLOGY AND TRUTH

While the market ideology diminishes moral restraint, technology creates new power centers. Power is magnified by the enormous levers put into the hands of people at key points.

People at the top of giant organizations in a modern society claim they don't have as much power as the broad public; they operate within very narrow constraints of the institutional organizations of which they are the heads. In a democratic society, they operate within the limitations of public opinion, and of powerful collective interest groups; within a free market economy they claim to be governed by the choices of the scatter of consumers in the public. Perhaps there is a half truth in what they say. The presidents, executives, chief executive officers, boards of trustees, cabinets, and owners have a more limited power than it appears to outsiders. But they have a lot still. Nielsen Ratings make Fred Silvermans jump; but Fred Silvermans decide which programs the public can turn on and off. Maybe the distribution of power as between the mass and the elite, bound together, interacting, is a chicken-and-egg arrangement. Technology makes it a very big Chicken and a very big Egg.

TECHNOLOGY AND VALUES

Modern economic growth is, of course, correlated with the growth of the technological wonders of the modern world, and with the science of which that technology is an application. That modern technology has, indeed, brought human goods is beyond dispute. Those of us who write critically of it do so with the dictating machines and electric typewriters it provides. The articles are then reproduced by its Xerox and printing machines. We go back and forth to our offices in its automobiles and expect to go to its hospitals with its modern technical marvels of medicine to have our lives prolonged in order to write more articles critical of modern technology. Jacques Ellul's *The Technological Society* is available in a small town bookstore in a paperback version; making it available required a whole series of those technological actions on which the book is an unremitting attack. Modern technology is at the core of that economic growth that has raised life expectancy, met basic human needs, and made life more interesting for vast millions of the human race. That deeper human life for which this chapter is a plea, as over against the tendency of modern technology, is nevertheless in part made possible, at least for larger numbers of men and women than in the past, by the very technology that we here criticize.

So, once again, the point is not to reject what is undeniably good, but to point to its overextension and its intrinsic bias. Technology is not simply a neutral instrument that can be used now this way and now that, according to our choice. It has its own warp in it. I refer not to this technique or that machine, but to the entire complex—not only machines and techniques, multiplied, but also social organization, bureaucracies made possible by techniques like the telephone and the Xerox machine—and behind both the machinery and the bureaucracy, a habit of thinking.

The "technical" outlook has its own strong undertow on the modern world. It begins with a sharp sorting of purposes, a predetermination of the end to be achieved, a narrowing of the goal. This narrowing is done for the purpose of devising the most efficient means; Robert Merton has described technological society as a complex of standardized means to predetermined ends. Among the ends and purposes of life, some are sorted out and fixed. The ingenuity of man is turned to the devising of the means to achieve these fixed ends most efficiently, with the least cost. These are single, linear, and narrow goals.

It is, of course, part of the genius of modernity that it "predetermined" such ends. It has shown what it is about. Where other civilizations dreamed of, prayed over, speculated on, made myths and poems about what was to be found on the other side of the river, practical modern man built a bridge and crossed over to see. We are practical-minded, knowing what we are going to do and doing it. By fixing the goal, by honing the definition of it and the means to it, the machine man attains that goal beyond the dreams of merely tool-making man: there follow all the wonders and comforts of the modern world.

But if, in the original analysis and identifying of his purposes, he left something out, or if it is in the nature of higher values not to be separable from other values, or if it is in their nature to change and to blend, then the machine age is in trouble. As indeed it is. Having crossed to the other side of the river, practical modern men now ask, "Why was it we wanted to be here?"

All of us in our practical tasks need to focus on the goal. From Ben Franklin to today's self-help manual comes the message: keep your eye on the ball. Don't be distracted by a butterfly or a cloud. Keep your eye on the hammer hitting the head of the nail.

There is an old saying that a pickpocket who looks at a saint sees his pockets. He has focused on his role in the world. But he may leave out something important. Such a focus prolonged and underscored makes it difficult to shift. A conscientious life-long pickpocket is going to have a hard time describing what *sainthood* is, even though he may dimly sense that there is something more to life than picking pockets. He isn't going to be very interested in or good at describing what those other aspects of life are—whatever else there is to a human being than pockets to be picked.

I have a dentist who is a utopian. Not too long ago he looked at my teeth and found them not to be perfect and made a model of them. Here were my teeth smiling or grimacing back at me in a clay form. He proposed that this set of teeth be brought to perfection by means of braces worn for eight years and manufactured by him at great cost. I pictured going home to my family and telling my children that they would have to forego their college education in order to pay for daddy's braces. I asked him how my overbite compared to the rest of the human species. If one went out onto the street and stopped the people going by, how many would have an overbite like mine or worse? His answer: At least eight out of ten. Nevertheless, perfection of teeth being his objective, he proposed that I submit my dental equipment to this attainment. Up to a point, this focus of the dentist (and the focus of the pickpocket were we to approve his line of work) is admirable. Beyond a point it is not.

The focused goal is achieved by standardized means. The process hardens into a machine—including a social machine—that can achieve the narrowly defined results more efficiently—in a manner untouched by human hands. What we do as the machine age develops is to fix and take for granted and narrow the end in order to standardize the means and attain the results more "efficiently." We seek the one best way to accomplish the desired result: the efficient way, the way with the least cost for the most benefit. Having found it, we repeat it, and repeat it, and build the repetition into institutions: the machine, the system, the assembly line, the giant office, the "repetitious emphasis" of advertising jingles, dinning the brand name into your head whether you want to know it or not.

Labor is divided so that the parts of it may be done repeatedly and, therefore presumably, more efficiently: Charlie Chaplin bending and bending his wrist. Repetition and standardization will persist, in particular at the lower levels; they must to some degree, because they are part of the genius of this kind of society. They are essential to its many benefits: multiple Chevettes, multiple hamburgers, multiple pay checks from the payroll bureaucracy.

Technique—skill of all kinds—has its gentle satisfaction for the human spirit. It is a pleasure to understand well and speak properly, exactly, about the art and practice of some interesting human activity: running pass formations, marking up a bill, trading stocks, or editing a manuscript. But that satisfaction begins to fade when the skill is given over to the machine.

A tool, an extension of human powers, may be used for multiplicity of purposes: a hammer, a knife, a pair of scissors. A machine, on the other hand, performs much more efficiently a much smaller range of actions. Characteristically, it can be used for fewer goals performed in a more standardized fashion. It comes to be independent of human powers in achieving the narrowed results. The thermostat can't exactly put another log on the fire, or poke it to make the flames more lovely to the eye, or open the windows to the south to catch the breeze; but it can much more efficiently than you or I adapt the machinery that is heating and cooling the house to a desired set temperature.

When Adam Smith gives his famous description of the pin factory, he is telling us not about the enormously important idea of the free market—but about the related and even more important institutional arrangement, the division of labor.

I have seen a small manufactory . . . where ten men only were employed . . . and where some of them consequently performed two or three distinct . . . operations. Those ten persons . . . could make among them upwards of forty-eight thousand pins in a day. But if they had all wrought separately and independently . . . they could certainly not each of them have made twenty, perhaps not one pin in a day. . . . The division of labour . . . so far as it can be introduced, occasions, in every art, a proportionable increase of the productive powers of labour.[20]

In this most famous example the focus is on *production*—the amount turned out—not on the satisfactions and values in the work itself. But this is a *small* manufactory, where communal feeling may still apply, and share in the increased return. When the principle in this famous passage is applied broadside, in huge impersonal settings, we make a skewed society: more pins, less happiness in work. The hand of abundance attends to our subtlest desire as consumer but not to our needs, our safety, our happiness as workers (because that all appears at a cost).

The great principle of the division of labor—not only on the assembly line but also in the modern bureaucracy, the giant social organization, everywhere in our specialized society—has at its base the same separating, narrowing, fixing of ends to be achieved in order to be more efficient that, carried too far, turns into a human disadvantage.

We are describing not only factories, offices, and machines but also a set of values and a belief system. The struggle with traditionalism in modernizing countries is, of course, a struggle not only to introduce machines and social arrangements but also to get rid of the belief systems that stand in the way of this utilitarian narrowing of human powers. We cannot build the efficient pin factory that Smith describes if two or three of the pin makers object to the making of pins because of a religious commandment; or refuse to show up on particular days because they would rather fish or look at the clouds or observe an ancient religious custom; or if they go by nature's time and not by the (so to speak) real time (Mumford explains that the clock is the original device of modern technology);[21] or if they decide on the basis of a revelation or of an independent judgment that pins are undesirable. The premise of the pin factory is that pins are good and more pins are better than fewer pins. And the expectation is that each of the workers performing his little function believes that or is willing to acquiesce in order for the wealth of nations to be increased. Rapid economic growth requires a system of belief. In the recent period of disillusionment, some have leaped clear outside the Western attitude to a different outlook.

The "ends" and the "means" are not as easily separated in life as they are in thought. "Means," so called, do not retain their subordinate place. Human beings who devote their energies to presumably instrumental activities make them an end in themselves: the army, the business corporation, the government bureau, the data processing center. Our affections follow our attention. The operation itself takes on an independent value. There is a withdrawal of attention and positive feeling from the original purpose and a transfer to the means.

No one who knew this country at the height of its love affair with the automobile would believe that the primary focus of devotion was simply convenient movement from one place to another. The automobile people had a revealing, very subordinate and dismissive, category of plainer vehicles kept in the rear of showrooms for old-maid schoolteachers: "transportation" cars. So much for them. In one nice moment in the motion picture "Dr. Strangelove," the American general played by George C. Scott rejoices in the proficiency of the American bombing force, even though the result will be to blow up the world.

"Efficiency" and "practicality" are regularly listed as "values" characteristically held by Americans, independently of what one may be efficiently or practically doing, rather in the way that "success" is admired, to the astonishment of a foreign observer like Charles Dickens, without regard to the field or activity in which success is achieved.

We all experience having our own minds altered by a long-term change in focus. The man who has been in the Executive Branch, disdainful of the compromising dilatoriness in Congress, watches in wonderment as his mind slowly alters its chemistry to include a greater sympathy for representative government when he goes to work on Capitol Hill. The hard-news, shoe-leather reporter who rejects the "punditry" of the "deep-think" armchair columnists finds, if he is given a column or editorials to write, that this other activity is harder and worthier than he thought. His mind shifts: a man's mind is shaped by what he actually spends his days doing, and by the value those around him place on it. A reduced and narrow concentration upon particular limited objects diminishes the capacity to deal with the broad, unreduced remainder of human existence. The pickpocket is so boring when he tries to talk about sainthood that one is eager to get him back on the subject of pockets and picking them.

The "operator" resists questioning and criticism if it falls outside the domain to which his operation commits him. The human will, having resolved its questions, reached a decision, and moved to action, does not welcome the reopening of original questions. That is true of all of us, in our practical activities, including moral practice: after a certain point, and with reference to a certain set of activities, the decision is made. Decision and action require that such a closure be reached. And that characteristic of purposive human action is institutionalized in the operational society. In that shoe factory, the person on the line who puts in the laces depends upon the man who punches the holes: the latter is not expected to be exploring new values in sandals, throwing off the assembly line. One can't produce, produce, produce, if one has doubts about what is being produced. Smith's famous pin factory would not have functioned well if several of the workers doing their particular functions had decided pins were a bad idea or that there was a better way to produce them.

A man I worked with and admired in city administration, a mayor who was a great urban developer, used to say that "critics build nothing"—a very American economic-technological sentiment. In a sense it's true. But critics may keep something from being built that shouldn't be. Perhaps the MacDonell Douglas Corporation needed critics of the DC-10. In the midst of the Vietnam War a general asked Vice President Hubert Humphrey a rather direct question. "Mr. Vice President," he said, "just what is our American MISSION in Vietnam?" "That's a good question," said Humphrey.

I served as an alderman with the mayor I mentioned, voting for a lot of redevelopment projects that I wish I had opposed. The city of New Haven would be better off if the Oak Street Connector didn't split the town physically and socially, creating an "Other Side of the Tracks" situation that wasn't there before. And I was proudest of something that didn't get built exactly because we were critics: a connector to the interstate going north to Hartford, which connector would have swept down through East Rock Park, straightening out the Mill River—as the state highway people showed it to us on their "rendering"—destroying the trees, removing the muskrats, and eliminating a natural area within the city boundaries. I'm proud to go back and see that truncated limb, that exit broken off at the shoulder—that something that was never built because of critics.

If your focus is narrow, your capacity to criticize may atrophy. The story is told about the salesman who remarked, after watching Willie Loman's tribulation in *Death of a Salesman*, "Well, New England never was very good territory." It is necessary, continually, to be in touch with the true ends of living—therefore, continually to criticize, examine, reform. The critical function of the human mind is not just negative—stopping the East Rock connector, the DC-10, Thalidomide, or the Vietnam War—immensely valuable though such functions are. Criticism is also evaluation: making explicit the true ends of living, the spark, the imagination, to guide and enliven life, as well as resist its horrors.

Ironically, the "rationality" of technical reason turns irrational by being stuck in grooves. Since it must take static objectives for granted, and static circumstances, it gets caught when conditions change as they always do. The know-how people, like French generals in the familiar criticism, are always fighting the last war. The externalized, repetitious precision of technical knowledge carried too far fails in part because of its virtues: it is too closely linked to those objects and conditions it has mastered. Tied to those particulars, it falls down when faced with new ones. The productive machinery develops a life of its own. It develops momentum and moves inexorably in its own direction.

Once a machine has been constructed for its narrow tasks, it can do them by its own motion. Bureaucracies constructed for a presumed rational end persist beyond the memory of whatever that end originally was; people go each day to the office to shuffle the paper back and forth and answer the phone calls and memos without a memory of what the whole enterprise was originally supposed to do.

Why does this happen? Because questioning and redirection are at odds with the rudimentary form of technical reason, when it has sprung free from moral reason. Once the show is rolling and we have a stake in it, we want it to keep rolling even if it's ridiculous. No man or woman wants the source of his or her own livelihood to be abolished, however much he or she may harbor somewhere in a secret heart a notion that what he or she is doing doesn't make any sense. I want my job to continue; I want to rise in the hierarchy of the organization; I want my salary to increase; I want the number of people working under my direction to be larger; I want my prestige to be greater; I want my reputation to increase; I want all those things whether or not the activity with which I am associated really serves any purpose. Men fight with each other to be the top person in the chocolate drop division of a candy company with all the intensity of emperors fighting over the Eurasian heartland. And making chocolate drops is a great deal more beneficial than many other activities that we fight to carry on.

Modern society achieves certain kinds of human objectives more effectively than others: it serves the limited, tangible, discrete, the making of commodities. Other values of human living—aesthetic, intellectual, religious, moral, the joys and satisfaction of sociability and human companionship, all the values that cannot be made into something like commodities—are not so well served. They are harder to identify and fulfill; on the other hand, generally speaking, they are more important.

Once the machine, the organization, or the belief is shaped by calculating market-measured utilities, one cannot go back and reshape society in order to recover what has been set aside. One of the comedies of our modern world is the effort to do so; to plaster on, to add on, to recover what has been lost by the machinery of an advanced consumer society, as when big impersonal firms try to "personalize" their service.

The larger vision or belief must guide and discipline the techniques of the material order. The story of our modern economy is that it has distended its connection with fundamental human values and has shaped itself in a way that makes it difficult to restore a solid link to them.

We are all aware of the side effects of inventions and new technology; new social developments and ideas have the same. And they accumulate.

The automobile, intended for speedier transportation, brought with it the carnage on the highway that makes each holiday weekend as destructive as a major battle in a war; brought the end of the Sunday night church service and of courting in the parlor; brought the drive-in theatre, the drive-in ice cream joint, the drive-in bank, the drive-in life, leading to the drive-in funeral parlor; accelerated the centrifugal movement of the population to the suburbs and the exurbs; filled the Los Angeles air with smog; tangled each major city's life in the unbelievable concrete spaghetti of freeways. The airplane led not only to even faster transportation but also to crashes of the DC-10 at O'Hare Airport and the growth of Phoenix and the Sun Belt; to conferences with visitors who have had lunch in London and supper in New York; to shuttle diplomacy; to the deterioration of the railroad. The telephone made possible not only point-to-point conversation across miles but also the growth of modern bureaucracy. (There were bureaucracies going back to ancient Egypt, but our modern form is unthinkable without telephones.)

Television led not only to 80 million people being able to watch the President resign, in living color, but has also led to the decline of magazines and of books; to the new form of the pseudo-events; to the decline of literacy; to the cult of the personality in its new form. The new medical technology creates dilemmas about death and dying that didn't exist before we had it: whether to keep people alive by machinery; x-rays for diagnosis not only save lives but also can kill them. The patient may die, but not undiagnosed. The computer helped create the credit card society. And so on. And on.

Through all these inventions and machines and a hundred others there runs a threat of technological imperative: if you can make it, you do. And this is followed by the second rule: once you have built it, you then have to use it. It can't be undone. A university builds an elaborate state-wide system of terminals for television and computer relationships among its campuses; it then has to generate the courses and conferences that will make use of this expensive machinery.

I remember a constituent saying to me as we argued about the East Rock Connector: "You can't stop progress." Well, in the first place, that begs the question of what is progress (I thought it was regress not progress); and it took for granted that it couldn't be stopped. Happily, in that instance it was. But the undercurrent of society is as she stated it.

It *is* possible for a society to make discriminate judgments about technology if it has its values clear. In *Giving Up the Gun: Japan's Reversion to the Sword*,[22] Noel Perrin tells the story of Japan's rejection of firearms, even after they had developed and used an advanced form of firearms in the early days of gunpowder. They stopped progress. They even went backward—back to the sword. For reasons of samurai tradition, and aesthetic commitment to the sword, and a sense of how human beings gracefully engaged in warfare—and also geopolitics—Japan was able, for three centuries before Commodore Perry landed, to reject an "advance" in technology that it was fully capable of producing and using. Moreover, Mr. Perrin makes clear that *other* forms of technological development continued, even though there had been this conscious rejection of one particular possibility in the field of weapons. The Japanese were *selective*: it can be done.

But that, usually, is not our way. And the effects of technological development are not simply those of this one and that one taken separately. Society is shaped by all of them put together. By machines and techniques—and by the form of social organization they engender. Machinery and bureaucracy encourage the outlook that we have called technical reason: task oriented, fact oriented, pragmatic, limited, unspeculative, not holistic, linear—the mind of modern man.

All these technological "advances" have their side effects; all of them taken together have an accumulation of side effects. The energy problem, for example, is a symbol of the accumulation of an enormous set of energy-using devices: automobiles, air conditioning units, airplanes, computers, the elevators you don't want to be trapped in during a black-out, at least not without picking the people you would be trapped with. We find ourselves now unable to believe that the flow of energy upon which the apparatus of our society depends might be in real danger.

We human beings are always social selves, with the community of which we are a part constitutive of the self. Modern technology increases the links of the self to the society. Our technological development dramatizes our increased interdependence. As Karl Mannheim[23] says, a wreck on a stagecoach line didn't affect much of anybody except the people on the stagecoach; the wreck of the railroad, he said, or of the DC-10, we may now say, has widening effects far beyond those immediately affected.

If you think that the consumer's choice in the market is free and sovereign in the way some pictures of the operation of the economy suggest, try banishing the television set from your home if you have a ten-year-old child. What you will discover is that the children with whom your child plays tell him about television programs and the products advertised thereupon and his appetite for this fruit increases in the measure that it is forbidden.

The mass production world moves in enormously powerful grooves. Since Los Angeles has the automobile as its premise, you cannot live there without one, however abominable you may find the internal combustion engine to be.

The immense increase in life's possibilities that growth and abundance bring brings also the impossibility (or difficulty) of retreat. I'd like to live the simple life, but I can't afford it.

The expansion of life's possibilities shuts off other possibilities, too. David Halberstam tells the story, in his book *The Powers That Be*, of William Paley's effort to keep Bill Moyers at CBS—Moyers, symbol of the Ed Murrow tradition nearest to Paley's heart, emblem of CBS superiority. What would it take to keep Moyers? A regular news show in prime time. Paley's head sinks. "I'm sorry, Bill," the chairman said, "I can't do it anymore. The minute is worth too much now." He couldn't do what he *most* believed in because the entertainment–commercial network had made each minute too profitable to allow it. Reading about Paley's sad condition one is reminded of the old American gag: what good is happiness? You can't buy money with it. What good is the heart's desire? You can't get prime time rates for it.

For the rest of us, the dilemmas are simpler but related. It is not just that you can't live the life of a family on a family farm if inexorable economic logic of riches creates factories in the field; you can't live a family life in a little brown house in the middle of the city either. People who like to walk, or to ride a train, or to ride to work on a bus, cannot do it in Los Angeles; and one fears that economic growth means the Los Angelization of the world.

And so the efficient society sometimes curiously comes to this kind of thing. After the near accident at Three Mile Island, the Nuclear Regulatory Commission referred to what it called "the hassle factor": room full of flashing lights, sounding alarms and sometime contradictory instruments, all demanding attention, which the operators inevitably faced during an emergency. The Commission described the situation of a busy control room, of people standing on each others' feet, of red lights that are green except when they are blue, indicators fourteen feet off the ground and behind you. It sounds familiar, even to people who have no connection with nuclear power plants.

MEDIUM-SIZED IS BEAUTIFUL

I have stated several times that the "true ends of living" may be ill served by an indiscriminate economic growth, or by the spirit accompanying it. Let me now approach the matter from that side, with some remarks about several important areas of living.

Work: From "To Labor Is to Pray" to "A Hard Day at the Office"

A modern, advanced economy, prosperous and growing each year, provides jobs for the great mass of its citizenry, and a reward for those jobs that is beyond the dreams of such small-scale avarice as there may have been in premodern civilizations. Russian observers in the old days, shown the parking lot at the Ford Motor Company, could not believe that all those automobiles belonged to workers—to members of the putative proletariat.

Economic growth is a necessity, in some form, where the population is growing and the work force increases. To provide jobs for everybody who wants to work is a commitment that runs through modern American life—as yet unfulfilled—and has its unassailable human base. One wishes the nation were committed even more firmly than the Employment Act of 1946 committed us to full employment. Nevertheless, a modern society has, at the same time, a problem of another kind with work: what jobs mean and are and represent. A joking Hoosier sign says: "I love my job. It's the work I hate."

At the top of the list of "productive" values that make for economic growth there stands the "work ethic," the "protestant ethic," the Calvinist and then the Yankee virtues: work; avoid idleness; avoid frivolity; order your time; improve yourself. These virtues Americans have had in good measure, and they have helped to make us a people of plenty. The virtues are genuine, and many people of the world wish they had them the way Americans and some other industrious folk do, or did. But they are virtues only to a point and in a context.

Once upon a time, monks would say "to labor is to pray." The "work ethic" of modern advanced industrial nations has moved a distance away from that belief. Already in the American colonies there was, among the Puritan settlers, reference to another religious statement about work, which, although drawn from the New Testament, had a different edge to it: "He who does not work, neither shall he eat." This sentence, unlike the earlier one, has its echoes right down to contemporary attitudes. It was a sentence used in the colonies under the pressure of dire necessity; its equivalent has come to be used today for invidious class purposes.

The quotations take their modern turn with this one from Benjamin Franklin: "God helps them that helps themselves." That idea is a long way from "to labor is to pray," in which the working human being is doing a service to God, instead of the other way around.

The original quotation I have given implies that work is an intrinsic good, an end in itself—or, rather, that it has its immediate relationship to the transcendent and, therefore, is not merely instrumental, to be measured by its productivity. Work is intrinsic or constitutive of human life, and not to be tested solely by some other end (production or success). I mentioned Schumacher's presentation of a Buddhist understanding of "right livelihood," in which the point about work is its meaning in the life of the worker. If we look not to the monks, Christian or Buddhist, but to Sigmund Freud we find the same thing: work paired with love at the center of life.

To be sure, work is a function in society. The monks carefully copied manuscripts and chronicles which historians 500 years later would use to understand the time; they grew grapes and made wine which would one day be drunk presumably with pleasure by fellow human beings probably on the whole of lesser spiritual stature than themselves. We do work to serve the ends of our fellow human beings and of the community of which we are a part. But that instrumental aspect is not always primary, and certainly should not be exclusive.

As the logic of productivity takes over, work changes. It is changed to underline the product and the reward. Work comes to be exclusively measured by its *product*, which, in turn, is defined by the market. And it becomes more sharply individual and linked to a person's "success" and "failure," which, in turn, is taken to define his worth.

When Max Weber examined the relationship of the Protestant ethic to a capitalist ethos, he turned to American materials for his example[24]—specifically to Benjamin Franklin and the injunctions that have burned their way into the memory of every American. Be frugal; avoid idleness; use time well: fools make feasts and wise men eat them. Every American knows, though he may not know where it comes from, what early to bed and early to rise will make him. Weber summarized the ethic that Protestantism, helped along by capitalism, came to recommend: "Time is money." One's day, one's life, one's time is to be measured out against its market-measured productivity.

The work ethic changed into the ethic of individual enterprise and success, and then into an ethic of *jobs*. The value is not "work" in and of itself but work as it is defined in the market, producing tangible results, and serving an individual's material interests. The value of such "work" is defined and rewarded by the market: a job ethic.

There is also an *enterprise* ethic in which one proves oneself by initiatives taken. One buys up the land where the city will spread, or builds a better mouse trap, or sees the possibilities in fast food hamburgers or magazines that have more pictures than text: imagination, cunning, opportunism, shrewd practical observation, calculating self-interest, and all the qualities that accompany "enterprise"—which is not the same thing as "work."

It is now altogether possible to have a job and not do much work. Because the work ethic defines itself by the transactions of the market, the very important work of homemaker and mother is not counted as "work." The woman who is paid a monthly check by the American Broadcasting Company for composing moronic jingles is doing work, whereas the woman who is raising a family of five children is not. Most of us in the middle class know situations in which the husband has a job without work and the wife works without a job. A man goes off to an office where he sits behind a desk doing nothing, phoning his buddies to make dates for lunch and squash, while his wife at home drives the children to school, the dentist, the music lessons, and their friends' houses; makes the meals, cleans the house, writes the checks, deals with the man who fixes the leaking roof. He gets a check. She doesn't. He "works." She doesn't. He counts in the GNP. She doesn't.

When work gets turned around and made into an instrument to other ends—to money, success, and even to one's sense of worth—individuals feel a necessity to prove themselves by an upwardly mobile career, as a kind of test of manhood or personhood. Market earnings or position in a bureaucracy are the measure of the person: the money you have; the size of the house; all those people who work under you; the size of the budget of the outfit you administer—these are the tests of your worth. It is invidious: getting ahead means getting ahead of somebody. Room at the top means others are at the bottom. The underside of the American success ethic was dramatized by the numbers of people in the depression, men trying to support their families, who couldn't because of the enormous social catastrophe—one quarter of the work force thrown out of work by a major depression—who, nevertheless, blamed themselves for their inability to support their families.

The other side of the ethic that defines work in terms of success, money, and the market is its lack of sympathy for those who are not "successful." The rich and generous American people, though willing to do many acts of charity, have not been able to create a system of welfare for the least advantaged that is consonant with our wealth and our fundamental generosity. We are hamstrung by our ideology. As is often said, many of the elements of the welfare state created by Bismarck in the late nineteenth century in Germany have not yet been achieved in the United States. We now have the world's first *minority* poor, all the more difficult to aid because they are a minority and lack even the prestige of numbers. That shows that there is something inhumane in the way we have understood our relations to each other and to work. When you hear the man who has risen to the top from one suspender saying contentiously of those who have not: "I worked; why can't *they*"; you hear the echo of a distorted system of values. It exalts the merely tangible success and atrophies the impulses of human community, fellow-feeling, and sympathy.

Leisure: Full Morocco Territory

I know that the phrase, "The Quality of Life," which has been widely used now for almost a decade, is intended to convey some of the points made in these pages; its ubiquity is a sign that these points have some currency. At the same time, though, it seems a little suspect. Might it not be co-opted or assimilated into the modern machine, with no serious criticisms of it? (That's one of the marks of the modern commercial apparatus: notice how the recent, youthful-romantic interest in "natural" foods is picked up by the big sellers; but, of course, there are a thousand examples.) Can one not imagine "Quality of Life" as a heading in well-designed letters over an institutional advertisement in a slick magazine, with a picture of some beautiful person in his study with his excellent cigars? I remember Robert Hutchins telling the story about the time when he was moving—the Fund for Republic and he were moving—to Santa Barbara, California, and creating the Center for the Study of Democratic Institutions. Hutchins, also chairman of the editorial board of Encyclopedia Brittanica, had a courtesy visit from the head salesman in the region. To make conversation, Hutchins asked how the Encyclopedia was received in that area. The head salesman's memorable reply referred to the Brittanica's most expensive binding: "Doctor," he said, "this is full Morocco territory."

One may fear that "Quality of Life" may mean Full Morocco Territory: a more refined hedonism, a more discriminating materialism. "Quality" is a word badly damaged by advertising, and "Life" may mean what one reads about in the "Living" sections of newspapers. Hirsch quotes from a book called *The Harried Leisure Class*, making a slightly different point, picturing the modern man "drinking Brazilian coffee, smoking a Dutch cigar, sipping a French cognac, reading *The New York Times*, listening to a Brandenburg Concerto and entertaining his Swedish wife—all at the same time, with varying degrees of success."[25] The life of quality sounds like hard work.

I do not mean so much to question a phrase as to break open the differences it may paper over. Aesthetic values may be co-opted by the privately purchased "Quality of Life" ethic, for those who can afford the security guards to protect it. Prosperity helps the arts to find their patrons, yet aesthetic considerations are not part of the central spirit of the dynamism of modern abundance. Often they are at odds with the imperative to produce. The people who oppose billboards on the highway because they prefer to look at trees seemed an enormous nuisance, and were described scornfully as "aesthetes" by the late Senator Bob Kerr of Oklahoma. The same word appeared, in the same tone of voice, in the debate about the East Rock Connector and the aesthetic values of the natural environment of East Rock Park. A great deal of the appreciation of the beauties of the world in which we live draws upon decisions made by many other people. Very few of us have control of the environment of our day-to-day surroundings sufficient to make them what we want them to be from an aesthetic point of view; nobody, happily, controls the whole of his or her environment. Therefore we depend upon the decisions of others and upon what the economists calls "externalities"; neither imperative of production nor the logic of the market is inclined to be very sensitive to these values, and often finds them an impediment.

The same may apply to the goods of the mind—education, curiosity, intellectual satisfaction. By their nature, they aren't well served by the production ethic. One is reminded of the story of General Eisenhower when he was President of Columbia University. Told that the Philosophy Department had no metaphysician, he is said to have responded: "Procure one!" Metaphysics doesn't fit the procurement system the way material goods do.

Among the genuine goods of human living there is a young person or an older person using the capacity to understand, to perceive and conceive, to enjoy the shock of recognition when the truth about our life is stated truthfully, forcefully, with insight.

The undertow of a society intent upon maximum production pulls against the intrinsic worth of the intellectual virtues and tries to adapt them to its own native instrumentalism: what good is it to take a course in English literature? What good is it to take philosophy? The vocationalism that afflicts higher education at the present moment is but a more intense form of a continuing problem. The true goods of liberal arts—of understanding, appreciation, criticism, perception, creativity—are crowded by the piston thrusts of a productivity that can be measured. When the human soul, feeling a lack, seeks them again, the modern society is clumsy in its efforts to recover them. For Aristotle contemplation was the highest human activity—one could scarcely think of a more unAmerican, uneconomic, unmodern, idea than that.

Communication: The Voice You Hear May Be Your Own

There are no human persons without community, and there is no community without communication. Modernity has provided us with a remarkable means of communicating with each other (maybe) across a great part of the globe. But what is this "mass" communicating?

No doubt, somebody writing a paper with the outlook of this chapter said the same thing I'm going to say when confronted with that first major mass production technological innovation, the printing press. Probably, somebody then wrote that that mechanical means represented a great loss from the humane crafting of written documents by quill pens on parchment. Technology often looks better in retrospect than it does to its critical contemporaries. People writing a hundred years ago attacked the noisy, disruptive, dirty railroad train; today, most of us would like to bring back the railroad. Despite what I will say in a moment, there will come a time when many, including intellectuals and social critics, will look back with affection and nostalgia on today's TV. It has already happened with radio and movies. One also has to concede that modern means of mass communication do have the capacity to make life more interesting and to make it possible for people to share in the life of the community in a way that was not possible before. In housing for the elderly, in nursing homes, in a hospital, in a lonely cabin, in an isolated apartment, people can share in the Pope's visit to Poland and the Inauguration of a President and the hearings of the House Judiciary Committee, as well as in the Hollywood Squares. Nevertheless, there is something fundamentally wrong with American commercial television (in its present form: new technology itself may improve it. We'll see.).

Human communication is a peculiarly inappropriate activity to be subject to the operation of the market and of modern technology's repetitious machinery. But we have a central system of communication—our "mass" communication—that is built on those foundations. It includes ironic reversals of two kinds: First, the producer creates demand. The system's foundation is demand creation, advertising, which Potter describes as the representative institution of abundance.[26] But *creating* demand—the producer persuading the consumer to buy something he wouldn't have bought—surely fits oddly with the theories of the market economy. Second, this demand-creation supports a system of mass communication, governed by a search for enormous numbers. The enormous pressure of competition for huge numbers presses the material into formulas with quick flashes of the most elemental appeals to an audience: material that will snatch the attention without engaging the mind. The material communicated on American network television is bait to attract an audience for the real message, which is the commercial.

Advertising and its subpart, television, are enormously important segments of American society by any measure: 98 percent of the households have a set, more than have bathtubs, watched 6½ hours a day, on the average; nightly audiences of 20, 30, 40 million people—audiences on extraordinary occasions (Super Bowls, Presidential resignations) of 90, 100, 110 million. Expenditures of the advertising industry, in all phases, for 1977 were $38 billion, for television

$7.6 billion, of which $576 million plugged toiletries; $279 million, laundry soaps, cleansers, and polishes; $104 million, pet products. Advertising and mass communication are overwhelmingly larger than the institutions that have traditionally been the bearers of the symbols, values, and meaning in the society: the churches, the schools, and the arts.

The difference from these traditional institutional sources of value and meaning goes beyond the difference in size. There is a difference in kind. A difference in motive. The modern mass communication media built around advertising represent the central example of the reversal of true purpose I have been describing.

In the case of each of the other institutions I have mentioned, there is some antecedent content which they brought to the audience, the congregation, or the students. There was content, message, a standard. The religious communities have the traditional norms of the message of their beliefs; an educational system has a content that has to be taught to students whether they want to learn it or not; art has at least the individual artist's vision, in the modern period, and in an older period a larger tradition of artistic expression which may not win any immediate response from the public to which, nevertheless, the artist is committed. But modern mass communication has nothing antecedent to or independent of the attraction of the audience; attracting the audience is the entire purpose of the enterprise. It is, therefore, essentially empty. It is not communicating in order to carry a message, but creating a message in order to attract an audience. When Thoreau asked, on the invention of the telegraph, whether Maine had anything to say to Texas, he certainly asked a relevant question. As the communication has developed beyond the mere point-to-point of the telegraph and the telephone on down to a modern, 50 million person nightly television undertaking, it is exactly the case that California and New York have nothing to say to the rest of the country—nothing that should interfere with their getting the country to watch, except, of course: buy dog food.

The defense that the industry people use is, in fact, an indictment: they give the public what it wants. They give the public what it wants at its most undiscriminating and superficial level in order to attain the largest numbers. (Human beings in their depth and reality need and want something that is not the same as that which will momentarily attract their attention.)

The case of mass communications is particularly important because what we are talking about is the ethos of the society. How does one alter the ethos of the society? How do leaders, teachers, heroes, and examples get a purchase upon the mind and conscience of the nation in order to shape it? The age of celebrity, publicity, and mass communications drives out communications of a more substantial sort that would go against the main trend of the easiest ways to attract an audience.

Friendship, Sociability, Communal Feeling: The Crowd Still Lonely

If you think about what is truly of value in living, almost anyone would name links to others: to friends, to comrades in shared enterprises, to a small group of people who know each other and share each other's woes and pleasures. When the calculations of costs and benefits or the commercial exchange of the market enter, the true but fragile goods here described are put in peril. Part of the pathos of an advanced commercial society is the effort falsely to reimpose what has been driven out by the pressures of powerful economic forces: to "personalize," to dial-a-friend, to imply a chummy atmosphere by the name or advertising of a firm. All of this false personalization and false sociability is an offense against the real thing.

Now, it is not only true that friendship and sociability are not compatible with the forces that make for economic abundance (high technology, technical rationality, the market and commodity focus of life); it is further true that the highly advanced economies with a heavy emphasis on economic growth endanger and make difficult friendship and sociability. Fred Hirsch, in *The Social Limits to Growth*, notes this fact and mentions one cause: "the increasing premium that people have put on their time . . . has eroded sociability." And, "this erosion has been deepened by the influence of advertising and the self-interest ethos of the market. . . ."[27] I remember a Washington reporter, high powered, saying of a high government official, that being a friend of his was worth his time because—and then he specified the particular utilities of this friendship for his career. It did seem to me a rather tarnished view of friendship.

This point plainly could be carried further and dramatized by reference to romance, marriage, and sexual relations. Hirsch deals with the destructive role of commerce in that realm. If you would list all of the human interconnections that make life true and real—of parent to child, of husband to wife, of friend to friend, of neighbor to neighbor, of the group at the Sunday evening church service or the other group at the pub, of a group of companions in their companionship, you would have covered a very large part of the most valuable elements of our life. The ways of thought developed under the impulse of the economy of the market and advanced technology don't suit them. Calculation of personal advantage in tangible terms; commerce; utility in the simplest sense . . . that's not appropriate to friendship, love, family, and group loyalty. I remember arguing on the corner of York Street and Chapel in New Haven with a very bright Yale undergraduate. I said that, after all, he wouldn't apply a cost-benefit analysis to his relationship with his girlfriend. He answered instantly: yes, he certainly would. I went away depressed.

That this friendship and sociability is a fundamental human good is so evident it needs no elaboration. It is featured not only in the high terms of the Christian religion's emphasis on "love" but, at a lower degree of temperature, in Aristotle's chapter on Friendship, and in Montaigne's description of his friendship with La Boètie: "If anyone should ask why I loved him it could only be answered: because it was he; because it was I; it was by some secret appointment of Heaven."[28] The secret appointments of Heaven don't fit very well with cost accounting.

HUMANISM IS STILL AVAILABLE

Growing numbers of people perceive a faulty connection between the material apparatus for a "better" life and its actual achievement. The economic and social machinery often miss the mark; overshoot it; overelaborate the effort to achieve it. (Barry Commoner wrote about a nuclear power plant that it surely is a complicated way to boil water.)[29] Most important, the apparatus makes its own demands. Because you have the machine you do things to use it, you do what it requires. Means determine ends. It need not be that way. But, for it to be otherwise, there must be a deeper consciousness of profounder purposes in life.

The problem is mostly in the realm of *belief*, and not of the material base of living, per se, or even of the technology, machines, and giant organizations of an advanced economy—although it isn't easy to make the distinction. The growing economy, technology, machines, and giant organizations would be less hazardous if we believed in them less.

Of course, as this elaborate modern apparatus develops an imperative, it encourages certain beliefs and values, and discourages others; and it develops, partly because of these beliefs and values, its own hazardous inclinations. But one need not share the notion of a modern monster presented, for example, by Jacques Ellul and Herbert Marcuse. Nor need one adopt any economic or technological determinism. To the contrary, the point is human freedom and responsibility. The modern society of Big Production needs, more than another society, a conscious awareness of the human goods which it inclines to override.

The phenomena discussed in this chapter tend to assume that the ends of living may be taken for granted; that there isn't much need to articulate or examine or affirm in explicit belief the meaning and purpose of life on this earth. All of that kind of thing can be assumed. Assuming it, taking it for granted, means that a set of answers about the ends of human living are, in fact, bootlegged into the society and held de facto, even by people who nominally hold otherwise. One creates the kind of institutionalized and taken-for-granted selfishness, hedonism, materialism, and instrumentalism that are the implicit working creed of the modern commercial world.

There is a long history of criticism of this modern industrial civilization, a criticism that emphasizes also its impersonality and its fragmenting of life. That long tradition, a minor voice in the civilization of economic growth, is now becoming louder. In the late 1960s it burst out with a roar, and also with an error. In that period, the antagonists to the modern world—returners to the soil, greeners of America, the ''counterculture,'' residents of communes, the further reaches of the youth movement and the new left—feeling that there was something wrong (feeling is the right word), indulged in a sweeping rejection of the civilization that has been created around modern economic growth. That sweeping rejection was the opposite side of the uncritical endorsement of economic and technological ''progress'' and of indiscriminate growth. The two reinforced each other. Both reflected a lack of roots in an intellectual and moral order.

But the roots of an understanding that would put economic growth in its proper place are not so hard to find. They are available not only in Schumacher's perhaps rather romanticized Buddhism but also in the classical and Biblical foundations of Western society, as amended by the democratic ideas of the Enlightment and the social democratic movement of the nineteenth century. Our immense, modern, economic boom came out of that Western humanism. The profounder understandings of that Western humanism are still available as a corrective to the present excesses.

NOTES

1. E. F. Schumacher, *Small is Beautiful: Economics as if People Mattered* (New York: Harper and Row Perennial Library, 1973), p. 48.

2. David M. Potter, *People of Plenty: Economic Abundance and the American Character* (Chicago: University of Chicago Press, 1968).

3. Ibid., p. 86.

4. Ibid., p. 85, note 7.

5. Ibid., p. 88.

6. Ibid., p. 122.

7. Ibid., p. 134.

8. Ibid., p. 89.

9. Fred Hirsch, *The Social Limits to Growth* (Cambridge: Harvard University Press, 1976), p. 7.

10. Charles C. Moskos, Jr., ''Why They Fight: U.S. Combat Soldiers in Vietnam,'' in *The White Majority: Between Poverty and Affluence*, ed. Louise Kapp Howe (New York: Vintage Books, 1970) pp. 88-9.

11. Schumacher, *Small is Beautiful*, pp. 53-62.

12. Ibid., p. 54.

13. Ibid., p. 55.

14. Charles E. Lindblom, *Politics and Markets* (New York: Basic Books, 1977).

15. Paul A. Samuelson, *Economics: An Introductory Analysis* (New York: McGraw Hill, 1961), pp. 8-9.

16. John M. Clark, *Economic Institutions and Human Welfare* (New York: Alfred A. Knopf, 1957), p. 25.

17. Schumacher, *Small is Beautiful*, p. 69.

18. James Madison, *Federalist Papers*, no. 51.

19. Douglas Jay, *The Socialist Case* (London: Faber and Faber, 1937).

20. Adam Smith, *The Wealth of Nations* (New York: P. F. Collier & Son, 1905), 1: 44.

21. Lewis Munford, *Technics and Civilization* (New York: Harcourt Brace, 1963).

22. Noel Perrin, *Giving up the Gun: Japan's Reversion to the Sword, 1543-1879* (Boston: D. R. Godine, 1979).

23. Karl Mannheim, *Man and Society in an Age of Reconstruction: Studies in Modern Social Structure* (New York: Harcourt, Brace and World.)

24. David Halberstam, *The Powers That Be* (New York: Alfred A. Knopf, 1979), p. 734.

25. Hirsch, *The Social Limits to Growth*, p. 73.

26. Potter, *People of Plenty*, pp. 172-7.

27. Hirsch, *The Social Limits to Growth*, p. 84.

28. Montaigne, "Of Friendship," in *Essays*, vol. 1, chapter 27.

29. Barry Commoner, *The Politics of Energy* (New York: Alfred A. Knopf, 1979).

Chapter 4

THE ROAD TO QUALITATIVE GROWTH

Robert Hamrin

The 1960s are already seen historically as the decade of growth. It now appears likely that the 1970s will be looked upon by future historians as the decade of debate about growth. Much of this debate has been reflected in the first three Woodlands Conferences.

Prior to 1970, the value of growth was rarely seriously questioned. Growth was discussed or analyzed solely in terms of how to maximize the growth rate. The first overt challenge to this growth ethic came on Earth Day in April 1970. The simple message, carried by the media in the U.S. and other industrial countries, was that growth may not be all good—there may be some serious adverse by-products (costs) that at least partially offset the benefits. The debate had begun.

Full–scale debate was launched two years later with the publication of *The Limits to Growth*. This computer-based study, sponsored by the Club of Rome, frontally assaulted the fundamental operating assumptions of the developed nations that growth of both the economy and population is good: industrialization, having brought us material abundance, is good; production inputs are, and will always be, available (if not domestically, then readily through imports); and the environment is a convenient, free, and abundant receptacle for our wastes.

Into this complacent world came the staggering hypothesis that the world system would suffer catastrophic collapse before the year 2100 unless drastic movement was begun immediately toward reaching a state of global equilibrium defined by a set of minimum requirements that: (1) capital plant and population be held at a constant size; (2) all input and output rates—births, deaths, investment, depreciation—be kept to a minimum; and (3) levels of capital and population, as well as the ratio of the two, be set in accordance with the values of society. Though much critiqued (and justifiably so, since it was a highly aggregated, relatively crude global model), the study did spread the idea of physical limits to growth and spawned, between 1972 and 1976, an outpouring of literature and the establishment of institutes in numerous countries to study what Aurelio Paccei, founder of the Club of Rome, had dubbed the "world problematique."

The first Woodlands Conference in October 1975 focused primarily on these physical limits to growth, but also raised the central topic of the next phase of the debate from 1975 to 1979—the social limits to growth. This was essentially a shift in focus of analysis from the supply to the demand side. The essential point of the social limits argument is that, increasingly, the demands of individuals in the affluent countries are changing from material goods to nonmaterial forms of consumption.

Whatever the limits might actually be, the issue raised the question of possible alternatives to growth. This became the theme of the Second Woodlands Conference in 1977, which ended in much frustration because no clear alternatives were put forth. This "negative lesson," however, was extremely important, for it began the constructive thought process that has led to the positive, realistic, and necessary theme of the Third Woodlands Conference—how to achieve managed, sustainable growth.

This theme is not a break from the limits theme but, on the contrary, a logical extension of it. It assumes that growth will continue but that, because of both physical and social limits, it will likely be slower and it will certainly have a radically altered composition. Thus, the way we grow and the end purposes of growth will be the crucial considerations in the managed, qualitative growth era of the 1980s and beyond.

One of the most profound insights to emerge from the growth debate is that economic growth is not solely economic but is properly viewed as an irreversible economic, physical, social, and cultural process in which changes in noneconomic factors are as critical as the economic factors. The latter should no longer be viewed as isolated phenomena but as part and parcel of broader socio-cultural-economic trends. A holistic perspective is required, to examine the total structure, including not only material resources but also the social and cultural underpinnings of society. Only thus can one attain a balanced understanding of the broader reality which will determine the rates and patterns of future growth.

This examination concludes that the United States is currently in the midst of a profound transformation of its underlying economic structure, accompanied by an historic shift in values, attitudes, and priorities. At the same time, a revamped international economy requires a new role of the United States. These changes require that the U.S. adopt the goal of qualitative growth to replace its former allegiance to undifferentiated quantitative growth.

THE NATURE OF THE TRANSFORMATION

The intensity of the growth debate and its rapidly changing focus should be viewed as a history of attempts to comprehend, first, the fact that the United States is in the midst of a time of epochal change; and, second, the nature of that transformation. Our era has been described variously by a number of leading historians and social critics as: "a great transition in the state of the human race"; "a time between civilizations"; "one of the great discontinuities in human history"; "comparable in the history of Western civilization to the Renaissance and Reformation."[1]

An important element in this broad transition, whatever one calls it, is the changing nature of the traditional sources of growth which made such positive contributions in the 1950s and 1960s. These have now departed onto new paths which are no longer so supportive of growth. In the area of natural resources, for example, the United States has moved from a "cowboy economy" to a position of serious resource constraints manifested in periodic energy crises. A part of the constraints is the no-longer-free use of the environment.

Another key factor in the transition is the shift in human resources. There will be a sharp slowdown in labor force growth, particularly in the latter half of the 1980s. Thus, the U.S. labor force will grow by 12 percent in the 1980s, compared to a 21 percent growth in the 1970s. Labor productivity, averaging 0.7 percent per annum in the 1970s, has grown at less than one-third the 2.3 percent annual rate of increase during the 1960s. The productivity of capital has also experienced a long-run decline. From 1947 to 1966, the output/capital ratio increased at an average rate of 0.5 percent per year, while from 1967-74 it decreased by an average 1.3 percent per year. The falling ratio is a key contributor to the general expectation that capital expenditures in the 1980s will likely be insufficient to generate enough capacity to meet the demands of the economy, and that bottlenecks and sporadic shortages will result.

Research and development (R & D) spending for basic and exploratory research, measured in price-adjusted dollars, has been on a plateau for ten years. Moreover, it has been argued that a state of "technological maturity" characterizes much of corporate America—that much of the innovative thrust of corporations seems to have dissipated as several vital postwar industries have reached a mature stage.

Just the changes in traditional sources of growth are bound to mean that the days of wide and handsome economic growth will not be seen in the 1980s. Since many of these same trends are being experienced by other OECD nations, they, too, will witness slower growth. The OECD recently concluded that, even under favorable assumptions, growth rates of output would not return to those witnessed in the 1960s, while unemployment levels and inflation rates would remain, on average, higher than in that period.

As important as these changes are, however, the United States is experiencing three much more fundamental, long-run trends which constitute the heart of the transformation: the structural transformation from an economy based in manufacturing and industry to one based in information, knowledge, and communications; the significant change in people's values, attitudes, and priorities regarding growth and material progress, the environment, and work; and the complete revamping of the international economy in which new actors on the world stage, wielding considerable power and largely "uncontrollable," have made it impossible for "domestic policies" to control U.S. economic activity.

The Information Economy

The United States first achieved service economy status in the early 1950s when, for the first time, more than half of the labor force was engaged in service occupations. Since then, the services sector has become increasingly entrenched. The number of employed service workers between 1950 and 1974 rose 90 percent versus only a 34 percent rise for goods-producing workers. Whereas, early in the century, only 3 of 10 workers were employed in service industries, by 1970 it was 6 of 10, by 1980, 7 of 10, and by 1985, it will likely be 8 of 10.

This general story is well known. What is not well known is that the dynamic heart of the evolution has been the increasing role of information and knowledge as a basic resource and the dramatic rise in the number of information workers. The statistics speak for themselves: by 1955, information workers had surpassed industrial workers as the dominant category in the labor force; by 1967, the total information activity—meaning the resources consumed in producing, processing, and distributing information goods and services—accounted for 46 percent of the GNP; currently, more than half of the workers in the United States are primarily engaged in generating, processing, distributing, analyzing, or otherwise handling information.[2]

These statistics indicate that the United States is in the midst of a structural transformation from an industrial-based economy to an information economy. Recent investment figures bear this out. Most of the basic or heavy industries of the past—such as steel, autos, and even chemicals—are experiencing a relative decline in importance which will continue into the 1980s. A major study of 2500 corporations in 37 industries found less and less new investment money going into such industries.[3] In general, low investment has shown up most clearly in manufacturing industries, especially the basic industries that largely produce capital goods. The best case in point is steel, where investment has fallen to a 6 percent rate of growth per year in the past ten years, compared to more than 10 percent for industry as a whole. The importance of this trend lies in the fact that five basic industries—steel, chemicals, paper, oil, and autos—account for almost one-third of the jobs and the value-added by manufacture.

Though these industries are experiencing a relative decline, they will still be very much with us. Thus, the transformation to an information economy does not mean that the society is less technologically-based or is moving "beyond" industry, but simply that the production/manufacturing industries become relatively less important as the motive force and generator of wealth. The characteristics of the central technologies and industrial forms become inherently different.

In sharp contrast to the maturity of basic industries, industries in the communications and information field are bringing about what is increasingly being referred to as an information revolution. At the heart of this revolution is the microprocessor (or computer-on-a-chip). It has been called the engine powering a second industrial revolution and the single invention most radically changing U.S. industry. It is said to be rewriting the economics of computing and that it eventually will become as basic to industry as steel. It can increase typing efficiency fourfold or more and has made possible the creation of electronic robots that can "see" moving parts on a conveyor belt, pick them up, and transfer them to another location. It is little wonder that microprocessor shipments have gone from 2.3 million in 1976 to an anticipated 100 million in 1980.

In general, the phenomenal growth of computer use is expected to continue. Through 1985, the installed population of computers and terminals is projected to grow at a compound annual rate of more than 30 percent. Because the price/performance ratio in electronics decreases dramatically, the information processing capability of the U.S. economy can be expected to increase by a compound annual rate greater than 100 percent.

Such dynamic activity is by no means confined solely to the United States. Electronics has been one of the most rapidly expanding sectors throughout the industrialized world in the last decade. Its tremendous future importance was emphasized in a 1979 OECD draft report (of the Interfutures Project) on the long-run future of its member countries:

> Through its links with data processing and telecommunications, the introduction of automation throughout industry, the changes which electronic office equipment is producing in service activities, and the actual services which it creates, the electronics complex will during the next quarter of a century be the main pole around which the productive structures of the advanced industrial societies will be reorganized.[4]

Because of such tremendous growth and the central role anticipated for electronics and information activity in general, the information revolution could bring about a socioeconomic transformation as far-reaching in its ultimate effects as the industrial revolution, one which would certainly alter profoundly the pattern of growth. The two revolutions, however, are quite different. Whereas the industrial revolution made available and employed vast amounts of mechanical energy, the information-electronics revolution is extremely sparing of energy and materials. Much of the industrial technology was crude, with only a modest scientific or theoretical base. The information revolution, however, as the product of the most advanced science, technology and management, represents one of the greatest intellectual achievements of mankind.

The impact of the shift to an information economy will be pervasive—it will be seen in how and what we produce, how we transact business, how we pay, how news is gathered and spread, where we work and what kind of work we do, and how we communicate. Not restricted to just the computer or communications industries, the information revolution at minimum embraces banking, insurance, transportation, health, education, communications, entertainment, and manufacturing.

Given this wide range of impacts in so many different industries, there can be little doubt that the key economic variables—productivity, investment, consumption, employment, prices, and trade—will be affected. The three most distinctly impacted are employment, trade, and productivity.

Both the quality and distribution of employment will be significantly affected. Regarding quality, it has been argued that revolutionary change in the office of the future could redefine how the largest single part of our working population works and how it functions. It should be conceived, according to Paul Strassman, not just as a means of overcoming existing technological limitations, but as "a restructuring in the thinking and working methods of the professional, managerial and administrative occupational categories."[5]

Electronics will cause a widespread change in employment patterns in such establishments as banks, schools, offices, newspapers, the postal service, factories, retail outlets, and communications industries. For example, retail store automation systems bringing point-of-sale terminals to retail and grocery stores will enable automatic checkout and inventory management with less need of labor for these functions. Overall, most government economists and industry executives maintain that, despite such changes, expanding opportunities in information-related activities will more than offset jobs lost to more productive electronic equipment. Thus, advances in electronics should have very little impact on structural employment.

The single most critical question concerning the employment effects is whether the potential for future growth of the information work force is limited. Are the private and public bureaucracies becoming glutted with information workers so that no more can be easily absorbed? If the information work force is indeed saturated, this will have profound impact on employment prospects in the 1980s. If not only agriculture and manufacturing but also the information sector no longer are providing dynamic growth in employment opportunities, where will the new jobs be generated? The jury on this most fundamental question is still out.

It is likely, as Daniel Bell argues, that in the 1980s traditional, routinized manufacturing, such as the textile, shipbuilding, steel, shoe, and small consumer appliances industries will be "drawn out" of the advanced industrial countries and become centered in the new tier of rapidly developing countries, including

Brazil, Mexico, South Korea, Taiwan, Singapore, Algeria, and Nigeria. The response of the advanced industrial countries, as Bell sees it, "will be either protectionism and the disruption of the world economy or the development of a 'comparative advantage' in essentially the electronic and advanced technological and science-based industries that are the feature of a post-industrial society."[6]

Japan has recognized this, and the powerful Ministry of International Trade and Industry has declared that the information industry is the strategic industry of the future. Ministry officials see the industry not just as a source of export earnings but as the "nervous system of the future economy, the technology that will help us cope with the limits to industrialization, such as pollution and finite resources."[7]

The United States government has not engaged in such systematic, long-range thinking and planning, even though the United States is currently out front in terms of the technical sophistication and the manufacturing distribution base of its telecommunications industries, and even though the stakes are high. It has been estimated that $1.6 trillion will be invested in telecommunications outside the U.S. by the year 2000 to establish a global information network in which information will be transmitted and stored electronically.

There is a great opportunity for the United States to exert a significant leadership role in the information field. To do so, the government must abandon its adversary relationship with industry in the telecommunications field. The government should begin to consider positive initiatives which would help American industry retain both its technological leadership and its commanding share of the export market in the information-electronics field.

The potential is more than economic. The magnificently productive U.S. industrial machine has brought material goods not only to Americans but to many of the world's people in undreamed-of abundance. Tomorrow's challenge is even greater. Information and related technology could be directed to helping people achieve their full human potential once material needs have been satisfied, a goal more nebulous and complex than supplying material goods, but likely to be more exciting and rewarding.

New Values and Priorities

The transformation is so profound because it consists of more than change in the underlying economic structure, fundamental as that may be. Even more important, perhaps, is the fact that the values, attitudes, and priorities of the American people are undergoing substantial change. The change was authoritatively advertised in President Carter's televised speech in July 1979 in which he spoke of a "crisis of confidence" and a "fundamental malaise." Daniel Yankelovich in Chapter 1 succinctly summarizes the source of this malaise.

In short, the American people are extremely confused right now. The old "givens" are crumbling, and there is nothing to take their place. America, in the past, had as a central goal the pursuit of growth through hard work and the application of new technologies. Growth is no longer an adequate goal; fulfilling work is being sought and new technologies are being increasingly questioned, controlled, and sometimes rejected. Such a state of confusion ultimately demands a resolution. What seems to be needed is a new goal of qualitative growth as a major part of that resolution; of this, more below.

Historical experience demonstrates the critical nature of such changes in values. The rise of the Protestant ethic was a primary force behind the remarkable expansion of commerce and advance in technology in Europe and America beginning in the eighteenth century. Far more than a religious doctrine, it encompassed new views on the purpose of work, the virtues of material progress, and man's dominion over his natural environment which were fundamentally different from the previous era and provided a fertile environment for the institutions of capitalism and the industrial revolution.

It is precisely the values regarding these three areas—material progress, the environment, and work—that are undergoing the most serious change today. Certainly, the origin of these changes lies in the civil rights, environment, and antiwar movements of the 1960s which shook people out of their complacency, causing them to seriously question previous "givens."

Such questioning and the activities that followed it, not only persuasive speeches by established leaders, have been the origin of most of the recent U.S. policy shifts in areas such as minority rights, the status of women, and consumer and environmental protection. The OECD Interfutures study emphasized both the value change underway and the interdependence of values, growth, and structure:

> Values are changing radically, even if the changes are slow and of the greatest possible variety. The legitimacy of growth as a goal in itself is questioned by some, and is running afoul of structural rigidities. Influenced by growth, changes in structure are accelerating or curbing certain trends in values. The complex relationships between these three poles make any linear view of development untenable.[8]

Certainly, growth as a goal in itself has come under serious questioning in the United States. Recent surveys and changes in life-style provide evidence that the American people are increasingly skeptical about the nation's capacity for unlimited economic growth, are becoming wary of the benefits unlimited economic growth is supposed to bring, and are beginning to place a higher priority on improving human and social relationships than on raising the material standard of living. This suggests a significant turnabout from the values and lifestyles of the first 25 years after World War II.

Studies by the University of Michigan's Survey Research Center have documented the change in people's priorities during the last ten years. To the average family, the following priorities have gained in importance at the expense of an increase in the level of consumption: a secure job; continuous income even in case of sickness, disability, and old age; appropriate forms of job and career; safety in one's home and on the street; and neighborhoods that do not deteriorate as the result of rapid changes in the human or physical environment. Two leading researchers at the Center conclude:

> These new desires are more than a passing fad. They compete with the pursuit of traditional ways of progress and growth. Much thinking and debate will be required to strike a balance between an expansion of production that ignores the new priorities and an attempt to create a humane economy regardless of its consequences for production.[9]

Another element of the new view regarding growth and material progress is the growing recognition of the fact that scientific and technological advance is a double-edged sword that threatens man and his environment while, at the same time, offering the tempting possibility of a better future. Perhaps the most dramatic examples of changed public attitudes regarding technology and the impact they can have were the halt in production of the ABM system and the SST in the early 1970s. This would have been inconceivable in the 1950s.[10]

As knowledge of the delicate interwebbing of the ecological system has spread, people have demanded wide-ranging measures to protect the natural environment. As the carcinogenic nature of some industrial products and processes has been recognized, the demand for other protective measures has risen. Current apprehension about the ultimate physical and social consequences of nuclear power technology may retard a development which, until recently, was expected to yield a cornucopia of benefits. Research into genetic manipulations also has aroused new reservations about the independence of science.

Additional evidence that changes in values have already played a significant role is seen in the fact that three of the four types of public investment which rapidly are rising, and will continue to rise, stem from value changes which occurred in the 1970s: environmental, health and safety, mass transit, and energy.

Changed attitudes regarding growth are also manifested in the no-growth or controlled growth measures enacted by local communities. Santa Barbara, California, which since 1975 has consistently discouraged new industry and residential growth, is a typical example of a very widespread movement. The traditional approach is to control land use, but other methods, such as setting population ceilings or imposing moratoriums on construction until services are provided, are also used. Localism also connotes the fact that conflicts very often arise between the interests of a local area and the national or regional good. For instance, there are no new deep ports on the East coast of the United States and no refineries have been built there for a decade.

What the foregoing makes clear is that the god of growth has not brought full contentment. The vast majority of Americans had acquired an abundance of material things yet remained vexed, not only by the immediate noneconomic problems in their individual personal lives, but by larger issues that extended outward to embrace their whole society and forward into the future. When an increasing number now think of the quality of life, it is not simply a vision of an overflowing cornucopia of personal material luxuries, but something less tangible and yet more important—job satisfaction, good schools, responsive government, and fulfilling leisure activities. Most Americans have "enough," but many are beginning to think that this kind of enough is still not adequate.

Gardiner Ackley has offered one of the most intriguing statements on the relationship between affluence, value changes, and economic growth:

> I doubt that externally imposed limits to growth will require a new ethic and life–style: rather, I suspect, an ever–increasing abundance will generate an ethic or life–style which will cause economic growth to slow and ultimately to cease. And as that happens, we may not even recognize it because the question of growth will have become so uninteresting.[11]

Recent evidence suggests that such a new ethic or life–style is emerging. It has been described as a move toward "voluntary simplicity"—an effort to live life within a new balance between inner and outer growth. The five values which seem to lie at the heart of this emerging way of life are material simplicity, human scale, self-determination, ecological awareness, and personal growth. It is estimated that currently about 4 to 5 million adults fully and wholeheartedly live a life of voluntary simplicity, while partial adherents may include up to 8 to 10 million others.[12] It has been argued that perhaps as much as half of the adult population sympathizes with many values associated with voluntary simplicity but do not presently act on this sympathy. These "sympathizers" will play a pivotal role because their numbers are so large that, if only a small percentage become acting adherents, this will represent a major shift in American life-style and values.

So widespread and fundamental is the shift in values and priorities that virtually all growth experts—including both optimist Herman Kahn and Jay Forrester, the founding father of the physical limits school—now acknowledge that the "social limits to growth" represented by the new ethic are of greater importance than physical limits in determining growth prospects. However, the two types of limits are closely related for much of the social change has occurred in response to perceptions of physical limits.

The change in values has been most clearly manifested in the environmental movement and charter of the 1970s. The resulting environmental control standards have brought about resources. In 1978, business invested $7.1 billion (current dollars) in pollution control equipment and total pollution control expenditures required by federal regulations were $22.5 billion. Such federally induced expenditures are estimated by the Council on Environmental Quality to total $361 billion in the 1977-86 decade. Faced with such heavy costs in a period of high inflation and general economic difficulties, one may expect to see the public backing away from strong environmental values. Yet this has not occurred. A comprehensive National Environmental Survey conducted in late 1978 found that more than half the people think protecting the environment is so important that "continuing improvements must be made regardless of cost."[13] Contrary to the white, upper–middle class image of environmental concern, black support (55 percent) was virtually identical to that of whites (54 percent) and 49 percent of those with very low incomes (under $6000) chose this option. By three to one, the respondents favored paying higher prices to protect the environment over paying lower prices but putting up with more air and water pollution.

Attitudes regarding work—both the content of the job at the workplace and how the job fits into one's life—are also beginning to change radically. In a 1973 survey of college students which asked about major influences on their choice of a job or career, "challenge of the job" was endorsed by 77 percent, "ability to make a meaningful contribution" by 72 percent, and "ability to express yourself" by 68 percent. The general adult population responded similarly in Detroit Area Surveys. When respondents were asked in 1971 what they would most prefer in a job, the top two choices were "the work is important and gives a feeling of accomplishment" and "chances for advancement."

Yet not enough challenging, meaningful jobs are being created to match the rapidly growing demand. The result is increasing job dissatisfaction. The Survey Research Center noted an appreciable drop in overall job satisfaction between 1973 and 1977 which was pervasive, affecting virtually all demographic and occupational classes tested.

A major cause of job dissatisfaction and increasing demands for improvements in the quality of working life is underemployment. It affects all workers but particularly the highly educated. Recent surveys have indicated that over one-third of American workers consider that they are already educationally "overqualified" for their jobs. These workers have significantly lower levels of job satisfaction than other workers. Moreover, the problem will become worse in the future. The Bureau of Labor Statistics in 1978 projected that about 10.4 million college graduates will enter the civilian job market during the years 1976 to 1985, but that only 7.7 million job openings requiring a college education will be available. The increasing job dissatisfaction of the highly educated holds considerable potential for increasing the social malaise. On the positive side, it should give new stimulus to efforts to reconstitute jobs and the work environment to enhance employee satisfaction.

There has also been a major change in attitude regarding the role women should play in the work force. This is reflected in the full-force entry of women into the labor market since the 1950s. The great increase in the women's labor force participation rate (from 33 percent in 1950 to 48.4 percent in 1977) has been an exceptional and very influential force on the economy as well as many other aspects of American life. In particular, it has become increasingly accepted that a mother can work without being uncaring or raising derelict children. Hence, the proportion of married women in the labor force with school-age children nearly doubled between 1950 and 1976, while the participation rate of those with preschool children nearly tripled. The culmination of these trends was that, in 1976, the percentage of mothers who chose to participate in the labor force was almost equal to the proportion (nearly 50 percent) of all women who were working. Clearly, this large-scale entry of women into the labor force has been a major part of the movement toward more flexible, fulfilling, and nondiscriminating employment.

The Global Context

The international economy at the close of the 1970s is entirely different in character from that of 1970 and would have been undreamed of in 1960. All nations have become highly interdependent, but this has affected the United States in the 1970s more than most because of its relative isolation from the world economy throughout most of its history. Americans became vividly aware of their new interdependence in 1972 when bad weather in the Ukraine led to higher bread prices at home, and in 1973 and 1974 when the actions of a group of small countries contributed to a quadrupling of oil prices in a period of a few months.

Trade historically has been a very small proportion of the GNP as the United States was among the world's most self–sufficient countries. This is changing rapidly and radically. In 1971, the value of exports plus imports was less than $90 billion in an economy just over $1 trillion. By 1977, the value of exports and imports had tripled to $270 billion while the GNP had fallen short of doubling ($1.9 trillion). By the late 1970s, leading U.S. firms had so greatly increased their stake in overseas production that the output of their foreign subsidiaries and branches had grown to over one-quarter of the firms' U.S. production. Thus, with upwards of $350 billion of assets held by Americans in foreign countries and around $275 billion held by foreigners in the United States, there can be no doubt that the United States is irretrievably linked to the world economy.

The nature of this interdependence goes well beyond such trade or financial figures. It is best understood through examination of five new factors with considerable power and influence on the world stage in the 1970s: the OPEC oil cartel; the Third World's demand for a New International Economic Order; multinational corporations; floating exchange rates; and the Eurodollar market. Each of these factors is largely "uncontrollable"—that is, no single national economic institution or policy can substantially alter their global actions or effects. The result is a significant diminution in the effectiveness of "domestic" policies on economic activity; this amounts to a reduction in a nation's sovereignty over its economy.

The impact of interdependence has been and will continue to be felt primarily in the energy arena. OPEC is the dominant actor; its 1973-74 quadrupling of oil prices triggered three fundamental changes: a sharp slowing in world economic growth, a slowing in the rise of world trade, and an increase in the productivity of labor relative to capital that is strengthening the economies of the more advanced developing countries.

Two basic lessons have been learned from this experience. The first is that the energy crisis is essentially an oil crisis. Virtually all recent studies conclude that the production of oil will peak in the last two decades of this century. The United States is right in the midst of the oil crisis. In 1977, it bought $45 billion of oil from abroad, the prime factor underlying the $27 billion trade deficit that year. An immediate result of the huge outflow of dollars was the dramatic decline of the dollar's value in international money markets. Regarding projected domestic oil production, there is little optimism. Most estimates show production levels in the 1990s lower than current levels. It can be safely concluded that the chief dependency problem for the United States in the 1980s will be retaining access to secure sources of petroleum and natural gas. This relates to the second basic lesson—political realities (which encompass personal ego, institutional power, conflict and relationships among different nations) have more influence on the international flow of oil than economic factors.

The oil crisis highlights a very basic element of the U.S. domestic transformation. The "cowboy economy" had already been replaced by one in which there are important resource constraints on growth. Due to increased competition from other nations, the United States can no longer rely on readily available and relatively inexpensive imports as it has in the past. A recent U.N. study directed by Wassily Leontief concluded that the world was expected to consume, during the last 30 years of this century, from three to four times as many minerals as have been consumed throughout the whole period of civilization.

The United States is now experiencing a serious race between technological progress and the depletion of its resource endowment, or even availability through imports. To the present, technology has been the clear and easy victor. It is not possible, however, to guarantee which force will be more powerful in the future. Certainly, new technologies will be applied which increase efficiency in extracting and using resources. But whether these technologies will be able to offset the hard fact that, because of continuing resource depletion, given inputs of labor, capital, energy, and other factor inputs will yield less output of energy and resources is a major unknown. The other primary consideration is what effect will the information revolution and relative decline of the heavy industries have on energy-resource demand? Conceivably, changes at this broadest structural level could have a much more energy-resource conserving effect than all the direct conservation measures combined.

To the degree that the transition to an information economy does not take place to the fullest possible extent, two conclusions can be reached. First, the real cost of mineral resources is likely to increase substantially over the next few decades. This is due both to "imperfections" in market systems and institutional arrangements as well as the fact that many of the real costs involved in present mineral production and consumption will be paid in the future. Second, U.S. mineral import dependency, already high, will likely be higher in the year 2000.

There are other important new factors in international economics. OPEC actions have been at the leading edge of Third World efforts to engineer a new international economic order. These countries not only recognize the enormous disparity between their incomes and respective rates of growth of the developed and the developing countries, but they are determined to see them remedied. They want a share of the developed countries' growth, technology, wealth, status, and power.

Their argument basically runs as follows. The advanced nations have exhausted their supplies of many of the most important natural resources needed for an industrialized society and are entering into increased competition for the remaining supplies in the Third World countries. Continually increasing consumption by the industrial nations, which shortens supplies and raises prices, is now shutting off growth possibilities for the developing countries. In short, they argue that the industrial nations hold all of the cards in the present international division of labor, and that it is impossible for many developing countries to begin to catch up with the industrial world unless a dramatic transformation of the international economy takes place.

The multinational corporations will have an important voice in influencing these relationships. In the first half of the 1970s, the multinationals grew at an annual rate of 10 percent, twice that of the world's economies taken as a whole. It is expected that, in 1980, their sales will constitute 16 percent of the Gross World Product. Large size *per se*, however, is not the most significant fact. The multinationals enjoy a freedom of action denied to smaller companies bound by the laws of a single nation. They can juggle their profits by arbitrary pricing arrangements that magnify the profitability of operations in low-tax areas and minimize it in high-tax areas. They can escape a country's labor standards by transferring production to plants located where wages or health and safety requirements are lower. They can wreak havoc with pollution standards and control by transferring pollution generating production to countries without such standards. They can ally their economic power with the political or military power of the host government, as in the case of IT&T in Chile, or deploy their considerable economic power in ways that are contrary to the political designs of their home countries. The multinationals are the key to stabilizing the global economy for they have considerable potential for raising living standards or for destroying them, for enhancing global awareness or for imposing their own profit-driven order.

This international climate has been much affected also by new financial factors. In 1973, the stable fixed exchange rate system yielded to a system of continually changing, transient relationships among currencies which significantly changed the face of the world economy. One has only to look at the headlong slide of the dollar in 1977 and 1978 to see the great uncertainty and effects floating exchange rates have wrought.

The massive Eurocurrency pool—estimated at between $400-$500 billion—is another of the new forces in the international economy that is out of reach of any single nation's control mechanisms. Around 80 percent of this pool is Eurodollars—dollars deposited abroad. This credit system in the 1970s has swollen faster than anyone's ability to understand, contain, or manage it. It currently is 10 times larger than it was a decade ago, and four times bigger than it was as recently as 1972. This pool, lying outside of increasingly tough exchange controls, is available for anyone to trade in it or invest it. Primary borrowers are governments and businesses, while the principal investors are the Arab oil countries and the multinational corporations looking to turn a profit on their idle cash.

International commerce has become totally dependent on this market and its associated supranational banking system; multinational business could not operate in the world of floating exchange rates without it. This new banking order tremendously increases the efficiency of moving cash around the globe; but as currencies move freely from one country to another, they may stimulate inflation in one, undermine a currency's stability in another, and generally cause a much more volatile floating exchange rate than anticipated. At times, such movements force exchange rates up and down beyond anything justified by accepted economic criteria. It has been said that, in trying to control the gyrations, central bankers (who in 1977 had only a $220 billion reserve pool) have been in the "untenable position of chasing an elephant with butterfly nets."

THE CHALLENGE OF QUALITATIVE GROWTH

The United States is in the midst of a structural transformation in its economic base. Its domestic policies are being buffeted and often rendered ineffective by the machinations of a number of new, powerful, and uncontrollable actors on the world stage. Perhaps most seriously, the American public is undergoing a transformation in values, attitudes, and priorities which, in the present confused state, has left them with little confidence in institutions to solve the major problems and a belief that the future will be worse than the present or the past.

If all these conditions are true, it would be the height of foolishness to think that we can go on as usual, hoping to muddle through by tinkering with the old policy tools. President Carter has gone part of the way in recognizing a change in values is underway and has called for significant efforts in the energy arena which will entail some sacrifice and change in life-style. But this prescription, while necessary, is far from sufficient to cure the general malaise which has very deep roots.

Ironically, part of the crisis can be traced to the solving of the economic problem—defined by Keynes as the struggle for subsistence—in the United States. If one accepts Keynes' thinking that "we have been expressly evolved by nature—with all our impulses and deepest instincts—for the purpose of solving the economic problem,"[14] then we as a nation have been deprived of our previous central purpose.

The American public needs a positive vision of America in the 1980s—where we are headed as a nation and why, and what the concrete steps (or alternatives) are to get us there. The serious questioning regarding the value of economic growth as such, and the growing expression of diverse quality-of-life concerns indicates the type of vision that should be put forth by political leaders as a major challenge for America to meet in the 1980s: a vision of qualitative, selective growth to replace the present allegiance to quantitative, undifferentiated growth. A three-to-one majority in a Harris Poll endorsed the statement "The trouble with most leaders is that they don't understand people want better quality of almost everything they have rather than more quantity."

The new questions must ask about purposes: Growth for whom? Growth for what? A goal of an expanded GNP must not be automatically applauded before we ask: How will the growth be achieved and what end will it serve? Growth must come to be seen as a means to an end and not the end itself.

At a minimum, we must move in the 1980s toward an economic growth that is restrained by sound environmental–physical principles and which is both sustainable and productive of desirable human ends. Redirecting future economic growth into nonpolluting and resource-conserving channels will help minimize its disruptive effects upon the environment. The broader objective of policy must be to maximize national economic welfare, defined to take account of pollution, resource depletion, the quality of work, income distribution, and other aspects of welfare not reflected in the GNP as now calculated.

What does it mean in practical terms to pursue a qualitative growth goal? What specific actions and policies will be involved? The wide array of actions and policies that would be involved in pursuing qualitative growth can be grouped under two main guiding principles which relate to physical resources and human resources. In the physical realm, the actions and policies of corporations and governments should conform to a Conserver Society goal, one which requires the use of the minimum amount of resources deemed necessary to carry out an activity or produce a product (including growth) without resulting in undesirable side effects. The guiding principle for human resources should be total employment, wherein everyone desiring a job would have the opportunity of obtaining one that essentially satisfies his or her personal desires.

New corporate and government actions and policies are needed to further these two goals but they, in turn, must have a firm foundation in a significantly revised discipline of economics.

Rethinking Economics

Marginal changes in economic theory will not suffice to meet the challenge. Current theory and resulting policies (1) are inadequate to deal with the new economic realities and major economic problems, and (2) do not complement well the goal of qualitative growth.

The public today accurately discerns that the government continues to rely on more-of-the-same policies in the face of entirely new conditions. Much of the blame, however, belongs to economists, for economics has not come up with any major new breakthroughs to resolve the increasingly burdensome problems of our day. Lester Thurow has noted that the last decade has not witnessed a major or even lesser "gold strike," while E. H. Phelps Brown, in his Presidential address to the Royal Economics Society of England alluded to "the smallness of the contribution that the most conspicuous developments of economics in the last quarter of a century have made to the solution of the most pressing problems of the times."

The result is that economic policies continue to rely almost exclusively on the theory and tools promulgated by Keynes 40 years ago which were developed to combat depression conditions by stimulating demand. The hard fact today is that many of our economic problems, including inflation, emanate from the supply side and many are not even due to "economic" forces. Thus, changes in the money supply, interest rates, taxes, or government spending, though still needed to help shape aggregate economic activity, will not in and of themselves resolve stagflation and other economic ills.

Why is it that most economists cannot come to this fundamental realization? One reason is the short-run (three months to three years) focus of economics. Short-run equilibrium analysis is not conducive to dealing with long-run, structural changes. Another reason is that economics has not been firmly grounded in institutional-power realities. Economists on the whole have pursued rigorous quantification to consolidate the scientific status of the discipline at the expense of ignoring almost completely the economic impact of social institutions, political power relationships, and cultural attitudes and values. Economic theory assumes that economic actors will have fairly equal opportunity in the competitive arena. Today's world of conglomerates, multinationals, big labor unions, and massive government intervention belies this assumption. Prices and markets and their role in the economic process are often subordinate to the use and distribution of power. Thus, prices and wages may remain the same or rise despite declines in demand. Economic theory is similarly deficient vis-a-vis the new power exerted by the new international economic actors discussed earlier.

De Jouvenel's critiques of economists' divorcement from reality are matchless in witty succinctness. He claims that the word "angelic" characterizes the whole of modern economics for it has become "bodyless, immaterial," a far cry from what it was in the days of the classics, from Smith to Marx. Smith, for example, was interested in coal, noting that manufactures were sited in the neighborhood of coalfields because indoor work called for heating the work rooms. De Jouvenel calls this "concrete vision," but says "many contemporary economists would regard it as 'intellectual slumming.' " In de Jouvenel's view, there are two reasons for economists' divorcement from reality. First, they have a remarkable nimbleness of mind and "nimble minds delight in playing with abstractions." Second, economists, unlike other professionals, have a ready-made tool for abstracting from reality: money. "Carried on the wings of this instrument, they lose sight of reality, and thus the more the science is academized and formalized."[15]

Economists in the past century have also largely ignored the third factor of production—land, which, broadly defined, encompasses natural resources and the environment. It is only in recent years that economic textbooks have begun inserting material on the environment or energy and, even then, it is usually in special chapters near the end of the book along with other "miscellaneous" topics. As E. F. Schumacher commented, it is inherent in the methodology of current economics to ignore man's dependence on the natural world. In pursuing solely economic goals, economists today still pay little or no regard to evolutionary processes, to the dissipation of energy, or to the depletion of resources.

Finally, there is the problem of the nonquantifiable. Following the reasoning of Pythagoras, economists have assumed that what is not quantifiable does not exist or, at the minimum, is not very important. This has led to a definition of the economy that uses a set of quantifiable terms (almost exclusively "economic" variables) and leaves out all other factors, however essential. This makes possible the use of impressive mathematical models, but at the cost of a sharp restriction of vision—almost the entire social system is kept out of sight. Oskar Morgenstern has asserted that, instead of first identifying the economic problems and then trying to develop adequate tools of analysis, economists tend normally to think first of the mathematical tools available and only then of the problems—more often than not imaginary ones—to which they could be applied.

The main reason why economics as currently constituted and the goal of qualitative growth in the 1980s do not complement each other well is that economics, seeking scientific positivity, has dealt almost exclusively with what "is," shunning the question of what "ought" to be. This is a serious consequence of the false pretension that economics is value-free. Focusing on what "is" unfortunately brings with it an unconcern for all the inequities which define and sustain the status quo. It also means that economics has operated essentially at the know-how level, paying scant attention to the know–why questions.

Certainly, some leading economists have been exceptions to this general rule, including Marx, Veblen, and Galbraith. Lord Robbins' attitude, however, is much more representative when he states that inasmuch as the economist and the political philosopher have nothing whatsoever to do "where the true positive goods of life" are concerned, they "must bow and take their leave." The danger is that refusing to decide today what the good things of life are or are not could make things awkward in the future. We must remember that even the means chosen may oblige us to alter the ends themselves. At the extreme, such an approach leaves economics wide open to the tragedy experienced by Captain Ahab—the use of means which are rational for purposes which are insane. Surely, to be able to do more efficiently that which should not be done in the first place represents a perverse form of progress.

What it boils down to is that concern over quality involves difficult judgments and imposes responsibility, while concern over quantity involves merely arithmetical operations that give everyone the same answer and impose no responsibility. It is little wonder that there has been no great vision in economic thought in the past three or four decades comparable to the illuminating power of past chapters of economic thought. The most formidable advances in mathematics or statistics or philosophic method do not help us in the formulation of those gestalts or visions by which we "grasp" the problems of a society and set out in new directions.

Not only is the lack of goals a problem for the discipline of economics, but it also reinforces the lack of clear national goals. The nation's economic and social "goals" are left to be voiced primarily by the marketplace, yet the marketplace is a poor goal-setting mechanism, since the goals it establishes (1) reflect the buying power of the rich far more than the poor, (2) are distorted by powerful aggregates of massed corporate wealth, and (3) are without recognition of any end result of economic activity that does not bear a price tag.

These deficiencies of the standard Keynesian intellectual baggage which economists have carried around for the last 40 years mean the discipline simply cannot deal adequately with the new realities of the 1980s. Ironically, if Keynes were alive today, it is likely that he would be among the first to jettison many of his views. Just over a half century ago, he wrote the following passage to point out the intellectual impotence of the then prevailing economic orthodoxy, which seemed less and less capable of explaining what was happening to the economy and quite incapable of providing a coherent set of economic policies which policymakers and business leaders could look to with confidence: "We do not dance even yet to a new tune. But change is in the air. We hear but indistinctly what were once the clearest and most distinguishable voices which have ever instructed political mankind."[16] His words certainly ring true now, when the discipline of economics is once more in the midst of crisis and change is in the air.

Keynes' words about the inadequacy of the prevailing economic orthodoxy fit precisely Thomas Kuhn's classic discussion of paradigm change, wherein the passage of time brings anomalies that the existing theory can no longer explain. As the divergence between theory and observed reality increases, questions begin to arise and a new theory or explanation more consistent with reality eventually evolves. Certainly, today such anomalies and questions are rampant and the divergence is clear.

Keynes himself noted one of the major reasons behind the development of new paradigms: changing purposes or end goals of economic activity. Certainly, the new goal of qualitative growth could constitute a challenge to today's economists as great as the earlier challenge to achieve and sustain high rates of economic growth. For now, economists are being asked how to organize the economy so as to better accommodate the human needs for security, continuity, equity, and self-actualization within the limits imposed by the environment and the available natural and human resources. This does not call for more "changes at the margin" but a thorough restructuring of economics. It encompasses the theory, methodology, and policy tools, since all of these have developed in a growth-oriented environment.

The new economic paradigm to be developed must be holistic, which involves extending economic analysis in regard to scope of data, time, and space. It will have to consider many noneconomic factors and trends; it will have to develop a long-run analysis capability; and it will have to explicitly incorporate the workings of the global economy and subnational economies into its domestic, national considerations. The broad dimensions of needed change are direct, simple to understand, yet revolutionary; much more than a further sharpening of nuances, they demand reconceptualizing the entire problem—a most difficult task.

Among the noneconomic factors and trends which economics must incorporate, those in political science and the physical sciences are the most critical. Economics must become more firmly grounded in people's needs and aspirations in all their diversity, and then seek the most effective and equitable means to achieve them. When economics becomes so oriented, it will again deserve its original title—political economics.

The word "equitable" is the key, for the considerations of politics are centered very much on equity while economists talk in terms of efficiency, costs, and benefits—which point to two very different worlds. Equity and efficiency often conflict, or at least appear to do so, because neither is a free good, each tending to "cost" in terms of the other.

Such considerations will become increasingly important as we move into the slower growth of the 1980s, for slower growth will necessitate much more explicit attention to questions of equity. Periods of high growth help diffuse the equity issue since even those on the lower rungs of the ladder are able to climb a few rungs higher. If they are to continue climbing under slower growth, the people higher up on the ladder will find their pace correspondingly slowed.

Political economics will undoubtedly appear relatively loose and unstructured when compared with the artificial neatness of highly abstract economic or econometric models. However, it is only by attempting to put together theories that really fit functioning societies and exploring political-economic processes in depth that we will move toward a coherent analysis of economic growth and other social issues of our day.

Economics does take some account of physical factors, but in exactly the reverse relationship from what should be. Currently, nonphysical parameters such as technology, capital, preferences, and distribution of wealth and income determine the equilibrium to which physical variables must be accommodated. In the new theory, physical parameters, such as resource availability, net energy, a complex ecosystem and the laws of thermodynamics, must come first; and nonphysical variables must be brought into a sustaining equilibrium with the complex biophysical system.

Economics can no longer assume that resources are given or infinite. Economic policy alternatives must be evaluated as they relate to long-run ecological balance. In the 1980s, economic policies must be examined not only with respect to their impact upon such economic variables as investment, prices, employment, and income, but also upon the rate of resource depletion and environmental deterioration. This would reduce the risk of potentially disastrous disorder in the earth's systems of life support. Such a milestone in economic thought would lead to changes in the very definitions of output, income, wealth, productivity, cost, and profit.

If economics is to serve in the pursuit of qualitative growth, it also must become more explicitly normative. The adjective "explicitly" is used purposefully to indicate that, while most economists would shudder at the suggestion that their discipline all of a sudden will or should consider norms, the fact is that economics has always been implicitly normative. As far back as the "invisible hand" of Adam Smith, economics has assumed a normative model of how *Homo Economicus* is thought to act and, implicitly, of how he or she ought to act. Based on this view, for 200 years economists have avoided making economic decisions directly on the basis of moral considerations about the larger social good. As a result, we have (in the words of Oscar Wilde) economists who know the price of everything and the value of nothing.

This is a far cry from the view expressed in the first textbook of political economy that "the science of Political Economy bears a nearer resemblance to the science of morals and politics than to that of mathematics."[17] Economics did, indeed, begin as a branch of moral philosophy, and the ethical content was at least as important as the analytical up through Alfred Marshall and the burgeoning of marginal analysis. Theodore Roszak is thus merely calling for a return to economics' roots when he writes ". . . we need a nobler economics that is not afraid to discuss spirit and conscience, moral purpose and the meaning of life, an economics that aims to educate and elevate people, not merely to measure their low-grade behavior."[18]

Is it only dreaming to advocate a "nobler economics" in the service of qualitative growth? It would appear that Keynes, who was a solid blend of the great theoretician and the practical man of the world, did not think so. He saw such a possibility in the world of his grandchildren:

I see us free, therefore, to return to some of the most sure and certain principles of religion and traditional virtue—that avarice is a vice, that the exaction of usury is a misdemeanor, and the love of money is detestable, that those walk most truly in the paths of virtue and sane wisdom who take least thought for the morrow. We shall once more value ends above means and prefer the good to the useful. We shall honor those who can teach us how to pluck the hour and the day virtuously and well, the delightful people who are capable of taking direct enjoyment in things, the lilies of the field who toil not, neither do they spin.[19]

Rethinking Inflation and Productivity

In order to illustrate how these new approaches could be applied, let us consider the problem of inflation. Only when a long-run, holistic framework is applied will economists and policymakers be able to more fully understand the root causes of inflation and, thus, what approaches are most appropriate to lessen inflationary pressures.

To understand the inflation of recent years, analysis must begin with the underlying long-run forces shaping the patterns of economic activity which manifest themselves, among many other ways, in inflationary pressures. Thus, one should have a much better understanding of why inflation has been so virulent in recent years after looking at: the increase in real costs of resources; changing personal values and priorities; the local no-growth phenomenon; competition from abroad and increased global economic interdependence; the tremendous labor force influx; the increasing "nonproductive" demands on capital; the maturing of technology and of many basic industries; and the decade-long slowdown in productivity. The clear conclusion which emerges is that inflation is a complex, multi-faceted phenomenon which is not going to be significantly ameliorated by the one-shot solutions which have been so prevalent (a prime example of which is the call for a constitutionally-mandated balanced budget).

Although there is no one-shot inflation solution, it is the case that only by achieving a faster rate of productivity growth will the rate of inflation decline below 5 to 6 percent for more than brief periods of time. Productivity growth is the foundation upon which all other "inflation-fighting" policies must be constructed if they are to have an appreciable effect. Economists using a more holistic perspective would explore those elements of the structural transformation which might provide significant impetus to productivity increases. They would then use their more traditional analytical tools and models to see how these areas may be most effectively supported and encouraged.

A thorough examination of the shift underway to an information economy would reveal an increasingly antiinflationary, more energy/resource-conserving economy which carries exciting potential for major productivity gains. This is primarily due to computer advances and particularly computer-aided manufacturing (CAM). In manufacturing, computers have potential applications in controlling automatic assembly machines, automatic parts–handling robots (3500 are at work already), automatic warehousing and inventory systems, and computer-based management and factory scheduling systems. If computers were used to control all these steps, the Rand Corporation estimates that overall reductions in total manufacturing costs of two to four times are achievable.[20]

One feature of computer-based automation which helps to make such cost reductions possible is its flexibility. In the 1980s, computer-controlled automation could, for example, make small-lot manufacturing as economical as mass production is today. Nathan Cook of MIT predicts that computers and robots may reduce overall costs in small-lot manufacturing by 80 to 90 percent. This would be a most important development, for more than three-quarters of U.S. industrial output comes from batch manufacturing, a system currently plagued by long lead times, high in-process inventories, low machine utilization, and very little automation. The revolutionary feature of this process is that it feeds on itself. Thus, the introduction of computers into the manufacturing process has the potential for increasing productivity on a scale never before conceivable.

These are not fanciful imaginings for a far future time. Computers are already being put to many new uses in industry with positive productivity results. At McDonnell Douglas, a proprietary software system called CADD (computer-aided design and drafting) allows engineers to generate the design of the airframe and its parts in three dimensions without touching a pencil, and to make changes, store the images in the computer's memory bank, and retrieve them. Effect on productivity? The company's engineering vice-president claimed an engineer may be able to do four or five times as much work as before.

In late 1978, it was reported that a quiet revolution in machine tools is sweeping the metal-cutting industry. Computerized controls, which originally were found only in complex, multimillion dollar machining systems, are rapidly joining low-priced, simpler equipment such as milling machines and lathes in doing the bulk of U.S. metal-working in factories of all sizes. Some controls feature sensors and feedback loops that can boost productivity by automatically compensating for such mishaps as broken tools and set-up variations caused by temperature fluctuations.

Economists should also examine the potential for major productivity advances in services in the 1980s through what has been termed the "industrialization" of services. Basically, this means the adoption of better management practices—the prime mover historically behind increased productivity in the industrial sector. A graphic example is McDonald's which applies through management the same systematic modes of analysis, design, organization, and control that are commonplace in manufacturing. It is one of the "soft" technologies which essentially substitutes organized preplanned systems for individual service operatives. Other examples are preplanned vacation tours and production-line type income tax service.

Services can also be "industrialized" through hard technologies which substitute machinery, tools, or other tangible artifacts for people-intensive performance. Common examples are the consumer credit card, the automatic car wash, convenience home appliances, and automatic coil receptacles at toll booths. Examples of hybrid technologies which combine hard equipment with carefully planned industrial systems to bring efficiency, order, and speed to the service process are: development of unit trains and integral trains; computer-based, over-the-road routing which optimizes truck utilization and minimizes user cost; and limited service, fast, relatively low–priced repair facilities such as national muffler and transmission shops.

The changing composition of the labor force in the 1980s will also work to improve productivity advance, but this will occur independent of the efforts of economists. There is a complementary set of trends, however, which bodes well for productivity prospects in the 1980s that would receive more attention from economists guided by a qualitative growth norm since they center around developing a more flexible labor market. One trend involves quality of work life (QOWL). Productivity expert John Kendrick says that the main areas in which labor efficiency can be improved are those in which there are restrictive work practices, union work rules, or just plain lack of motivation and concern.[21] He argues that QOWL programs and other programs designed to stimulate worker cooperation and efforts to cut unit real costs must play an important role.

Flexitime has been the response to an increasing desire by workers to help determine their own work schedules. Productivity could be increased by more efficient utilization of labor through better matching of job schedules with job needs. Additionally, flexitime experience in Europe and to a more limited extent in the United States indicates that it tends to raise work morale and increase the work effort for each hour on the job. An alternative arrangement concerning the timing of jobs, the four-day workweek, has also been shown in recent studies to have a positive impact on productivity.

As more analysis becomes focused on these broader areas of potential productivity advancement, the nation should see a return to the healthy productivity growth rates of the first two decades after World War II, and a concomitant reduction in inflation. Similar positive effects in such areas as employment policy, tax policy, government spending, and trade should be experienced once a holistic perspective is widely adopted.

The Role of Market Forces and Corporations

Fortunately, many corporate activities and trends, driven by market forces, will help to significantly foster qualitative growth in the 1980s. Since the last two decades of this century will most certainly be a time of increasing resource scarcity and increasing real costs, the transformation to an information economy at this time is most fortuitous. Without this transformation, the United States would find it very difficult to generate a substantial share of its GNP from heavy, basic industries and to remain competitive in these industries in world markets.

Since this revolution is so sparing of energy and materials, it is virtually guaranteed an enduring impact. This point is supported in the President's 1976 Report on *National Growth and Development* which singled out computers and communications as vital growth industries now spawning an economics of abundance in our information resources rather than the economics of scarcity that tends to characterize energy and other natural resource sectors.

Thus, as information activity continues to generate an increasing share of the GNP at the expense of energy/resource-intensive basic industries, we will move toward a more energy/resource-conserving base and, hence, one that will also be more environmentally benign. Qualitative growth in the physical realm is given a substantial thrust by this transformation in the economic base.

In the workings of so broad a market force, corporations are rather passive participants. The question remains: Do they have an active role to play, perhaps even a responsibility, in helping to shape a qualitative growth future? The traditional perspective on corporations and social responsibility has been succinctly expressed by Milton Friedman, "There is one and only one social responsibility of business—to use its resources and engage in activities to increase its profits."[22]

An opposing viewpoint developed in the 1970s argues that corporations, as creatures of the state, have an explicit responsibility to conform to broad societal goals. This view has been expounded not only by those outside business and critical of its ways but by an increasing number of corporate leaders. Louis Lundborg, former chairman of the board of the Bank of America, argues strongly that corporate leadership should be responsive to environmental and social concerns: "Those in corporate life are going to be expected to do things for the good of society just to earn their franchise, their corporate right to exist."[23] He believes debating the issue of corporate social responsibility is nonsense: "Environmental and other social problems should get at least as much corporate attention as production, sales, and finance. The quality of life in its total meaning is, in the final reckoning, the only justification for any corporate activity."[24]

The likely response would be that this is all well and good in theory but, in practice, production, sales, and finance would then likely suffer. The notion that pollution control or resource conservation practices have to be a burden to the firm, ultimately lowering profits, is widespread but false. Indeed, the variety of resource conservation activities already underway by corporations quickly belies the argument. Environmental regulations have forced companies to engage in wholesale rethinking of their production processes and products. This often shocks them out of a rather inflexible production system and, thereby, provides the catalyst necessary for beneficial innovation to occur. For example, when the initial requirements of the Clean Air and Water Acts began to take effect, many industrial firms reevaluated their entire system in terms of product recovery and waste minimization. The process review led many to adopt innovations that led to greater industrial efficiency, improved fuel conservation, and profitable recycling. Thus, there have been bottom-line economic benefits as well as energy benefits.

The list of corporations which have attained such benefits in the United States is already large and continues to grow. Companies of all sizes and from a wide spectrum of industries are represented. Perhaps the most innovative and well-known program is the 3P program—Pollution Prevention Pays—of the 3M Corporation which rewards employees who can find more economical and less polluting ways to manufacture products. In the first three years, 37 3P projects yielded an estimated savings in compliance costs of about $17 million.

Recycling is another area that can yield a number of benefits: it contributes to the supply of materials; it reduces the volume of waste; and it usually results in energy, environmental, and process savings. The contributions to materials supplies and waste disposal are inherent in recycling. They always accrue (at least in principle) and are the main driving forces of recycling. The energy, environmental, and process savings are contingent on the efficiency of the recycling operation. There is a considerable amount of recycling being done at the present time. It is ironic that "junk" materials for recycling in the mid-1970s supplied the nation with 44 percent of its copper, 20 percent of its iron and steel, almost 50 percent of its lead, and about 20 percent of its paper. The recovery rate for copper is the best in the world, and the high rate for lead is especially remarkable since several hundred thousand tons of lead are devoted annually to dissipative uses, the largest being gasoline additives.

Certainly, much is already being done in the area of energy conservation as well. Industry officials in 1978 reported that energy conservation was about the most profitable use for their capital, causing companies to pour hundreds of millions of dollars into it. For instance, according to a 1978 *Business Week* article:

> Scientists at a half dozen other respected companies, including General Electric, are also sketching windmills these days. Elsewhere in the white-smocked reaches of the nation's corporate laboratories, researchers are trying to squeeze power from ocean heat, design hollow ball bearings to cut auto weight, and convert wood chips from logged-over forests into fuel oil. Such projects, implausible or even ludicrous just five years ago, are evidence of the most far-reaching transformation to hit company research labs in more than a decade. Energy worries are sparking a basic reorientation of corporate research and development.[25]

Furthermore, the rapidly rising spending in this area which is currently adding to total capital outlays will be increasingly important in the next few years. Dow Chemical, for instance, sees over $200 million of energy-saving investment opportunities in its Texas and Louisiana facilities alone, each project offering over 50 percent pretax return on investment.

Significant resource savings have already been achieved through materials substitution. In many cases, the driving force of materials substitution is the development of technology. A principal area in which this has been demonstrated is communication, where technology development has significantly reduced materials requirements through solid-state electronics, microfilms, microwave transmission, and commercial satellites. This has been referred to as functional or system substitution in which a completely new way is found to perform the function of a component or system. The transistor, for instance, requires perhaps one-millionth of the material needed to make the vacuum tube it replaces.

Another example of recent corporate resource conservation activity resulting solely from market forces has been the establishment of waste bourses (exchanges) in at least 14 states. Basically, the bourse brings a company with an industrial waste into contact with another company that can profitably use the waste. Wastes that might be a liability because of high disposal costs or possible damage to the environment can given the seller additional income and the buyer a cheaper source of raw material.

It is estimated that the United States generates about 344 million metric (wet) tons of industrial waste each year. If only 10 percent of this waste could be utilized or recycled, the waste exchange would prove its worth. Further exploration is needed of possible cooperation between exchanges and the best ways to reach small and medium sized industries.

Though corporations have already undertaken significant resource conservation activities, there is considerably more potential, particularly regarding recycling, energy conservation, and process change for pollution control purposes.

Though the amount of recycling is significant in a number of cases, it has remained on a plateau for over a decade. Despite highly publicized programs to increase aluminum recycling, the market share for recycled aluminum as a percentage of the nation's total metal consumption still stands at 20 percent, the same ratio as in 1964. Over the same period, copper has remained virtually static, lead and zinc are both down about 2 percent, while tin has fallen from a 29 percent market share to 20 percent. Thus, even for a relatively high–valued commodity like copper, we are losing 40 percent of the supply available for recycling.

It goes without saying that energy conservation efforts have just scratched the surface. Energy savings of over 50 percent could be made in the steel industry if older plants were gradually replaced by more efficient facilities. Technical advances in the traditional Hall aluminum refining process can reduce energy requirements by more than a fifth; Alcoa recently completed a major facility using a new chloride process that is expected to reduce energy needs by a third. Such changes in the paper and cement industries also have much potential. The most efficient paper-manufacturing technologies require 50 percent less fuel than other commonly used methods. In cement manufacturing, an average of 1.2 million Btu's is used to decompose enough limestone to produce a barrel of cement, while some European plants only use 550,000 Btu's per barrel due to the recapture of waste heat. In building construction, technical improvements can save 50 percent in office space requirements.

Another area of great potential for conservation is "cogeneration," the generating of electricity as a by-product of the process steam produced in industry. According to a study directed by Paul McCracken, U.S. industry by 1985 could meet approximately half its own electricity needs by this means, compared to about a seventh in 1977. This rate of cogeneration would save fuel equivalent to 2 to 3 million barrels of oil per day as well as investments of $20 to $50 billion.

Most conservation measures to date have been applied to existing structures or equipment. The really significant potential can be realized with the introduction of new facilities. In a 1978 survey, corporate officials predicted that replacement of entire plants or offices with more energy-efficient ones, while only a small part of spending so far, would be a significant factor in capital outlays a few years hence.

To date, only a small fraction of all corporations have undertaken process change to control (or prevent) pollution. Since those that have been pioneers in this area have experienced such significant economic and/or energy savings, it seems inevitable that market forces alone will dictate the expansion of this pollution control method.

As a general principle, corporations in the future are going to have to learn how to get more from less and then have the will to do it. Corporations will have to place greater emphasis on the development of production methods, technologies, and products that extend the life of renewable resources, maximize the sustainable yield of renewable resources, reduce energy and resource use per unit of output, eliminate or cut down on the creation of waste, allow for recycling, provide for greater longevity and durability, and not overburden the "public service" functions (such as waste assimilation) rendered by the natural environment. This is an overall strategy of stewardship compatible with the coming age of world resource scarcity.

What Government Can Do

We have become great in a material sense because of the lavish use of our resources, and we have just reason to be proud of our growth. But the time has come to inquire seriously what will happen when our forests are gone, when the coal, the iron, the oil, and the gas are exhausted, when the soils shall have been still further impoverished and washed into the streams, polluting the rivers, denuding the fields, and obstructing navigation. These questions do not relate only to the next century or to the next generation. One distinguishing characteristic of really civilized men is foresight; we have to, as a nation, exercise foresight for this nation in the future.[26]

These presidential words were not spoken by Jimmy Carter in 1979—though they easily would apply. They were spoken by Theodore Roosevelt in 1911 in his keynote remarks to a White House Conference on the Conservation of Natural Resources. Unfortunately, the nation has never heeded his advice. Indeed, much of the malaise afflicting Americans is due to their perception that the United States is walking backwards into the future. With no vision of what America could or should be like in the future, the nation drifts into uncertainty like a ship without a compass or even a clear destination. It is little wonder that Americans today keenly sense that their government leaders do not have a good idea of "whither we are tending."

Key federal decision makers not only fail in vision but they continue to view trends as isolated phenomena even though it is the mode of interaction among the major forces and trends influencing long-term growth that is usually critical in determining their ultimate impact. Tackling problems on such a piecemeal basis leads to policy "solutions" which are too narrowly based and often produce results which are favorable in one area but counterproductive in another. While a given environmental policy, an energy policy, and an urban policy may each be perfectly reasonable when judged on their individual merits, they may culminate in an untenable whole, a pattern of mutual paralysis.

The fact that the really critical issues before the country are not the immediate and isolated ones but the interrelated and long-range ones suggests the desirability of developing a new policy framework which would examine the interrelationships among problems and be anticipatory in nature. However, policy formulation cannot be forward-looking until clear national goals and priorities have been established to which policies can be directed and measurement to assess progress applied. If government policies are going to effectively govern the long–run forces influencing growth, a national growth policy framework will have to be developed which would fulfill three functions:

- establish national priorities and goals and measure progress toward achieving them;
- conduct on-going analysis of long-range trends influencing the socioeconomic system, and anticipate problems likely to arise; and
- guide the coordination and integration of policies.

The heart of a national growth policy process would be long-range analysis and goal setting. This would involve traversing a logical sequence of steps: specify broad, long-term policy objectives; consider alternative policies; set priorities among policies; lay out alternative plans; evaluate the consequences of the alternative plans; coordinate the study; and disseminate the results. Such a set of aims, together with an estimate of priorities, would serve as a fundamental foundation for planning throughout government and in the private sector. The United States would then have attained that which Churchill stressed: "Those who are possessed of a definite body of doctrine and of deeply rooted convictions upon it will be in a much better position to deal with the shifts and surprises of daily affairs than those who are merely taking short views, and indulging their natural impulses as they are evoked by what they read from day to day."[27]

The keynote to the success of a national growth policy would be its comprehensiveness. Such a policy would provide for harmonizing frequently conflicting aims within our broader social and environmental, as well as purely economic, objectives. Economic efficiency itself will have to be viewed more comprehensively. We must aim for the efficiency of the whole economic performance, so that it is measured by the degree to which the capabilities of the economy are being used and by how well the economy is able to cope with problems such as ecology and the quality of life.

Comprehensive national planning would also encompass regional and state concerns and policies. This could be accomplished in a number of ways: by incorporating into the national process existing regional and state policies; helping to resolve conflicts between levels of government; explicitly taking into account unique growth requirements and impacts on regions or states; and by anticipating trends and problems that could emerge at state and local levels and potentially interfere with achieving national growth and development objectives.

Finally, the growth policy process could not operate in a domestic vacuum. As shown earlier, interdependence is a hard reality which demands that there be closer international cooperation in economic policies. More broadly, in a world where not energy alone, nor Eurodollars nor multinational corporations proffer problems beyond the capacity of any nation to resolve, but also inflation, worker migration patterns, pollution, and scores of other issues which have taken on transnational dimensions, government "drift" must be replaced by purposive, coordinated policies.

Granted, the process of developing a national growth policy which is integrative, anticipatory, and goal-suggesting is much more of an art-form than a science. Some may, on that basis, question its validity. To them, Lord Keynes would gently suggest, "There will be no harm in making mild preparations for our destiny."

There should be no illusions that a national growth policy will solve all current problems or prevent major ones from arising in the future. It will, however, help us regain our confidence in our ability to manage our affairs, to deal with our problems, and to take charge of our future. Most importantly, it should help bring about a renewed sense of national unity and purpose.

Let us assume that a national growth policy process were in place. What could the government do to promote qualitative growth under the two guiding principles suggested earlier—developing a Conserver Society and achieving total employment.

Most Americans today would readily identify the notion of a Conserver Society with energy conservation. But it would encompass much broader notions as well regarding the conservation of all natural resources and raw materials and the transition from a "linear" economy in which extracted resources move through production processes and are ultimately disposed of as waste to a "circular" economy in which a substantial portion of output would not be eventually discarded but recycled for further production.

Certainly, energy conservation, which simply means cutting back on energy demand, must be an instrumental part of a Conserver Society. The government is finally recognizing this due to two convincing arguments. First, the argument that cutbacks in energy use would cripple economic growth has been conclusively debunked. A number of studies have shown that there has been no historic inviolable relationship between energy and GNP growth and, also, that energy growth in the future could be substantially cut without seriously affecting economic growth prospects.[28] Second, energy conservation has been shown to be economically wise, since it is generally cheaper than investment in new energy supplies. To the degree that clean air and water, and indeed land itself, have become increasingly scarce and costly goods, it makes both environmental and economic sense to make the conservation of energy and the recovery of waste a matter of the highest priority. Indeed, it is in the field of energy conservation that energy, environmental, and economic policy goals coincide and are attainable simultaneously.

Because energy has become such an urgent matter, it is likely that energy conservation will become an ingrained habit in American life. But what are the chances of the United States ever becoming a Conserver Society on a broader scale? Certainly, past experience does not augur well for the future. Though the United States has flirted with elements of a conservation ethic in the past (primarily in preserving land in the national park system and other federally protected lands), by and large a growth and resource–exploitation ethic has dominated decisions by business and government. The government has not had, and still does not have, a comprehensive policy concerning natural resources which seeks to use judiciously our declining supply of domestic natural resources and the supplies available to us from abroad.

The government's one-resource-at-a-time approach to natural resource problems continues to have serious adverse effects. The "national" policy has really "emerged" as the result of an accretion of numerous individual decisions. This after-the-fact policy takes on some of the contradictions accumulated over separate and disparate decisions. By continuing our present piecemeal materials policy, the government is increasing the likelihood of inducing shortages or prolonging those caused by natural or political events, here and abroad.

A central element of a new, comprehensive natural resources policy would be the establishment of economic incentives to turn our industrial structure toward less resource and energy-intensive technologies and production processes. Perhaps the area where national policy could have the most significant impact is in recycling.

Three categories of problems tend to inhibit the recycling of materials: the lack of adequate technology, economic and institutional issues, and problems of information and interpretation. Many of the technical processing problems have a common origin in contamination which typically characterizes secondary materials. The economic problem revolves about the fact that secondary materials often serve in a substitutional capacity. During periods of high demand, they are used to fill gaps in supply but they are first to feel the effects of reduced demand. Thus, the prices of many secondary materials are more volatile than those of their primary counterparts which represents a substantial deterrent to investment in recycling. There are also artificial barriers and disincentives to recycling such as freight rates and taxation which discriminate in favor of primary production. Finally, we lack reliable information on the magnitude of the pools of the major secondary materials so that we cannot determine the physically possible maximum production from secondary sources. The costs of greatly increased recycling are not known. The question of the optimum amount of recycling has different answers for different materials at different times, depending on the labor, capital, indirect materials, and energy required for recycling. The economic optimum for recycling is also affected by other materials conservation strategies such as the increased durability of products or substitutions in the materials from which they are made. The effects of such changes are not yet well understood. All these uncertainties support the view that recycling is one element in a complex materials supply system, and they point to the importance of a systems approach to materials conservation.

There are a number of specific steps government could take to correct these problems, particularly the economic and institutional ones. First, it can ensure that price reflects scarcity value and that economic neutrality is achieved as far as the production and use of primary (virgin) and secondary (recycled) materials are concerned. Ending discriminatory freight rates would be a minimal first step to encourage conservation of scarce resources and to provide appropriate signals for the market place to reflect true scarcity values.

Since extractive industries have been so heavily subsidized in the past, it seems appropriate that the government go beyond economic neutrality and provide subsidies or other economic incentives for recycling. The National Energy Act of 1978 did include a 10 percent tax credit on recycling equipment and established goals for increasing the use of recycled commodities. This was the federal government's first demonstration of a willingness to provide a viable means for industry to increase the recycling of solid-waste materials.

Many observers feel the government should take further steps, particularly in light of the fact that recycling could save significantly on energy use: one–fifth of the total U.S. energy budget is now spent on materials production. In the most extreme case, recycling 100 percent scrap would reduce energy output 47 percent for steel, 96 percent for aluminum, and from 88 to 95 percent for copper. Further steps the government could take to aid the development and modernization of the recycling industry would be research and development grants and incentives such as low-interest loans, price guarantees, and preference for recycled materials in government procurement.

One form of recycling which deserves special consideration by the federal government is the recycling of urban solid wastes. The institutional separation of energy generation by public utilities and waste disposal by municipalities must be seriously questioned from both an economic and energy perspective. Currently, the United States recovers only one percent of the energy potential of municipal solid waste; while Denmark, which integrates the two functions, recovers 60 percent. The energy recovery potential from our municipal solid waste is equivalent to 400,000 barrels of oil per day.

Mining this "urban ore" has begun. Five refuse-to-energy plants processing more than 500 tons daily are in operation. Industry experts expect that, within eight years, about 17 percent of the nation's garbage will be recovered in 20 large plants and 50 small ones (treating less than 500 tons daily). The federal government could facilitate further movement in this direction by making grants to competing municipalities to help them figure out how to switch from solid waste disposal to resource recovery in the manner best suited to their individual community needs. Such planning grants are essential because, before cities can justify making capital expenditures, they have to first figure out how to overcome the technical, marketing, financial, legal, and organizational barriers. Thus, the 40 or so resource recovery planning grants that EPA awarded in 1979 must be expanded in the future.

Also in the solid waste area, the federal government could consider eliminating its disincentives, in the form of tax and revenue-sharing policies, that have prevented communities and citizens from giving serious consideration to the adoption of local user fees for financing solid waste management. This would allow communities to make an unbiased decision about the economic and social merits of having citizens pay for that proportion of the municipal disposal services that they actually use. Quantity-based local user fees have the potential to stimulate both a reduction in the quantity of waste individuals put out for collection and disposal and an increase in source separation and resource recovery. In addition to providing a more equitable means of payment for waste collection and disposal than the current flat fee and local tax-financed systems, local user fees would also heighten citizen consciousness about both waste and resource recovery.

Although recycling is an essential feature of the Conserver Society, it, nevertheless, consumes energy and often causes severe pollution problems. An important additional goal, therefore, is to increase the effective life of manufactured products, i.e., to promote reuse rather than recycling. Though providing effective rewards for the manufacture of durable products is difficult, a number of simple steps can be considered such as the setting of mandatory minimum standards for durability, requirements for labeling of guaranteed and expected lifetime, provision of consumer information on lifetime costs, and, in some cases, testing by government agencies (analogous to existing tests for automobile fuel economy). Because automobiles impose such substantial costs in depleted resources and pollution, mandatory minimum standards for durability and sustained performance (including at least the engine and drive train) should be considered, backed by a federal testing and consumer information program.

In general, the free market imperfection of consumer choices based on insufficient or misleading information should be reduced by a vigorous government effort to ensure accurate advertising and full labeling of products. In certain instances, the government could choose to influence consumer choice by altering the cost of energy or resource-intensive goods or activities. An example would be imposing an excise tax on gas-guzzling cars. Or it could provide the public alternative services which effectively reduce demand for energy and raw materials. Mass transit, for example, is a public investment which helps reduce reliance on imported oil.

The concept of total employment, focusing as it does on providing satisfying jobs, is integrally related to the earlier discussion of a more flexible market. In fact, most of the actions that would help bring about total employment would be by companies, not government. Government can, however, provide a supportive framework. One possibility here would be the development of a national policy on quality of work life. In such a policy, government could generate some pressures on organizations to become aware of quality of work life problems, perhaps through legislation requiring management and labor to jointly engage in an internal process of surveying QOWL problems and making changes; stimulate the spread of QOWL innovations by supplementing the current market strategy with a program of tax incentives, limited exemptions from federal employment regulations, and continued direct government funding of QOWL demonstration projects; expand efforts to advance the state-of-the-art in QOWL; and undertake major efforts to innovate and improve QOWL in its own organizations.

More generally, the government should begin to do some creative thinking on how education, work, and leisure can be experienced as continuing strands running throughout life rather than in the three distinct "boxes" they now occupy. The challenge is to facilitate a new cyclic life pattern in which some of the time currently spent in school and retirement would be redistributed to the middle years of life in the form of extended periods away from work for leisure and education. Workers in a recent survey expressed a preference for such a cyclic plan over the linear life plan. Government should examine whether distributing income earning work time more evenly over the total life cycle would reduce the problems of poverty, ever higher welfare and student support costs, and the possible bankruptcy of Social Security resulting from the compression of work into ever fewer mid-life years which has created a large number of nonearning years at the extremes of life. Whatever such benefits may be, the increase in freedom and flexibility that such a "continuing strand" brings should make work and all of life more enjoyable.

Qualitative growth must encompass a concern not only for rates and patterns of growth but also for the beneficiaries of growth. Specifically, it must develop means to move from a situation in which nearly all new wealth generated goes to current wealthholders toward one in which such wealth would be distributed much more broadly, perhaps even to all Americans. For example, if a means could be devised by which all Americans could become owners of capital—an income-producing asset—this would alleviate much of the burden and inherent disgrace associated with "welfare state" income redistribution schemes of the past two decades in which hundreds of billions of dollars have had little effect on the distribution of income. The importance of such a goal was recognized by the Joint Economic Committee of Congress in 1976 through a recommendation in its Annual Report:

> To provide a realistic opportunity for more U.S. citizens to become owners of capital, and to provide an expanded source of equity financing for corporations, it should be made national policy to pursue the goal of broadened capital ownership. . . .

A logical asset to target for use in achieving the goal of broadened capital ownership is corporate stock. It is a very significant component of personal wealth, it can be easily and widely distributed, and it would be the most direct way for Americans to share in the growth of the economy.

A highly innovative plan has been suggested to enable all Americans to acquire significant capital holdings. Its barest outline is the following. Companies would finance all capital expenditures by issuance of capital stock made available to the plan. There would be a full pay-out of dividends except for necessary operating reserves, and the dividends would either be tax deductible or the corporate income tax could eventually be eliminated. Breaking with traditional economic theory and financial practices, individuals could acquire these stocks not from their past savings but based on the future earning power of corporations. In the plan, the "purchase" of the stock by a household would be financed through a commercial bank loan made on a nonrecourse self-liquidating basis. The borrower would have no personal liability to repay the loan; the yearly dividends on the stock would repay it. With full payout of dividends and no corporate income tax, it should be only 5-7 years before the stock would be fully and directly owned by the household.

This "full-blown" version entails such radical change in the economic-financial system that there would have to be very gradual movement toward it. Such movement at the experimental level is currently under consideration by the Alaskan state legislature. They are considering establishing a General Stock Ownership Corporation which would initially invest in Alaskan energy-development projects and distribute equal shares of stock to each Alaskan citizen. The plan was developed by Senator Mike Gravel of Alaska who, in 1978, obtained the necessary tax legislation support from the Senate Finance Committee.

The idea of broadened capital ownership was supported by Keynes. In 1940, when income distribution through labor alone was threatening to cause inflation, he recommended income distribution through capital ownership by the workers which would bring "an advance towards economic equality greater than any which we have made in recent times."[29]

It is quite incongruous that capital ownership has never been the subject of a national economic policy, since the mark of a healthy capitalist system should be direct participation in the system by the citizens through ownership of capital. Surely, now, qualitative growth can devise some means by which all Americans can get "a piece of the action."

We have just come through a unique period in U.S. history. The economy, drawing on a confluence of positive and often unique factors, was able to generate material wealth for most Americans which those living just 100 years ago could hardly have dreamed of. Yet, in a fundamental sense, the United States has been too preoccupied with the material side of the great socioeconomic revolution brought about by its scientific and technological accomplishments. We have long given lip service to the fact that man does not live by bread alone but have rarely believed it. Only recently has the general feeling arisen that affluence does not buy contentment or a sense of community.

It is now time for America to exercise a new kind of leadership, having introduced much of the world to the industrial era and its concomitant spirit of material acquisition. In the new information economy era, America is called to three tasks: (1) to rechannel the present narrow thrust for growth and its associated materialistic ethic; (2) to foster economic practices centered around stewardship toward nature and providing opportunities for individuals to realize their full potential; and (3) to make the possibilities of the future imaginatively concrete.

The words of Lincoln urge us on:

> The dogmas of the quiet past are inadequate to the stormy present. The occasion is piled high with difficulty, and we must arise with the occasion. As our case is new, so we must think anew and act anew.

NOTES

1. The four quoted statements were made, respectively, by Kenneth Boulding, *The Meaning of the Twentieth Century: The Great Transition* (New York: Harper and Row, 1965), pp. 1, 2 & 24; André Malraux, quoted in Paul Ress, "Malraux: The End of Civilization," *Time*, April 8, 1974, p. 34; Lester R. Brown, *In the Human Interest: A Strategy to Stabilize World Population* (New York: W. W. Norton, 1975), p. 172; and Arthur M. Johnson, "The Business of America—Does it Face a Period of Basic Change?" *The National Observer*, February 28, 1976.

2. These statistics are drawn from the seminal work on this topic done by Marc Porat, *The Information Economy: Definition and Measurement*, U.S. Department of Commerce, Office of Telecommunication, O.T. Special Publication 77-12(1), May 1977.

3. The survey was done for *Business Week* by Investors Management Sciences, Inc., and was highlighted in "The Slow Investment Economy," *Business Week*, October 17, 1977.

4. Organization for Economic Cooperation and Development, Draft Final Report, Part V, Interfutures Project, Paris, January 13, 1979, p. 4.

5. Paul Strassman, "Office of the Future," unpublished paper, Xerox Corporation, September 20, 1978.

6. Daniel Bell, "Communications Technology—For Better or Worse," *Harvard Business Review*, May-June 1979, p. 26.

7. Bro Uttal, "Japan's Big Push in Computers," *Fortune*, September 25, 1978, p. 72.

8. OECD, Draft Final Report, Part III, Interfutures Project, Paris, January 23, 1979, p. 150.

9. George Katona and Burkhard Strumpel, *A New Economic Era* (New York: Elsevier, 1978), p. 132.

10. Harlan Cleveland and Thomas W. Wilson, Jr., *Humangrowth: An Essay on Growth Values and Quality of Life* (New York: Aspen Institute for Humanistic Studies, 1978), pp. 20-21.

11. Gardiner Ackley, "Prospects for the U.S. Economy—Growth or Stagnation?" *Vital Speeches*, May 1974, p. 432.

12. Duane S. Elgin and Arnold Mitchell, "Voluntary Simplicity: Life–Style of the Future?" *The Futurist*, August 1977, p. 203.

13. Robert Cameron Mitchell, "The Public Speaks Again: A New Environmental Survey," *Resources*, September–November 1978.

14. John Maynard Keynes, "Economic Possibilities for Our Grandchildren," in *Essays in Persuasion* (New York: St. Martin's Press), 1930.

15. de Jouvenel, "An Open Letter to Walter Heller," unpublished.

16. John Maynard Keynes, *The End of Laissez-Faire* (New York: Irvington), 1926.

17. Thomas R. Malthus, *Principles of Political Economy* (London: Kelley, 1836), p. 1.

18. Theodore Roszak in "Introduction" to *Small is Beautiful*, edited by E. F. Schumacher (New York: Harper and Row, 1973).

19. Keynes, *Essays in Persuasion*.

20. The Rand Corporation, *Computer-Based Automation of Discrete Product Manufacture: A Preliminary Discussion of Feasibility and Impact*, R-1073-ARPA, Santa Monica, California, July 1974.

21. John Kendrick, "Productivity: A Program for Improvement," in *Special Study on Economic Change*, Hearings before the Joint Economic Committee, U.S. Congress, June 8, 9, 13 and 14, 1978.

22. *New York Times Magazine*, September 12, 1970.

23. Louis Lundborg, *Future Without Shock* (New York: W. W. Norton, 1974), p. 84.

24. Ibid.

25. *Business Week*, February 1978.

26. Theodore Roosevelt, "Opening Address by the President," in Roderick Nash, *The American Environment: Readings in the History of Conservation* (Menlo Park, California: Addison-Wesley, 1968), p. 50.

27. Winston S. Churchill, *The Gathering Storm* (Boston: Houghton-Mifflin, 1948), p. 210.

28. Examples of such studies are: Sam H. Schurr and Joel Darmstadter, "The Energy Connection," *Resources*, Fall 1976, pp. 1-2, 5-7; Joel Darmstadter, Joy Dunkerly, and Jack Alternman, "International Variations in Energy Use: Findings from a Comparative Study," *Annual Review of Energy*, 1978, pp. 201-24; John G. Myers, "Energy Conservation and Economic Growth—Are They Incompatible?" *The Conference Board Record*, July 1976; Institute for Energy Analysis, *U.S. Energy and Economic Growth, 1975-2010*, September 1976; Dale Jorgenson, "The Economic Impact of Policies to Reduce U.S. Energy Growth," Harvard Institute of Economic Research, Discussion Paper Number 644, August 1978; William Nordhaus, "What is the Tradeoff Between Energy Consumption and Real Income?" unpublished paper, Yale University, 1977.

29. John Maynard Keynes, *How to Pay for the War* (New York: Harcourt, Brace, Jovanovich, 1940).

Chapter 5

ENERGY: A TIME FOR CHOICES

Eric R. Zausner

When historians of the future look back at the period in which we live, they will no doubt view it as a time of transition in the use of energy sources. Our current prosperity is based in part on past availability and use of cheap and abundant oil and gas. But the use of conventional oil and gas has peaked in the United States, and will decline through the remainder of the century. A new energy future is at hand, but its precise nature is uncertain.

Many people question what role energy will play in the economy and the society of the future. How will its cost affect the growth and distribution of wealth? Will its availability permit growth and prosperity, or will its scarcity place limits on the progress of mankind? These questions are of great concern to the world's thinkers and planners; yet, one can only guess at the long-term answers.

If history foretells the future, the full transition from a petroleum-based economy to the next energy source will take several decades. For example, wood was the world's most abundantly used fuel in 1850. But even as wood use was at its peak, the use of coal was on the rise (see fig. 5.1). In turn, coal use peaked in the 1930s, and began to decline as cheap supplies of oil and gas became increasingly available in the late 1940s. Now we have passed the peak of the oil era. What lies ahead?

The transition in which we find ourselves is characterized by a growing dependence on non-U.S.-produced petroleum. In 1973-1974, the United States imported 36 to 37 percent of its oil—just over 6 million barrels per day—with only about 700,000 barrels per day coming from the Arab members of the Organization of Petroleum Exporting Countries (OPEC) (see fig. 5.2). In 1977,

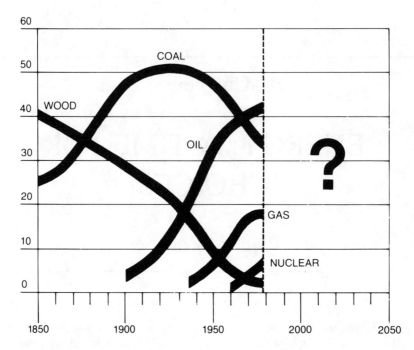

Figure 5.1. Over the years, energy usage has shifted from one source to another, driven primarily by abundance and cost.
Source: Energy Research and Development Administration (now the U.S. Department of Energy); *The Energy Research, Development, and Demonstration Plan*; 1977

we imported almost half of our oil—almost 9 million barrels per day—of which more than 3 million barrels came from Arab OPEC nations. The danger of our ever-increasing reliance on these uncertain supplies from a politically unstable region is obvious.

This growing import dependence is just one clear signal that our pattern of energy consumption—i.e., the mix of fuel types—is not consistent with our domestic energy resources, as shown in figure 5.3. We use oil and gas for about

Table 5.1.

	1973	1974	1975	1976	1977	1978
TOTAL CONSUMPTION	17.3	16.6	16.3	17.4	18.4	18.8
TOTAL IMPORTS	6.2	6.1	6.1	7.3	8.8	8.1
ARAB OPEC IMPORTS	714,700	752,500	1.38	2.42	3.18	2.92

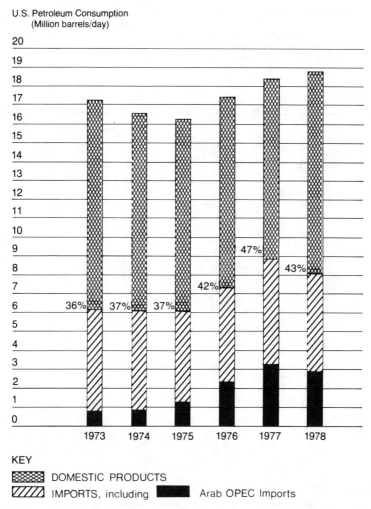

U.S. Petroleum Consumption
(Million barrels/day)

KEY

▨ DOMESTIC PRODUCTS

▨ IMPORTS, including ▮ Arab OPEC Imports

Figure 5.2. U.S. dependence on imported oil has increased dramatically in the last 5 years.
Source: U.S. Department of Energy; *Monthly Energy Review, Petroleum Section*; September 1979

75 percent of our energy needs; but these liquid and gaseous fuels represent only about 7 percent of our economically recoverable reserves. By contrast, coal accounts for only 19 percent of our energy consumption but constitutes 90 percent of all proven energy reserves in the United States. Ths imbalance predisposes an energy transition, for as our own—as well as the world's—oil and gas dwindle, we will be forced to turn to alternative fuels, whether we like it or not.

That transition can be characterized in several ways. First, in the new energy era, the price of energy will continue to rise. This increase began in 1973-1974,

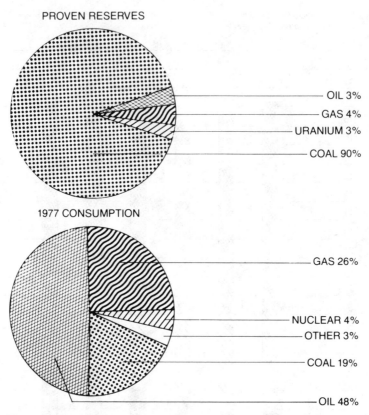

PROVEN RESERVES

OIL 3%
GAS 4%
URANIUM 3%
COAL 90%

1977 CONSUMPTION

GAS 26%
NUCLEAR 4%
OTHER 3%
COAL 19%
OIL 48%

Figure 5.3. The United States does not consume those types of fuel that are most abundant domestically.
Source: Energy Information Administration, U.S. Department of Energy; *Annual Report to Congress, 1978, Volume III*; 1979

when the OPEC nations raised the price of their oil from about $3 to $11 per barrel. Since that time, the price of oil has continued to increase in current if not in real dollars. The world oil price now hovers around $30 per barrel, with no end in sight for the upward spiral. In the few years that have elapsed since the oil embargo, the price of oil has changed from being a negligible economic cost to one of the most important economic indicators in the world. From 1962 through 1972, the real price of energy actually declined by about 1.4 percent each year. But from now until the end of the century, it is expected to outpace the rate of general inflation by 2 to 5 percent.

IMPACTS OF HIGHER ENERGY PRICES

The high price of oil has pushed U.S. payments for petroleum imports from

$4 billion in 1971 and $8 billion in 1973, to a staggering $40 billion in 1978. In 1979, with the surge in prices following the disruption of Iranian oil production, the total import bill could exceed $50 billion.

Slowed Growth

These higher prices will have a significant impact on the shape of our future. The world's economic growth rate has slackened since 1973, and most forecasts for the United States and world economies suggest slower growth than pre-1973 predictions indicated. Specifically, the average rate of growth over the next two decades is expected to remain below pre-1973 levels because of higher real energy costs, inflation, slower population growth, and slowing productivity increases.

Along with the overall economic growth rate, the energy demand projections have been revised downward. What was once considered a wildly optimistic—i.e., unrealistically low—energy demand forecast for 2000 is now viewed as suspiciously high. Specifically, in 1972, the conventional forecast of the U.S. demand for energy in 2000 was 160 quadrillion Btu (quads), while the pessimistic (i.e., high) forecast was for 190 quads. At that time, British physicist-philosopher Amory Lovins was earning the badge of "heretic" for forecasting a mere hundred and twenty-five quads of energy demand in 2000.

By 1975, however, as forecasters hastily replotted their charts in the wake of the OPEC price hike and the conservation it induced, Lovins' heretical forecast had become the conventional figure. Of course, by that time, Lovins, too, had issued a new forecast, this time of 90 quads. By 1978, the heretics had issued forecasts as low as 33 quads; the conventional forecast was pegged at 110 quads; and the heretically optimistic forecast of 1972—for a hundred and twenty-four quads—had, in six short years, become the pessimistically high prediction!

Search for Alternatives

The damaging effects of the OPEC oil embargo have led to a frantic search for other energy sources. With the need to find petroleum substitutes, a variety of fuel supply options—domestic crude, natural gas, coal, and nuclear energy—have been examined. All of these options, however, pose problems for which there are no easy solutions.

● *Domestic crude oil.* As good a substitute as any for imported oil, domestic crude is neither easy nor inexpensive to find or produce. Although overall domestic oil production may increase somewhat in absolute terms through the end of the century, as shown in figure 5.4, the sources of that production will shift. Production from "old" onshore and offshore reserves will decline rapidly, while increased production will originate in new onshore and offshore

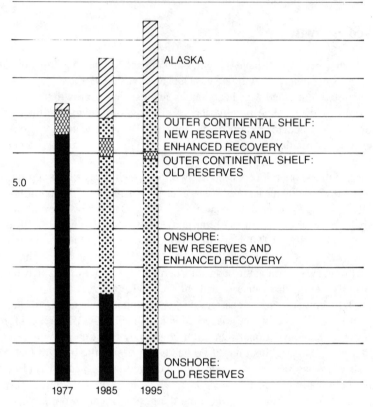

Crude Oil Production
(Million barrels/day)

10.0

ALASKA

OUTER CONTINENTAL SHELF:
NEW RESERVES AND
ENHANCED RECOVERY

OUTER CONTINENTAL SHELF:
OLD RESERVES

5.0

ONSHORE:
NEW RESERVES AND
ENHANCED RECOVERY

ONSHORE:
OLD RESERVES

1977   1985   1995

Figure 5.4. Domestic crude production in the remainder of this century will be characterized by
sharp drops in the production of old reserves and increased production of new sources.
Source: Booz·Allen & Hamilton estimates, 1979.

fields and through enhanced recovery techniques. Oil produced from these
sources will sell for the world market price, under the government's phased
price-deregulation program. But this will only slow the rate of decline in
production during the remainder of this century; indeed, the ratio of reserves
to production in the continental United States is 7.5:1, one of the lowest in
the world. At best, then, we are only "buying more time" for the transition
to other sources to take place.

● *Natural gas.* Under the price deregulation program of the 1978 Natural Gas
Policy Act, the cost of natural gas will climb. High demand for gas may bring
in supplemental gas supplies from Mexico and Canada and imported liquefied

natural gas. However, unless we can tap several other types of gas sources, such as geopressured methane and very deep gas resources, natural gas production will never again reach its early-1970s peak.

- *Coal.* Increased coal production and use, the keystone of the federal government's energy program, are constrained by a host of economic and regulatory restrictions—e.g., air quality standards, railroad rate regulations, and surface mining regulations. As a result, coal production is only expected to rise from 660 million tons a year in 1978 to less than 1 billion tons per year by 1985. Any slippage from even this modest goal could lead to a great increase in oil imports.

- *Nuclear energy.* Only a few decades ago, the energy oracles of the postwar era foresaw a future abundant with nuclear-powered electricity that would be "too cheap to meter." And, indeed, we began to use nuclear power for civilian energy production in the late 1950s. Today, nuclear power provides some 13 percent of our national electricity needs. But numerous problems have eroded public confidence in this technology; the crisis of confidence climaxed with the accident at Three Mile Island in March 1979. In addition, the unresolved problems of licensing and waste disposal continue to plague the industry. As a result, the utility industry's orders for new nuclear plants (versus coal–fired plants) have fallen from 103 nuclear plants in the period 1971 through 1977 to only 2 new plants in 1978. Furthermore, 39 of the original orders were cancelled. Despite these set-backs, nuclear energy's contribution to the nation's total electricity supply probably will rise from the current 13 percent to 20 percent by 1985, through contributions of plants now under construction.

In addition to these four energy sources, a number of other options—synthetic fuels, wood, solar, geothermal, and photovoltaic cells—could play increasingly important roles in the energy picture during the remainder of this century.

- *Synthetic fuels.* The cost of producing a barrel of oil equivalent from coal or shale is not yet economically competitive with the cost of even an imported barrel of oil. For example, the cost of obtaining oil from shale is estimated at $27-30/barrel (in current dollars) in the 5-year period, 1985-1990. Moreover, most of the synthetic fuels and oil shale technologies have some unresolved technical, environmental, and institutional problems. With shale oil production, for instance, there are numerous environmental issues, such as waste disposal and the availability of fresh water in the arid West where most of the oil shale exists. The institutional problems associated with shale oil production also are immense; one shale oil company identified more than 50 permits and approvals needed to initiate its proposed project.

- *Wood.* The use of wood as a heating and cooking fuel peaked in the United States about 100 years ago. Today, wood–burning stoves are making a tremendous comeback, especially in timber-rich but oil-poor areas like New England. Some 5 million American homes now use wood for all or part of their energy requirements.

- *Solar, geothermal, photovoltaics.* Although the extent to which the various renewable sources will alter U.S. energy consumption patterns is unknown at present, the government does forecast market penetration of all solar and geothermal technologies to double by the end of the century—from about 4.8 quads in 1978 to about 11 quads of primary energy by 2000, assuming oil prices reach $32 per barrel or more (in 2000 dollars). The technologies contributing to this figure would include passive and active solar heating/cooling systems, industrial/agricultural solar process heat, hydropower, and wind machines. The tremendous potential of solar photovoltaic systems, however, probably will not be realized until costs are brought down significantly in the next century.

Today's energy situation, then, will very likely be with us for the short term. What comes after that is somewhat uncertain, but we do know it will be characterized by high energy prices; continued dependence on petroleum imports; an economy dependent in the mid-term on oil and gas, coal, and nuclear power; and the emergence of some new, commercially viable energy alternatives.

IMPLICATIONS OF ENERGY PICTURE

The future energy situation has broad implications for the United States and its people. It will test politicians' willingness to make some difficult—and perhaps unpopular—decisions on the basis of hard-nosed trade-offs among the various "costs." It will dare us to change our foreign and domestic policies to allow us to reshape our international role. And it will force the American people to decide what actions they are willing to support and take to preserve our economic and social well-being.

Making Tough Trade-off Decisions

Since the oil embargo of 1973-1974, three presidents have struggled with the energy problem. President Nixon's ill-fated Project Independence was followed by President Ford's National Energy Act and President Carter's "Moral Equivalent of War." Each time, a comprehensive energy plan was proposed, only to be met by a recalcitrant Congress and years of protracted debate.

The problem, then, is not primarily technological or economic; it is institutional. By nature, the energy problem is complex, and no single, sweeping action or policy can deal effectively with the myriad ways in which we produce and consume energy. Conversely, any *comprehensive* policy has something in it for someone to oppose. Most of the less controversial, more politically acceptable solutions already have been proposed and enacted. We have legislated a 55-mile-per-hour speed limit, as well as various energy-related tax incentives. Congress has increased the budget for energy research, development, and demonstration. A few years ago, we even extended daylight savings time in a dubiously successful move to save just a little bit more energy. All of these legislative programs were approved because they involved only minimally increased federal outlays, or their effects were years away. *None* had severe impacts on the American people.

Every effective program exacts some "costs"—intensified regional development, environmental impacts, higher prices. For example, incremental oil or gas production involves tapping reserves in the frontier areas of Alaska or on the outer continental shelf. Yet, it is precisely those areas where environmental impacts are the greatest and where a limited regional population bears the development burdens. The impacts of regional coal development are similar; here the environmental burdens of surface mining and the associated impacts of the economic "boom" that follows increased mining operations are placed on the Western states and their populations. The coal itself, however, is used to generate electricity for urban Californians or to produce synthetic fuels for the Midwest.

In some ways, nuclear power plants epitomize this dilemma. The electricity generated by a typical nuclear power plant benefits an entire region, especially the metropolitan areas. Yet, the residents of the rural town or county in which the plant is located often oppose siting the facility in their area. Similarly, on the national level, national energy policies that will produce national benefits can be—and are—opposed because they adversely affect regional interests.

Potentially negative environmental impact is a second "cost" that must be considered. Every one of our feasible energy supply alternatives has at least some negative environmental effects. Even the extraction of natural gas—our cleanest-burning fuel—affects the environment. For oil, coal, and nuclear power, the environmental impacts are more severe and less politically palatable.

To a large extent, our political process has failed to make the necessary trade-offs between energy development and complete environmental protection. The examples are legion, but one of the most notorious will suffice. In several pieces of legislation enacted since the oil embargo, Congress has *mandated* oil- and gas–burning utilities, which had forsaken coal years ago because of stringent federal environmental standards on stack-gas emissions, to convert to coal. And, adding insult to the injury of coming full circle in little more than a decade, Congress failed to align the existing environmental standards with the new coal-conversion requirement. The result has been a tangled web of exemptions, and armies of lawyers on both sides. More importantly, almost no oil and gas have been saved from mandatory coal conversion in the more than five years the legislation has existed.

The final energy program "cost" *is* cost. The old axiom "there is no such thing as a free lunch" is painfully applicable to our energy solutions. For example, there is no more effective means of stimulating production or conservation than raising energy prices—but increasing consumer fuel prices in a period of massive inflation and dollar-value decline is one of the most politically disastrous actions a Congressman can take. In addition, funds needed for investment in tomorrow's energy projects could be generated in part by raising prices from yesterday's to today's levels—yet, any changes in federal petroleum and oil regulatory programs or state-set utility rates to generate such funds would face severe public backlash.

In and of themselves, these development, environmental, and financial costs would be politically acceptable were it not for the length of time between seeding and harvest—that is, the time needed to build oil and gas production facilities or nuclear power plants, or to replace existing building and transportation vehicle stock with a significant percentage of more energy-efficient models. Politicians are willing to pay significant costs to solve the energy problem as long as the benefits accrue quickly enough to be used in reelection campaigns. Unfortunately, many energy solutions bear little if any fruit for up to 10 years. (Perhaps "five elections" is a much more meaningful measure of the time involved.)

Faced with these realities, politicians are forced to respond to—rather than shape—public opinion. And therein lies the crux of the problem—because the public is in no better position than Congress to mold a cohesive, well-thought-out stand on energy priorities and solutions. On the contrary; people are numb to the implications of the energy problem, doubtful of the reality of the crisis, divided on who is at fault, and at odds concerning the types of programs needed to remedy the situation. As a result, there *is* no clear public consensus on which the political system can act.

Given this pessimistic assessment, are we doomed to failure in our effort to ensure a stable domestic energy future? I think not. The political process may move slowly, and actions may be piecemeal, and it may take years to forge effective policy; but economic forces *are* at work, and the political process *is* moving—and in the right direction.

Creating a New Foreign Policy Role

But while we wait for our political system to forge an effective national energy policy, our already heavy dependence on foreign oil grows. This comes at a time when world events foretell new supply disruptions. With the world oil market stretched as tight as a drum already, any disruption—from political or natural causes—will ripple through the energy distribution network, quickly precipitating shortages and higher prices.

As the debate on a domestic energy policy continues, we must deal with the realities of our foreign oil dependence by trying to stabilize the fragile world oil market, whose production and incremental capacity are controlled by a handful of countries. We can take steps to slow the rate of growth in demand and help find and develop new supplies. But such an undertaking implies some changes in both our foreign and domestic policies.

On the international level, to develop new supplies, we should sell to the Russians and the Chinese—at a fair price—the advanced petroleum exploration and development technology and expertise we possess. Making such technology available in no way would compromise our national security posture. But it would accelerate oil exploration and production in those countries, thereby adding to the world oil supply and helping stabilize the world oil market. Moreover, the more petroleum they produce the less likely they will become major importers; indeed, they may even become modest *ex*porters.

In addition to Russia and China, many of the developing countries are blessed with untapped oil and gas reserves. We should assist these countries in developing their natural resources by providing financial backing, needed technology, and trained personnel. To do so, however, government policy must be reshaped to stimulate rather than restrict foreign investments. Developing the Third World's natural resources addresses more than the immediate worldwide petroleum-supply crunch. It will help supply their own oil demand, which will grow explosively as these countries industrialize and raise their standards of living. In such an environment, the energy demand will shift from the wood and biomass used today to commercially traded fuels like oil or electricity. At the same time, however, we should encourage the development and use of decentralized technologies, which are especially well-suited to countries without fully developed energy infrastructures and large capital investment in facilities that use conventional fuels. To this end, we should redirect our research to developing technologies with important applications in developing countries—such as a solar stove rather than an improved solar pool heater, which is of little value in emerging nations.

Changes in Domestic Policy

On the domestic front, the production of large quantities of economically competitive oil and gas from coal or shale—i.e., synthetic fuels—would be a giant step toward capping world oil prices. At the moment, the necessary technologies are not in widespread commercial use, and the end products—the synfuels—are much more expensive than oil. But the sooner we get several plants on line—at whatever price—the sooner these fuel substitutes will act as a governor on Arab OPEC price increases.

Obtaining the necessary Congressional support for the development of synfuels is easier said than done, however. In 1975, President Ford proposed a loan guarantee program to finance the development of a synfuels "industry" capable of producing 1 million barrels per day of synthetic fuels. If Congress had not balked, we would be well on the way to reaching that target. Now history repeats itself. Several members of Congress and President Carter are proposing even larger programs to develop these technologies commercially. These programs are sound and should be approved—but they are not enough. If synfuels are to act as an effective cap on Arab-OPEC price increases, a significant synthetic fuel capability must be developed throughout the world, not just in the United States. Our program must be redirected, then, to provide this capacity in the world in partnership with other countries.

Preserving Economic and Social Well-Being

The implications of rising energy prices and sporadic supply shortages will undoubtedly affect our economic growth and social well-being. Domestic energy prices declined literally for decades, and low–cost energy fueled our economic growth. Now that energy is a much more costly factor in our economy, the unavoidable consequence will be slower growth in the coming decade—and it *is* price, and not lack of supply that will be the important constraint. But the major economic impact of higher energy prices will be the varying degree to which they adversely affect different income groups, regions, and industries in the country.

The nation's poor will be hardest hit because the percentage of their total income devoted to necessities such as food and fuel is much higher than that spent by Middle America for the same items. Already plagued by inflation, the economy will experience new pressures to expand income redistribution and transfer payment programs. Regions of the country blessed with nearby sources of conventional fuels, such as the Southwest, will not feel the pressures New England or the Mid-Atlantic states experience—and these differences will further hasten regional population migration and the movement of industry out of the traditional urban centers. And energy-intensive industries, such as aluminum smelting and petrochemicals, will feel the sting more than, say, the food processing industry. Moreover, American industry in general can no longer rely on cheap fuel as a means of offsetting higher U.S. labor costs to remain economically competitive in the world marketplace.

SOME HOPEFUL SIGNS

These sobering predictions are balanced, however, by several more optimistic signs. One promising prospect is that the world will respond to higher energy prices by using energy more efficiently and by developing economies based on indigenous, renewable energy resources. Already, American industry has made significant progress in increasing energy use efficiency. The Energy Policy and

Conservation Act of 1975 set voluntary conservation goals for the country's five most energy-intensive industries; these industries have come close to—and in some cases exceeded—their goals. The chemicals industry, for example, pledged to reduce energy consumption by 13 percent over 1972 levels, and by the end of 1977, energy use had been cut by almost 10 percent. The petroleum refining and coal industries' goal was to cut 1972 energy consumption by about 11 percent; they have exceeded that goal by almost 4 percent. In fact, since 1973, total U.S. industrial fuel use declined while output rose significantly. Other advances are being made in the transportation sector, which accounts for 27 percent of primary energy consumption in the United States. Barrels of oil equivalent consumed each year by the average American-made auto should drop by 25 percent by the end of the century, from 22 barrels to about 12 barrels.

Legislated improvements in building energy efficiency also will begin to be realized as building stock is replaced or renovated. As the United States moves into the twenty-first century, these improvements, along with increased use of passive and active solar energy systems for space and water heating systems, will make a dent in energy consumption.

Beyond these modest gains in the near term, there is the long-term promise of a future in which energy demand will be met primarily with renewable resources. By the beginning of the next century, the United States should be well on its way to an economy based on renewable resources, which can sustain a higher economic growth rate over the longer term. In the next century, the renewable energy systems that are now uneconomical in most applications—e.g., photovoltaics, wind systems, and biomass—will become economically competitive and help change energy consumption patterns. In addition, the next century could see the maturation of a domestic synthetic fuels industry, the growth of oil production in non-OPEC countries, and increased world trade in coal, liquefied natural gas, and other fuels as supplements to the renewable resources.

Such a range of energy options will enable Americans to maintain the high standard of living and the freedom of personal choice that have long epitomized this country. But having the luxury of a range of choices tomorrow necessitates making difficult choices today and developing many types of energy alternatives. If we do not begin today to build energy facilities that take 10 to 15 years to complete, for example, we may find ourselves with a severe supply shortage on our hands before the end of the century. If that occurs, it would be a time of curtailed mobility, restricted freedom, and reduced living standards for almost all of our people. It would be a time of dramatically different life-styles, forced on us rather than freely chosen. It would not be the America we know now—or the one we inherited from our forefathers to pass on, undiminished, to future generations.

What *will* we bequeath to our children and their children? Will it be the treasures of personal free choice and virtually unlimited economic opportunity given us? Or will they inherit the wind?

Part Two

MUTATIONS IN THE MEANING OF "WORK"

Chapter 6

GROWTH, EDUCATION, AND WORK

Willard Wirtz

Fitting the work and education[1] pieces into the mosaic of growth policy requires making some assumptions about the general pattern of the broader picture that is being put together, especially about how both growth and the policymaking function are perceived.

It seems reasonable to anticipate the use of two different vocabularies with the consequence of creating an illusion of disagreement even in the saying of the same thing. In one lexicon, "growth" connotes only, or at least primarily, the expansion of those economic activities that are reflected in the conventional measurement of the gross national product. As others use the term, however, "growth" applies properly and fairly, except for the influence of habit, to all forms of enhancement of the human well-being. Yet, no one will be contending for, or even accepting, a prospect of lessening in what the American society ought to mean.

What follows here will reflect an acceptance of American growth policy as having an essentially dual character, directed toward both the enlargement of individual human opportunity and the maintenance of a viable economic system. For two centuries, we have found it unnecessary to decide whether the third star of our national purpose, after life and liberty, is "the pursuit of happiness" that was written into the Declaration of Independence or "the protection of property" inscribed in the Constitution and the Bill of Rights. This magnificent, enduring ambivalence is part of the authentic and distinguishing national character.

The conference architects have advisedly focused attention on the necessity of "sustainable" growth. The life processes of mushrooms have no more relevance here than those of cancer cells. Taking the case for larger reliance on the growth potentials that lie in the development of the limitless "human resources," as a substitute for traditional reliance on the now diminishing supply of various critical "natural resources," will mean assuming the burden of persuasion that such a concept be self-sustaining. The conference will have to face, so far as the human and economic functions of education and work are concerned, the question—too unpalatable for the political forum—of whether there are going to be sufficient opportunities for those looking for them to use the talents a rapidly widening education is providing them. Yet, the Woodlands Conference charter, embodied in its program, permits and encourages the recognition that economic and political systems, including all institutions, are only agencies of the human purpose. They are means, not ends.

So education and work, and their more effective symbiosis, will be considered here in terms of their relationship to both the "human" or "individual" and the "economic" or "system" aspects of growth policy. What is said will reflect, however, a strong and clear bias against any use of the "growth" term that confines it to only the second of these aspects and leads to "no-growth" thinking; for *some form* of growth is probably essential to the vitality, surely to the meaningfulness, of either individual or institutional being. The only *ultimate* growth form consistent with underlying American ideals is in the expansion of people's net opportunity to make the highest and best use of the human experience. But this bias will permit no lesser regard to the principle of *sustainable* growth.

One other consideration has affected strongly the organization of these comments about education and work. Changes in social policy, in a society such as this one, are inevitably more accretive than drastic or dramatic; and this is true in only a slightly lesser degree of economic policy as well. Proceeding at the Conference from any assumption about the body politic's choosing, as a matter of logic, between one growth (or no-growth) policy and another would ignore the way this country decides what to do next. Our custom is to build on beachheads of experience that seem, if only by the measure of developed and perhaps inexplicable consensus, to be good ideas, and to abandon efforts that appear, by the same ineffable measure, to have failed.

So it has seemed appropriate here to concentrate less on any broad logic or theory about the relationship of education and work to growth than on the *experience* in the evolution of these two functions, and then to try to find, in this experience, possible keys to future prospects. To the extent that this means not coming squarely to grips with the central question of whether better management of these two functions constitutes in itself a significant and sustainable new growth potential can be returned to at the end—though with no implication that there is a clear answer to that question.

IN EXPERIENCE

Youth and Education

Although whatever is important to new growth policy in the relationship between education and work affects the entire scope of life's activities, it is in connection with young people's development that this relationship has received, so far, the most extensive consideration. This part of the record is proper preface to the rest. In looking back over the broad reach of two centuries, four underlying propositions regarding the interrelationship between education and work seem fairly identifiable.

First, the American decision to extend education way beyond the provision made for it anywhere else in the world is accepted as being at the base of the nation's 200 years of unparalleled social, political, and economic growth.

Second, the education of American youth has developed in a pattern of consistent responsiveness to the economy's constantly changing personnel needs—to the extent of affecting strongly the always widening decision about how many young people should receive how much of what kind of education. Other forces and considerations have entered into that decision making, but with no comparable degree of sustained influence. The idea of education's being only or even primarily preparation for work has been substantially and meaningfully rejected in principle. Consistently, however, education has accommodated in one way or another to the economy's changing "manpower" needs. That this has resulted in only minor strain among education's philosophers reflects the American instinct for taking steps and measures that serve both economic interests and human values sufficiently to prove generally acceptable at the time, though not always with a balance that hindsight has endorsed. It was typical that the Land Grant College Act of 1862 subsidized, for training in "agricultural and mechanical arts," institutions that became capitals of liberal arts education. In broader terms, the constantly increasing sophistication of the economy's production, service, and professional needs has always been sufficient stimulus to more and more liberal education, so that those who favored this for other reasons haven't had much of an issue.

Third, in interesting contrast to this history of intimately close *functional* relationship, the record has been one of virtually total *institutional* separatism. For reasons confirmed by what have been, on the whole, salutary results, education's stewards have held at arms' length any and all of those from "outside," including particularly the economic sector, who might attempt to influence the educational system unduly, or at least directly. The separatism was reflected in the insistence that even vocational education, after it was given significant impetus by federal legislation in 1917, be developed outside the general education system. Technically-oriented schools and colleges were set apart, and proprietary training institutions not recognized at all by the educational establishment. It wouldn't have occurred to anybody to think of the apprenticeship training programs as part of the educational system.

The fourth dominant strain in this record has been the assumption that the economy will always, except in occasional periods of cyclical distress, have a strong need and good use for as many young people as come along, with as much good education as they can get. It has been a critical corollary of this assumption that this need will be so strong that the transition from school to whatever comes after it will involve, in most cases, a relatively short passage beset by no more than transient difficulties.

It is important to an appreciation of the implications of these four developments or principles that their context was one of strongly dominant concentration on the relationship between the economy and the education of males. The work of most females, performed in the home and not paid for, didn't enter much into the thinking. So the story, until very recently, has been one of education and work (in the competitive market) developing as integrally interrelated functions, each dependent on and affected greatly by the other, but keeping such a distance between them institutionally that the common reference has been to "two worlds."

Yet all of that earlier history has now almost suddenly been brought into such question that to look for "experience" that will instruct for the future requires giving large account to what has happened so recently that it is still hard to evaluate.

One set of statistics reflect these developments in what may or may not be their basically significant form. The reported unemployment rate among 16-to-19-year olds who are looking for work is between 15 and 20 percent. For those in this group who are also black, it is over 40 percent. Almost half of all the unemployed in the country are under 25 years of age. The related but separate problem has developed that a good many young people with various kinds of college degrees can't find jobs that will use the talents those certificates attest.

The first reaction, when this situation became obvious in the early to middle 1960s, reflected the conditioning of those traditional underlying propositions. Something, it was assumed, must have gotten out of kilter in education's handling of its end of the relationship; there must be too many young people leaving school without adequate preparation. So a national campaign was instituted to reduce the high school drop-out rate. "To get a good job," the National Advertising Council posters read in the 1960s, "get a good education." The high school retention rate moved up. But so did the youth unemployment figures. Then the schools intensified their vocational education efforts and mounted various kinds of work-study and experiential learning programs. The national Commissioner of Education launched a "career education" initiative.

In apparent confirmation of the schools' responsibility for what was happening here, the country learned in the early 1970s that the average scores on students' college entrance examinations were dropping sharply.

This isn't the place for attempted detailed analysis of the maze of coincidence, correlation, cause, and consequence resulting from the virtually simultaneous declines, between 1960 and 1975, of (a) the high school drop-out rate, (b) the averages on the college entrance examinations, and (c) the percentage of employed upper teen-agers. There are obviously, however, implications for various kinds of growth and no-growth prospects in each of these declines and in the three braided together.

The most commonly accepted view, with probably the least basis, is that the key decline among the three is in the test scores. This is an involved question.[2] Suffice it to note two things here: that the administrators of these tests insist, despite people's apparent unwillingness to listen, that the statistics don't say what the public is reading into them; and that the test data themselves make it clear that most (though not all) of the decline is attributable either directly or indirectly to the fact that many young people who used to leave school earlier are now staying on, with the consequence that more academically borderline students are now taking the tests. As of 1960, about two-thirds of all young people were staying in high school until they got their diplomas; seven years later this had gone up to three-quarters, where it is apparently staying.

Is the lesson, then, that high schools don't educate three-fourths of all 17 and 18 year olds as well as they did two-thirds of them? But if this is true, is it a reflection of teachers' diminishing competence or of deterioration in the average quality of the students they are working with (at the 11th and 12th grade level, which means beyond the compulsory school attendance age)? Or are broader elements involved here?

Yet even as these questions on the education side are being debated widely and with considerable heat, and as state after state passes "minimal competency" testing measures in a probably short-sighted effort to cure causes by treating symptoms, quieter attention has been focused increasingly on the job side of this equation. Perhaps the most significant "experience" here, though it may lack the maturity to call it that yet, involves the questioning of that traditional assumption about the economy's continuing and constant desire and need for all the young people who will be coming along to receive all the education they can.

One factor here is an unquestionable increase in the competition for the opportunities available to young people. The *adult* unemployment rate has apparently leveled out at about twice what it was ten years ago. Employers are under statutory mandate to hire minority group members and women; and women are coming into the work force in vast numbers. Older people are staying on their jobs longer than they used to, again with recent legislative encouragement. The movement of illegal aliens into the country has become large enough to affect significantly the market for the kinds of jobs that have always been available to young people leaving school without having developed any particular skills. That so many machines now have the equivalent of a high school education or more, and work at less than living wages, complicates job seeking efforts of those who leave school early. As a consequence of these and other factors, the situation today is that many large manufacturing companies are simply not hiring anyone under the age of 20 or 21.

There is, however, another set of factors involved here. The jobs into which most young people, particularly boys with relatively limited formal education, used to move were unskilled, entry-level jobs in production industries. In the last two decades, however, the American economy has shifted so sharply from a concentration, in terms of number of employees, on production to a service emphasis that, now, two out of every three people in the work force are in service occupations. While there is nothing in the nature of employment in the private service sector that appears, on the face of it, to present youth with more serious problems than they encounter in trying to move into production sector jobs, the experience so far is that there are complications here.

What is implied for the future by the present experience of watching the youth unemployment figures continue to rise when there are "Help Wanted, Full Time or Part Time" signs in windows of most fast-food service establishments across the country (except, significantly, in central areas of large cities), when a great many other small retail and service establishments look almost desperately for help that students or recent school leavers could provide, and when trying to get help around the house or yard has become a national middle–class headache? There is a widespread feeling that the biggest reason for today's youth unemployment is that young people don't want to do the work that is available; that there are a lot more of them looking for jobs, at least for pay, than there are looking for work. Yet, if this is at least half true, it is an oversimplification. It bothers parents at least as much as it does their children that a lot of these lowest-level service sector jobs are what have come to be called "dead-end" jobs; they don't lead anywhere. For reasons not entirely clear, furthermore, a kind of stigma has been attached in this country to service jobs that carry much larger dignity in most parts of the world. Some suggest that there are elements here of rejection of work that was associated with the bigotry of slavery. Perhaps these attitudes can be traced in part, to rising expectations that accompany the spread of affluence and education. Whether it is this, or possibly some associated psychology, similar attitudes have emerged about even the craftsman's arts.

Although such conjecture hardly fits in a summarizing of "experience," one fact with clear relevance to the future emerges plainly from the recent record. It involves the traditional assumption regarding the passage from school to work. This transition used to be, for most young people, characteristically quick and relatively patterned. Most boys moved into the kinds of work their fathers were doing. The jobs were there and waiting, and a great many of them required little education. The need for more highly educated young people, and it was very largely males who were involved, seemed insatiable. Where girls were concerned, the situation was even plainer and simpler. When they left school, they would do what their mothers had done; the job requirements didn't include education, and the society's need and desire for homemakers and child bearers was simply taken for granted.

This has all changed so much in so short a period of time that it is hard to accept the reality or grasp the full significance of what has happened. Immediate moves from classrooms onto lifetime career ladders have become almost exceptions to the rule. Few boys follow their father's footsteps and most girls have decided to do, at least at first, some things very few of their mothers ever did. With new questions arising about how much the economy wants of what the school-leavers have, the transition period from school to permanent, long-range employment is now more likely to be years than the weeks or even days it used to be.

Perhaps the biggest question this still uncompleted chapter of experience raises is about the continuing validity of the historic institutional separatism between education and work. With an apparent dislocation between demand and supply having developed, with the evidence becoming increasingly clear that it isn't a matter that can be taken care of just by adjusting the educational valves, and with the passage now stretching out to cover what may be four or five years, it suddenly becomes critical that nobody is in charge. James E. Colman has summarized succinctly what may be the largest single lesson in this whole record: that so far as the movement of youth from school to employment is concerned, the fundamental problem is that a capitalist or market economy does not have a natural place for an intermediate status between full dependency (when young people are in school) and full productivity (when they are in the work force).

What instruction is there, then, to the formulation of future growth policy in the experience so far with the interrelating of education and work for youth? The most basic lesson is so obvious that it has been only mentioned here: that education *means* growth, as much to the individual as to the society and the economy. Yet what is most immediately relevant to growth's architects includes some sobering counsel based on recent experience. Serious growth questions are raised by what we are seeing in the transition period between youth and adulthood, and by our apparent inability to cope adequately with the situation.

Those conditioned most strongly by tradition will find only confirmation of the value, to individuals and the society and the economy alike, of young people getting adequate educations and developing full respect for the "work ethic." Others will see additional need now for reexamining the assumptions that the economy's need for youth is going to continue indefinitely to provide the demand to meet the supply, including adequate use of young people's rising educational attainment.

What seems clearest in all of this is that there has emerged a transitional period, involving what will frequently be three or four years of several millions of young people's lives, which demands some new form of institutional response.

The traditional separatism of the structures within which education and work are administered has come into sharp question. Whether the proper consideration is of young people's individual growth during this period or of their preparation for future contributions to the economy, or of both of these, the current situation—with nobody in charge—makes less and less sense.

Adulthood and Work

The story of work in America has been so well and fully told so recently that it remains only to try to identify quickly those pieces of this record that appear most relevant to an inquiry into the prospects for developing new dimensions of economic, societal, and human growth. Although the particular focus here is on experience in interrelating work and education, this has to be put in a broader context.

Proceeding again from the briefest possible boxing of the compass of the familiar and obvious, four generalizations provide adequate context for noting some more particular elements in the record.

First, it is a basic proposition of conventional growth theory that one critical key to continuing expansion of the economy lies in increasing productivity. The American experience has been one of unparalleled rise in productivity levels—until recently.

A second commonplace is that American economic history has been characterized by a unique commitment to the concept of the individual's *right* to work and to the ideal of full employment. Failures in particular respects and from time to time to keep practice up to these precepts haven't diminished the importance of these cornerstones of national purpose. The acceptance of a prospect of backing up or even standing still, consciously cutting back the right to work or deliberately whittling down the concept of full employment, would violate an authentic strain in the national character.

Third, there is a time-proven procedure for developing and improving the terms and conditions of American employment. It involves a complex and virtually indefinable interplay among the exercise of managerial responsibility, decision making by private collective bargaining, the lawmaking of public legislatures, and the judgments of courts. The procedure is essentially pluralistic. It has worked, at least by comparison with most other models, remarkably well. Although this procedure has been used primarily to develop the concept of labor as a unit of production, it has developed in the past 40 to 45 years a considerable capacity for recognizing work as a human value.

Fourth, so far as the interrelationship of work and education are concerned, the training of the work force has been recognized as an element of production and, in that sense, of growth. There has also been an increasing tendency over the years to try to isolate, so that it may be better administered, the on-the-job training element that inheres in almost all work experience. In the sense, however, of any combining of formal, institutionalized, external education with employment, only very recent experience can be identified; the operative tradition has been that youth is for education, maturity for work, and that the one ends when the other starts.

Turning to more particular experience bearing on what appear to loom today as growth/no-growth issues and prospects, several developments are appropriately noted not as involving any combining of work and education in themselves but as relating in one way or another to the possibilities of such combination. The relationship may lie either in the employment time factor that affects the education possibility, or in the suggestion of broader common denominators relating to the balance between economic and human growth emphasis.

The first of these developments came when the country faced squarely the disparity between the number of people needing jobs and the number of work places immediately available. The workweek averaged, at the turn of the present century, 60 hours. Its reduction to 40 hours didn't just happen. One of organized labor's first effective collective bargaining demands was for a shortened workweek; and then, in 1935, the Congress included in the Fair Labor Standards Act the time-and-a-half penalty or premium for "overtime" pay. The central purpose of that legislation was to promote a broader sharing of the available work opportunities.

There has been experience, too, with the shortening of the work-year and the work-life. The development of relatively standard patterns of two to four or more weeks of paid vacation and ten or twelve paid holidays involved little or no conscious or deliberate purpose to spread the work. It has had, nevertheless, some of that effect. Similarly, the practice of retirement at age 65, now takes on new significance because of the recent statutory change (in a period of higher-than-usual unemployment) prohibiting mandatory retirement, in general, before age 70.

In the related area of part-time employment, the American experience until very recently reflects considerable negativism, based in part on what are considered the additional fringe benefit costs of such employment and on what are perhaps exaggerations of managerial complications in handling it.

There is a similar likelihood of possibilities such as flextime work scheduling. Although consideration of such possibilities is frequently lumped with that of part-time employment, under the general heading of "alternative work schedules," some of the implications of more flexible work scheduling are substantially different, particularly so far as their growth implications are concerned. Flextime work scheduling has developed much less so far in this country than in a number of others, and the prevailing management attitudes toward it reflect something less than enthusiasm.

Additional developments with clear growth implications, but with only indirect bearing on the education/work potential, involve the new emphasis being given to "work enrichment" and "quality of working life" initiatives. The converse of these developments is reflected in increasingly common suggestions that "a new breed of American worker" is emerging, along with a significant enlargement of previous concepts of worker rights (in the technical sense of their being enforceable).

There may or may not be sufficient justification for noting this area of developing experience only in passing here. This area, furthermore, remains currently at such a protean stage that generalization regarding its possible implications would be as likely to prove wrong as right. Yet, the growth implications of these developments, including their posing of the question of what kind of growth is to be considered, obviously warrant careful thought.

Turning to experience with respect to the actual combining of work and education at the adult level, the fairest summation would perhaps be that "adult education" and "lifelong learning" are still phrases looking for clear content.

The National Center for Education Statistics (NCES) reported in its 1975 Triennial Survey of Adult Education that approximately 17 million people in this country over the age of 16 engaged that year in some kind of formal educational program other than full-time high school or college attendance. The definition is obviously critical. Other estimates, using different definitions, have been as high as 32 or even 60 million participants. The NCES 1975 Survey data show an over-all participation rate of individuals over 16 in non-full-time educational activities at between 11 and 12 percent. The largest concentration of participants was in the 25-to-44 age group; about half were women. (This represents a significant comparative increase in women's participation rate compared with 1969 data; and subsequent reports indicate that there are now substantially more women than men engaged in these activities.) The average participation rate was about twice as high among individuals with annual family incomes above $10,000; and, similarly, among white as compared with other groups. The participation rates rose progressively among individuals with different levels of previous educational attainment: from 2 percent among those with eight or fewer years of school, to 12 percent among those who stopped at the twelfth grade, to 27 percent among those who finished college. Two segments of the NCES data bear more directly on the education/work relationship. The participation rate was highest (15 percent) among those working full-time, and sharply lower (about 8 percent) among individuals working less than 15 hours per week. Although the definitional problem precludes statistical precision, the survey results indicate a much stronger comparative emphasis on job-related than on more general courses.

Although the NCES survey provides a fair suggestion of the rough outlines of the adult education situation in this country four years ago, it affords only limited insight into the critical character of this development. Very little of the available data differentiates between the various kinds of courses that are being taken or between relatively casual and more intensive participation in these programs. Nor is it possible or useful here to try to depict that variety of patterns of "nontraditional learning" that are currently emerging, involving more and more programs of various kinds at one educational level or another. They include not only new curricular offerings but significant recent attention to equivalency credentialing at both secondary and postsecondary levels. This is a belated recognition that learning outside the school house and after leaving it has values at least comparable to those flowing from more formal pedagogy.

Instead of attempting to assemble the still scattered evidence of what is happening in this broad area, it seems appropriate to emphasize two developments that represent institutional and procedural initiatives that may give the interrelating of work and education the ways and means to realize its potential. One involves the remarkable expansion in the past twenty years of the community college system. The other, comparable only in terms of its possible future significance, has to do with the relatively little recognized "tuition-aid" program.

In the 1960-61 academic year, reported enrollments in what were then commonly called junior colleges totaled about 750,000. The comparable figure for 1978-79 is over 4.3 million. These enrollees are by no means all adults; the majority of them are young people attending these colleges on the traditional and conventional basis. The median age of community college students, however, is 27 years, which compares with a figure of 20.7 for other types of colleges and universities. About 49 percent of full-time and 87 percent of part-time community college enrollees are combining this education with either full-time or substantial part-time employment (including the care of children at home). The community college development has its critics as well as its enthusiasts, its weaknesses as well as its strengths. There is real question about whether this system has been fit into the rest of the secondary and postsecondary system as well as it might have, or as effectively as may prove possible in the future. It probably represents, nevertheless, the most significant institutional development in American education in this century. Vast new possibilities have been opened for people to return to education after they have gone to work, either in the conventionally recognized labor market or at homemaking and family raising.

The potentially comparable "tuition-aid" development involves employers' covering part or all of the costs of their employees' participation in formal educational or training courses outside the employing establishment and (in most cases) in conventional educational institutions. A group of recent studies, including two conducted by the National Manpower Institute with the encouragement and support of the National Institute of Education (part of the Department of Health, Education and Welfare), indicate that between 10 and 15 million employees in this country are now covered by tuition-aid plans set up either by employers unilaterally or under collective bragaining agreements. Although the financial commitment involved here cannot be reliably determined, it is in the billions of dollars. Such figures, however, abuse the truth; for it turns out that only a small fraction of employees who have these entitlements are using them. The best data available indicate that the average use figure is between 3 and 5 percent. Yet this, in turn, *under*states the prospect here. Most of these plans are so new that arrangements for implementing and administering them have not been developed; and, in many cases, most covered employees have only limited knowledge of how they operate. Current in-depth studies of the results in companies making special efforts to implement such plans indicate that use rates in the 10 to 25 percent range (with significant variations between employees grouped in terms of skill and income levels) can be expected.

The significance of these tuition-aid programs, and of the broader worker–education systems being developed in more and more employment establishments, obviously depends on a number of details; whether, just as one example, the educational entitlements extend to non-job-related courses. There is still critical question, too, about the extent to which the use of these plans will mean, on the one hand, significant training and educational opportunities for those who had the least formal preparation before they entered employment, or, on the other, principally more education and training for those already most advantaged. The programs may or may not take on substantial significance in terms of enriching the work and adult experience for large numbers of employees. The tuition-aid development is, however, significant as a suggestion of the possibilities already opened up in actual experience for taking the age limits off formal education.

Perhaps the fairest summary, then, of actual experience so far with the interrelating of education and work at the adult level is that the record is more of suggested than of confirmed new growth potentials. It appears, though, a fair and proper sensing of the broader record of what is happening in American work-life that there is a considerable yeast at work here, with what are, in all likelihood, significant growth implications. If going to work still means for most Americans that education is over, there are now several million each year who are going back to school in ways that have career meaning. Adult education has now found two institutionalized procedures that make it something a good deal more than the grab-bag business it was a decade or two ago. Nor is it properly left out that these education/work initiatives bear more than coincidental relationship to such developments as alternative work scheduling and job enrichment programs and whatever is properly inferred from increased emphases on the quality of working life and broadening concepts of worker rights. Whether "growth" is conceived of in economic system terms or as including any expansion of individual opportunities for constructive and satisfying individual endeavor, these developments warrant serious notice.

IN PROSPECT

Purporting to find the elements of new growth policy in a hurried scanning of 200 years of education and work in America risks the dangers of a selective reading of what is in the record. Such an exercise could be considered little more than looking at a Rorschach ink blot and finding there reflections of one's own preconceptions and predilections. Even if it is true of growth policy, as Mr. Justice Holmes once remarked of the law, that its life lies less in logic than in experience, individual observers of that experience necessarily view it from particular perspectives.

A considerable basis emerges here, nevertheless, for identifying some new directions that education/work developments in this country appear to be taking. Their growth implications unquestionably depend on how the growth concept is perceived. It is, perhaps, best to identify the interpretations of these developments as essentially subjective and to put them in the form of personal conclusions about where they might be leading. The prospect, like the record, warrants separate attention to the periods of youth and adulthood.

School and the Transition to Work

The basic assumption remains that the most significance youth's education has to growth in its economic sense lies, quite simply, in the effectiveness with which students are prepared for work. This neither diminishes nor demeans education's broader purpose of developing individual capacities for human growth. Yet, when counsel is taken of experience as it is reflected in consensus, the commonest judgment today is clearly that young people are being inadequately prepared for what lies ahead of them.

To the extent that this may seem to imply that the answers to this problem—and therefore to new growth elements—lie in measures that can be taken *within the educational system itself*, quite a bit more remains to be said. If everything that is commonly suggested—better grounding in the basics, sterner academic discipline, fuller familiarization with what work means, better counseling and guidance, some rebalancing of general and more technical or vocational training—is accepted as being *programmatically* sound, the question remains and has emerged more clearly from recent experience, of whether the education work *processes* are properly responsive to current and prospective needs. The single central suggestion from the record is that the more effective interrelating of closely coordinated functions depends increasingly on neutralizing the effects of their classical institutional separatism.

Where the conventional approach has been to consider *programs* that can be devised and administered within the educational system to make it serve career and work needs better, the suggestion here is that primary emphasis be placed on the development—especially at the local community level—of new *collaborative processes*. This suggestion is pressed most strongly in connection with the recently emergent need to develop new ways of dealing with the difficulties that have developed in the transition between school and work.

Emphasis should be on the prospects for developing such collaborative processes primarily at the local level. What were previously two separate federal executive departments charged with responsibility for education and work and the development of human resources have become three. In most state government structures, there are from five to seven independent agencies with responsibilities in these areas. Interdepartmental task forces and coordinating committees have become exercises in futility. The record in federal and state capitals is one of increasing divergence of special interest institutional development rather than the combining of executive units or legislative committees along lines of functional interrelationships. "Turf protection" is the accepted name of the game large bureaucracies play with each other.

The comparatively recent emergence in a good many cities around the country of education/work councils, or industry-education-labor committees, encourages the thought that there is more promise of effective collaboration at the local community level, where the people whose lives are affected by whatever public programs are worked out know each other. It is easier for individuals than for organizations or agencies or institutions to work together.

In a community or neighborhood, furthermore, there is the opportunity for actual *participation* by individual citizens in the handling of their own affairs—which may be the only antidote to the present, almost total, disillusionment with *representative* government. If it is true, as there is reason to believe, that a potential working majority of Americans are developing, as individuals, a new commitment to quality-of-life values, this bears directly on some of the education/work human growth potentials. Such a commitment will be expressed more effectively and faster through people's direct participation in local community affairs than by relying on its filtering up to either federal or state capitols through the hardening arteries of conventional representative politics.

Neither experience nor logic warrants any expectation that new institutions will simply emerge in some unidentifiable way at the local level to find and exercise authority in areas already occupied by other institutions. It is a different matter to anticipate the development of local groups, including representatives of various public and private agencies and organizations, that will be in a position to consider and then influence significantly the implementation of various possible kinds of education/work initiatives.

The suggestion seems appropriately and even necessarily made here that a common education/work information base be developed as an essential condition for collaborative process in the education/work area. As a society grows, it relies increasingly on statistical measurements of both the societal and the human condition. What we do is affected more and more by what we measure and how we measure it.

The operational significance of a common education/work information base is obvious so far as the school-to-work problem is concerned. There are measurements made today, though with few details, of the number of youth who are in school and the number who are employed and unemployed. But the computers making these two sets of measurements use, literally, different languages; so the education and employment data cannot be correlated. The definition of unemployment applied to youth is substantially the one devised in 1941, with the adult work situation primarily in mind, to provide an economic indicator of the country's business affairs. The measurement taken each month tells little about the actual condition of the young people involved, about their critical characteristics, or about the causes of their unemployment. Nothing is asked and therefore nothing reported about job vacancies or other opportunities. And because this employment/unemployment measurement is readily available, the problem of youth in transition is perceived as an unemployment problem—with the implication that the cure for it is going to be found in more jobs. Both the problem and the right answers clearly go beyond that.

On the education side, public attention was focused four or five years ago on declining national averages on standardized college entrance examinations. These are unquestionably serious but when they are looked at closely and carefully, mean some things quite different from what most people have taken them to mean. The public response has been, almost ironically, to set up "minimal competency testing" programs that may well diminish the schools' contribution to both human and economic growth—unless equal emphasis is given to "maximal competency testing," for excellence.

Much of what is important in the education/work area involves transitions in people's lives and there is particular loss here from having so few "longitudinal" measurements of what affects the human condition. Annual statistical snapshots are taken of a particular condition and year-to-year comparisons then made, without recognizing that the pictures are of substantially different groups of people. This doesn't matter when only the over-all national economic situation is being reported. It results, however, in no reliable picture whatsoever of what is happening, or why it is happening, in particular individuals' lives. The information that is needed most can only be gathered by following groups of identified individuals through various patterns of interrelated education and work experiences to determine their effect—or determine if there is any effect on each other. The longitudinal study of the Department of Health, Education and Welfare of the Class of 1972 confirms, along with the three or four earlier studies of this kind, the value of this kind of measurement.

What is needed most of all is information about the particular local situations in which these problems develop and in which they will have to be resolved.

National data have only limited significance. It is at least possible to hope that the 1977 Congressional initiative in setting up the Occupational Information Coordinating Committee system, with committees being established in each state, will lead to improvement in this situation.

In broader terms, reaching beyond the school-to-work transition, the implementation of a dual growth concept—recognizing the importance and the potential of both economic and human growth—will be greatly influenced by whether there is further development of "social indicators" to complement the "economic indicators" we now live by. If we mean what we are professing increasingly about a larger interest in the quality of life, we need measures of whatever makes up the quality of life.

Extensive banks of such social indicators have been developed in most of the countries with which we compare ourselves. Two modest but, nevertheless, impressive sets of such measurements have been made and published in the past five years by the federal Office of Management and Budget, though only as reflections of the steadfast personal commitment of three or four staff members to this idea. The prevalent notion that significant measurement of qualitative accomplishment is impossible proves to be a myth when the effort is honestly made.

Specific proposals have been made, and consistently ignored, for the development of indexes of Net Economic Welfare, of Net National Accomplishment, even of Net National Satisfactions—to complement in various respects the measurement of the Gross National Product. No such indexes could possibly be precise or exact. But they would tell much more than is presently known about how we are doing as human beings rather than as units of production. It will be hard to break thinking about growth out of conventional economic limits until we can see reports on social and human growth (or no-growth) in at least something like the way we are shown, month by month, how the economy is doing.

New collaborative processes and revised forms of measurement obviously involve only the framework within which more meaningful growth policy can be developed and more effective programs administered. It remains to try to identify, in what is a considerable recent activity in this area, those initiatives that appear to suggest the most significant new growth potentials in the better handling of youth's education and their transition to whatever their careers are to be. Brief mention of several of these types of activity will have to suffice as suggestion of its broader range.

Continuing attention is being given various types of programs for better familiarizing students, especially at the high school level, with what different kinds of work mean and involve; and efforts are being renewed to establish a more rational relationship between vocational and general education. Although career guidance and counseling programs are still seriously underdeveloped, there is recognition in most schools and communities now that the improvement of this function will probably do more than anything else to link education and work together effectively. It isn't enough that the two *systems* be matched up in terms of over-all supply and demand; the critical linkage has to be established on a case-by-case basis, in individual people's lives. The importance of the career counseling and guidance function (not only at the youth but also at the adult stage) reflects a central element in integrating the economic and the individual human aspects of a dual growth policy. It can't be left to a comparative handful of counselors in the high schools, and is not being recognized as a broader community function.

The principal current emphasis in the educational system is on improving young people's growth potential (in the dual sense) by a firmer grounding in "the basics." This catch-phrase probably reflects as much concern about youth's attitudes and mores as about their ability to read and write and do arithmetic. While the criticism of the schools implied here takes too little account of the underlying reasons for the decline in academic test scores over the past fifteen years, there can't be much question, in growth analysis terms, about the critical importance of raising the general educational standards.

Perhaps it is more a subjective judgment than anything clearly identifiable in "experience" that the closely interactive roles of schools, parents, and community are being recognized as a critical component of youth's preparation. When we say "education" we think of what goes on in school houses. We ought to be talking and thinking in terms of "learning." What may be most important in considering the relevance to growth of the relationship between "education and work" is the recognition that the matter is one of "learning and growth," and that the schools are only partners in the learning enterprise.

A Youth/Community Compact

The imperative need for the development of new collaborative approaches in the youth education/work area emerges most plainly in connection with the transition between school and employment. Various aspects and implications of this situation have already been referred to here. In short, millions of young people are going through what is often an extended period of time without the individual resources to meet its trials, and a critical vacuum of institutional responsibility exists. Lying between the "worlds of education and work," the transitional area has become a nobody's land where the costs of mounting casualties are falling on individuals and the community alike, with increasingly serious consequences. The situation, involving clear growth implications by any definition, demands both the designing of new education/work programs and the establishment of new collaborative processes to implement them.

There is an emerging recognition of community responsibility for assuring young people some form of continuing, constructive preparatory opportunity, including one combination or another of education and training and work and service, between the ages of 16 and 20. This is done now, at least in effect, for those who stay on in high school and continue into college. As a matter of both equity and community interest, a comparable opportunity must be afforded those whose circumstances commend or even dictate following a different course.

This is not to suggest that laws be passed guaranteeing financial support during this critical period. Indeed, there is increasing question about the principle underlying legislation designed solely to assure young people job opportunities. It is, perhaps, becoming a fallacy that every young person should be *either* in school *or* at work. The better answer would appear to lie in a good many cases in providing the opportunity to combine *both* of these activities, broadened out to include job training and service as well.

It appears worth considering the possibilities in what might be called a Community/Youth Compact—based on "best effort" commitments running both ways. The local community's undertaking, probably through some kind of education/work council, could start from making arrangements to take up, on a case-by-case basis, the request for assistance of any young person asking for it. There would be provision for counseling and guidance going substantially beyond what is presently available to most young people in this age range. That counseling would include, where this was possible, the applicants' parents or parent and a teacher familiar with the individual. The organization in charge would develop a Community Youth Opportunity Inventory—not just of available jobs in the traditional sense, but of public service and various training opportunities as well. This would include making particular efforts to line up, so that they would be more easily identifiable, opportunities—both full-time and part-time—in the private service sector. Special attention would be paid to giving what are presently considered "dead-end" jobs the meaning and value they were once recognized as having. This Youth Opportunity Inventory would *not* include jobs that could be expected to be filled by adults. The relationship of such a program to military service, and whether such service should be considered part of the opportunity inventory, would require careful attention.

The expectation would be that arrangements would be made to provide individuals with a combination of different kinds of opportunities, some of them paying wages, some not. An established operating principle would be that the program be administered to prevent its encouraging anyone to leave school; every effort would be made in the other direction. As much attention would be given to the youth's prospects at the end of his or her Compact program as to its initial or interim stages. The commitment on the other side would be, turning the old phrase just a little, a fair day's performance for a fair day's opportunity. It would mean staying on a reasonable basis with any reasonable assignment.

This suggestion reflects the evidence that there are today a substantial number of unused youth opportunities available in most communities, at least outside the decayed sections of the large cities. It also assumes federal youth support programs, of some kind, probably with an emphasis on public service opportunities but certainly in a form permitting substantial local community flexibility in using the available funds. The proposal assumes, too, considerable local volunteer help in making the program work. It assumes, perhaps most critically, a collaborative policy development and administrative process initiated at the local community level.

Two other points should be noted in connection with the youth aspects of the education/work outlook.

The suggestion is made with increasing frequency that there is a demographic answer to concerns about the current youth unemployment situation. As the baby-boom-bulge passes on up the age-line, this argument goes, the number of young people looking for work will diminish sufficiently to remove the present pressures. There is even talk of "manpower shortages" at the entry level, developing between 1985 and 1990. This comes within the assignment of another Conference panel, and is properly left for consideration there. Suffice it to note here the possible minority view that this prospect is greatly exaggerated.

No mention has been made here of another familiar suggestion: that the proper accommodation of education to whatever diminution there may be in the economy's need for well-educated young people is to provide fewer of them with so much education. The omission in these comments of adequate treatment of the youth *under*employment problem weakens whatever justification there might otherwise be for summary dismissal of this suggestion. There is more than a little reason, including substantial and relevant experience in Sweden and several other countries, for looking at it seriously. So it will have to stand as essentially a personal persuasion that the rationing of educational opportunity cuts directly across the grain of the American character—and represents a no-growth policy, in every sense of the term, that this country is not prepared to adopt.

Education as an Element of Work-Life

The prospect of education's becoming a significant element in American work-life emerges most clearly from recognition of the predictable influence of three contextual developments noted earlier: the changes already taking place in traditional work scheduling, the increasing percentage of employed women, and the prospective increase in the number of middle-aged workers. In interestingly different ways, all three of these developments also relate to an analysis of growth policy.

For various reasons, it appears unlikely that serious consideration will be given in the short-term to any work-sharing program patterned on the 1935 legislative decision to shorten the work week. There is substantial movement, however, toward "alternative work scheduling" of one kind or another, with particular importance attaching to the increased reliance on part-time workers and on various "flexitime" arrangements. Because the conventional fixed 8:00-to-5:00, Monday-through Friday work schedule has represented a built-in obstacle to working adults' undertaking significant educational programs, these more flexible work scheduling arrangements will have an obvious liberating effect so far as adult education is concerned.

The single most significant current development in American work-life involves the entry into the "labor market" (a despicable term, attractive only to economists, for it brackets workers with commodities) of large numbers of women. Presently holding about 40 percent of all paid jobs, entitled by law to a larger share of them, and armed with both the equity and the political and economic strength to enforce their demands, women are about to play their hand—with a variety of implications so far as education/work and broader growth policies are concerned.

Any lingering opposition to part-time and flexitime work scheduling will almost certainly yield to women's insistence that their right to combine career motherhood with career something else is compromised by limiting them to full-time, fixed-schedule jobs. Nor will men accept more confining arrangements. What has been initiated as a women's issue will become a human issue, with the consequence that part-time or flexitime arrangements will be used not only to permit women to be with their children but to combine work with going back to school—and to permit men, incidentally, to do the same things.

It can be only a little time until adequate provisions are worked out for supplemental child care. The argument at the moment is whether the costs of this are to be borne by parents, consumers (through employers), or taxpayers (through the government). The costs will almost certainly be shared. This, too, started out as a women's issue; but men are obviously going to reject the suggestion that their children's care is any more their wives' interest than their own. Child care, too, is an education-related as well as a work-related issue.

Women are already enlarging the demand for "equal pay for equal work" to include "comparable pay for work of comparable value"—which is, again, potentially as much a male as a female workers' issue, and a high octane human and economic growth issue as well. Its implications are limitless, including the risks of leaving such issues to courts because neither private decision makers nor legislators come to grips with them.

Women coming in larger numbers into the work-force has a direct effect on the worker education prospect. Their circumstances are more likely to create stronger desires and needs both for refurbishing employment skills they had started to develop earlier and for training for career changes. Available data indicate that, while more men than women were engaging in one form or another of adult education (principally job related) ten years ago, these figures evened out in about 1975, and now show a sharply increasing percentage of women engaging in this kind of activity.

The baby-boom repercussions in this area are harder to identify clearly but may have the largest effect of all on the adult, or worker, education prospect. There will develop, in about ten years, a heavy concentration of the work force in comparatively top level jobs, which will mean an intensification of the "no place to go" frustration. The individuals involved will have had, in their youth, more education than the work generation that preceded them; and the evidence is clear that those with considerable education are most inclined to renew and extend it as adults. Given more flexible work schedules, and the likely development of "leave of absence" programs in this country paralleling those already established in Western Europe, there will be strong tendencies among those in this group to look for substantial and meaningful opportunities to find new outlets for their interests and perhaps new careers by going back to school.

In the context of these predictable developments, the already established community college and tuition-aid beachheads will take on additional significance. They represent, in effect, an institutional framework and a working mechanism that afford considerable implementive promise and encouragement.

The principle underlying the community college program, that identifying an educational institution much more closely with the immediate community in which it lies gives it significant strength, appears certain to receive serious consideration throughout the educational system. Education, from the primary through the post-secondary levels, is today seriously in need of more supporters and allies, partly as a consequence of the demographic fact that a smaller and smaller proportion of voters and taxpayers have school-age children. It will make quite a difference if the adult citizens in a community are given new reason to think of education, even after their children have finished school, as something in which they themselves have a direct self–interest.

It would be a mistake, though, to suggest that the wider opening of school doors is only, or even primarily, a counsel of economic and political necessity. The learning process involved originally, in the home and before the schools came along, an intergenerational relationship. There has been earlier reference here to the values that lie in recognizing the educational process that goes on in the schools as only one part of the broader learning process that takes place in the home, the schools, and the whole community. A considerable affirmative argument can be made in terms of education's highest ideals and pedagogy's purest principles that a mix of adults and youth in a classroom gives the teacher a superior opportunity to be constructive.

The tuition-aid initiative may well prove as important in the next two decades as the community college expansion already has to the new growth potential of the education/work relationship. The extraordinary proliferation of tuition-aid offerings of one kind or another during the past five years, both in unilateral management arrangements and in collective bargaining agreements, leaves little question that the *concept* of including an element of employer-supported worker education in the employment relationship has been established. This is significantly not a matter of just another employee fringe benefit. There is increasing consciousness of the huge expense employers incur in providing one form or another of employee training; estimates range from $30 to $100 billion a year.

The decision being made is that some substantial part of this money will be better spent in providing employees with educational and training opportunities outside the enterprise. Another way of saying this, with perhaps different implications, is that American management feels that some part of what the educational system can provide, with respect to the education and training of employees, will be more effectively and economically provided after individuals have established an employment base. This carries an advised warning against looking at the tuition-aid development uncritically just because adult education has an attractive ring to it. It will have to be asked whether movement in this direction implies a narrowing of the over-all educational emphasis, more on technical or skill training and less on liberal arts. Related deep-rooted questions remain: How much of the decision about what the society spends on education should be made (a) by the public and (b) by employers? Similarly, what are the comparative implications of the education bill being paid by taxpayers or by the consumers of goods and resources produced by private enterprise?

This remains, nevertheless, a prospect with infinite promise. Though the employee participation rates in most tuition-aid programs are today very low, the experience in situations where they are being aggressively administered is that this use-rate moves up toward 25 percent. The indications are that this will be affected strongly by such administrative details as whether full information is provided employees regarding the program, whether counseling and guidance elements are included, and whether arrangements are made to reimburse employees for the tuition costs (and other out-of-pocket expenses) at the time they have to be paid.

The clearly indicated future direction of these programs is away from their earlier limitation to job-related courses. More and more of the plans now cover courses in "the basics" for employees who left school without getting them; and, similarly, of courses leading to college degrees in the liberal arts and humanities as well as in more technically-oriented areas. Additional promise lies in the sophisticated arrangements being made in some companies to combine on-the-job training experiences in the plant or office with external courses in broader subject matters. There will probably be broader development of programs that include provisions covering employees' spouses and children, usually with some kind of matching-of-contributions arrangement.

An interesting extension of the tuition-aid concept is suggested by a recent statutory enactment in France which involves a mandated set-aside by employers of 2 percent of net profits to be used for employee education and training; any part of it not so used is paid the government as a tax. It seems more likely, though, that the American application of the underlying principle here will be kept within the private sector.

The National Manpower Institute studies already referred to disclose general endorsement of the tuition-aid idea both by employees who see here substantial prospects for widening their opportunities and by employers who already find evidence of consequent reduction in absenteeism and turnover rates, easier attraction of more sought-after job applicants, and rising productivity figures. These plans now provide adult education with a new working mechanism. A new set of "bridges between the worlds of work and of education," running the other way and carrying adults, is now in place.

Education and Non-Work

Separate note is properly made of another related potential that lies in combining educational and training opportunity with non-work, with unemployment.

There is particular reason in today's discouraging economic circumstances for considering in this country what Western Europeans call a counter-cyclical education and training policy. Its adoption here would involve only minor changes in present unemployment compensation administrative practice but would constitute significant recognition of a principle that offers considerable promise. What is required is the recognition that periods of higher-than-usual unemployment afford special opportunities for training and education that can then be put to use when economic circumstances improve.

Two steps could be taken in this direction. One would involve readjustment in the present unemployment insurance practice that affirmatively discourages an unemployed person's using his or her involuntary down-time for training or education. The law conditions unemployment insurance payments on the individual's being, from week to week, available for work. This means that an unemployed person, not likely to be reemployed with existing skills, can receive income support for up to a year but cannot take the steps necessary to fit himself or herself for another job. A different approach would be to have Employment Service counselors identify those recession-unemployed who are not expected to become reemployed and waive the "available for work" test so they can enroll in educational or training courses.

It would be even better to apply the education and training approach before workers are severed from their place of employment. Rather than laying workers off for six months or so, employers would put them in education and training status and pay them stipends instead of wages, with the government paying a portion of the cost. The workers would not be thrown onto unemployment lines, the employers would not have to rehire and train a work force, and skills would be upgraded rather than permitted to deteriorate.

Perhaps, instead of thinking in terms as pretentious as "counter-cyclical education and training policy," we should be reminded of the initiatives that were instituted in various communities in this country at the depth of the depression in the 1930s. An editorial in the August 1933 issue of the *Ladies' Home Journal* summarized the story:

> The free classes for adults in vocational and cultural subjects that have been set up by the New York State Temporary Emergency Relief Administration are double–edged

in their opportunities and value, for the teachers are trained brain workers who are out of jobs and the students are grown-ups who are fitting themselves for better things in better times. . . . In the first five months 30,000 adults enrolled, necessitating the employment of 1,000 men and women as teachers. . . . Many thousands of people are enriched or benefited economically. . . . It is to be hoped that so successful an experiment in adult education will not be allowed to pass with the emergency.

Forty-six years later might be about time to return to the lesson that periods of excessive unemployment afford particular opportunities for people to be enriched *and* benefited economically by renewing their education.

CONCLUSION

As was indicated at the outset, the conscious (but only partly successful) effort to concentrate narrowly here on the education/work relationship has had two promptings. One is respect for the conference programmers' design, leaving the broader architecture of future growth policy to the plenary sessions. The other, more basic consideration is the persuasion that American policymaking develop more from the enlargement of particular beachheads that appear promising by the measure of felt values than from the application of any articulated logic or strategy.

What emerges here is the conclusion that learning and working are strongly felt values in enough people's consciousness that the national consensus will support increasing the effectiveness of these functions and enlarging the common opportunity to engage in them. The specific corollary is that better integration of education and work, particularly by breaking down some of the walls of their institutional separatism, will facilitate these purposes.

Fitting this perhaps too obvious conclusion and its corollary into "sustainable growth policy" tempts exasperation—particularly if "growth" is defined solely in terms of economists' tired equations in calculating the conventional gross national product. This puts the answer before the question, which is whether that archaic formula is consistent with the dominant national purpose and the true human and societal potential. To get to the same place by defining "sustainability" in those same terms doesn't lessen the offense.

Though perhaps less clearly, a human growth theory is, in some ways, equally circular and self-serving—at least to the extent that it, too, suggests an accepted national logic, some syllogism by which the rightness of particular courses of action may be proved. It is not clear from experience that a working majority of the democracy's membership will always proceed from a calculation of what will maximize the common opportunity to make the highest and best use of the human experience. And if obsession with a selected notion of sustainability is the economists' pervasive error, the pen-on-paper humanists' hamartia is a disdain for ways and means.

If the temptation is strong to ask what is gained by trying to fit any particular decision into "growth" policy concept, one policy or another or even a combination of several, there is the reminder again that *some form* of growing is probably an essential element of the vitality and meaning of being. This can't be proved either. It just seems right. So it *is* argued here that a more effective symbiosis of education and work will contribute to providing this vitality and meaning. And this contribution will more than support itself.

Specifically and illustratively, there is the prospect of a continued extension and intensification of young people's education in subject matters having career and vocational relevance. This includes, in the first place, reading and writing and arithmetic. Although other countries have resorted, in circumstances comparable to ours, to a rationing of educational opportunity, we will reject that course. This enlarged education will mean, by any open–minded definition, growth. If it leaves, as it does, the question of whether there are going to be jobs to use this larger talent bank, we will eventually face up to that question. There is obviously plenty that needs doing. The use issue is more political than economic.

The single, largest education/work problem right now involves the handling of the nobody's land through which young people move from one "world" to another, not because of any theory of growth policy but, rather, because of immediate imperatives, the collaborative processes be established, and *then* the programs and the administration this transitional area demands. This will mean the constructive use of preparatory purposes, instead of the waste, of what have become several years of several million young people's lives. This will be growth by any definition. So far as sustainability is concerned, the dollars and cents costs will be less than those presently paid for permitting rot and decay in this area.

Part of the reason we have been slow in this country about building education into the adult work-life is that too much reliance has been placed on advocacy of "adult education" and "life-long learning" in policy terms alone. Effective catalysts have now emerged with the changes taking place in work scheduling, in the increasing participation of women in the paid "labor market," and in the changing age and educational attainment characteristics of the labor force. The setting up of community colleges is as much a stimulus to adult education as a response to any developed demand. Tuition-aid plans proliferate suddenly, not as applications of a national policy supporting life-long learning, but because employers find them better ways to spend training budgets, and workers find new opportunities to increase their skills, their career mobility, and their broader human satisfactions.

To attempt to analyze these developments in growth terms of one kind or another seems futile because they so obviously involve growth, however it is defined. Education and work *mean* growth, for both the individual and the system, in human and economic terms alike. Nor is there any real question whether the better integration of the two functions would enlarge their potential.

Yet, if all of this is so obvious as to seem almost to mock its exploration, the key question at Woodlands remains. An economy that has been living beyond its means in terms of some of the natural resources available to it is seriously concerned about the prospective overdrafts on this form of capital. Growth has been defined, contrary to underlying national ideals, exclusively in terms of what can be produced from that particular capital base. What we really want to know regarding education and work is whether improved handling of these functions, particularly their better integration, can provide a human resource base that will be an alternative to the diminishing natural resource base. The question is legitimately posed with emphasis on the necessity of the system's being self-sustaining.

To face squarely the issue of whether a better integration of education and work can be expected to fill this developing resource gap is to recognize the impossibility, given only present understanding, of answering this question. We don't have a system of accounts or accounting that permits computing the bottom-line effect in either conventional or alternative forms, so the "sustainable" character of enlarged reliance on the development of human resources as the supply of natural resources dwindles. We need this understanding badly. It is what the Third Woodlands Conference was all about.

NOTES

1. The Third Woodlands Conference program called for the inclusion here of leisure, along with work and education. The approach taken here, however, of emphasizing what may be identified as *experience* with the interlinkage of these functions, permits including so little about leisure that it is dealt with only incidentally rather than as a coordinate function—which the future may well prove it to be.

2. Whatever I understand, or don't understand, about the drop in the test scores is reflected in the College Entrance Examination Board's Advisory Panel report on it, issued in 1977 under the title *On Further Examination*.

Chapter 7

WORK IN AN ERA OF SLOW ECONOMIC GROWTH

James O'Toole

WITHOUT KAHN OR SCHUMACHER

I have set this discussion of work in the broadest possible social, political, economic, and technological context. No doubt many readers will be uncomfortable with such a sweeping approach. They will prefer that I had "stuck with the subject" and narrowed my focus to something "manageable," such as flextime or industrial democracy. But that would have been bad analysis, for it would not have helped to identify the problem at hand. In fact, a major problem is that, in all the current discussions about work and growth, *no* real problem is identified. My major aim in this chapter is, thus, to leave the reader with convincing arguments for why it is necessary to change the existing philosophy and organization of work.

At the most abstract level, the issue of what kind of work America wants in the future is synonymous with the issue of what kind of society America wants. Thus, it is at the high plane of inquiry about social justice that I have chosen to begin my analysis. Jean François Revel, the French social critic, argues that the search for a just society must proceed "without Marx or Jesus."[1] It might be said that, as a futurist, I am suggesting that the search must also proceed without Herman Kahn or Fritz Schumacher. Kahn's overweening optimism about the future is rooted in the established discipline of economics as it is practiced in America today. Kahn believes that if America just keeps doing what it is doing, things will turn out fine in the end.[2] This belief must be weighed against the one clear lesson history offers: failure to change to meet environmental exigencies is invariably fatal.

The late E. F. Schumacher's notion is just the opposite of Kahn's: he preached that for America to survive it must change *everything*—its technology, values, corporate economy, and political system.[3] This pessimism must be weighed against the costs involved: Is it at all realistic to assume that America can or should write off its many hundreds of billions of dollars of investment in "inappropriate" technology, large corporations, and the infrastructure that supports it all?

What is clearly needed is an astringent dash of realism to cut through Kahn's syrupy sanguinity and a splash of cold water to arouse us from Schumacher's fanciful and naive Buddhist reveries. After all, this is basically a pragmatic nation. Eventually, the consensus roar that "something must be done" will drown out Kahn's jeremiad, "resist all change." This is also a country with strong, competing interest groups who, in the long run, will ignore Schumacher's plaintive sermon that "small is beautiful," because they will perceive what he offers as a threat to all they have worked to achieve.

One wonders why it is that, as the signs of economic decline mount, the majority of "experts" denies the reality of the situation, and the minority prescribes a pill too bitter to swallow? Discussions of declining growth have a fabulous air about them, a chimerical unsubstantiality. On the one hand, they generate things like Laffer curves, and on the other hand, "voluntary simplicity." The dichotomy among the experts plays like a fantastic double bill: "La Dolce Vita," and "Last Year at Marienbad"—while the entire audience smokes the cannabis weed.

Things are going badly. In our desperate confusion we are willing to listen to every snake-oil salesman who comes along. Because we have no idea how to distinguish a good idea from a bad one, we run through them all, embracing each for a day or so, then discarding them all. No matter that one or two might have proved appropriate had we held on a bit longer, we had no *guide* to tell which to hold.

Our first shortcoming, then, is that we cannot identify the problem at hand. Is it job dissatisfaction? Is it the need to get back to craftwork to save energy? It is nothing so simple. In these pages I suggest the problem is that *America has an inappropriate philosophy and organization of work for an era of declining growth.* The current system breeds demands for more money, benefits, and entitlements from work; while, at the same time, it ensures that less and less effort will be expended at work and that workers will assume less and less responsibility for the quality and quantity of their work. This was not a noticeable problem when there was adequate economic growth to cover up the process, or when the United States was atop the world politically and economically. But times have changed, and Americans simply cannot go on demanding more and more from work while producing less and less. We can now feel the breath of our competitors on our necks, and soon we shall be seeing their heels.

One then asks what an appropriate philosophy and organization of work would look like in an era of declining growth. Can't America just do a better job making its technological and organizational choices the way economists say this *must* be done, that is, by getting the optimal mix of land, labor, and capital to maximize efficiency and then grow our way out of our pickle? No, we find that the world has changed. Energy availability obviates such growth. Moreover, social values have changed; today, the appropriate structure and function of work must be judged by multiple criteria, of which growth and efficiency are only two among many. Thus, America must look at the new constraints, identify its options, and choose a philosophy and organization of work that fits with reality. No pipe dreams need be offered—what is needed are alternatives that are squarely rooted in what is practical and "doable" and, most important, which address the real and immediate problem of declining growth.

To address this question, I do not attempt to design a utopia, invent a new social contract, or offer any other radical "solution" to the problems of work. Instead, I identify what is actually happening in American work places in terms of experiments with fringe benefits, incentive systems, and working conditions. I then suggest a set of performance criteria for evaluating how appropriate these various experiments would be if they were applied more generally throughout American industry. While the ultimate question is broad—i.e., Will the alternatives to the status quo create greater social justice in the hard times ahead?—no monolithic response is advanced. Work in America is far too complex, and the needs of American workers too diverse, for such simplicity.

In order to impose at least some limits to this inquiry (and to demonstrate *some* bounds to my arrogance), I confine my analysis to the United States. While most readers will, no doubt, desire even further limitations, a few ardent internationalists will find what I have done to be outmodedly "nationalistic." They will wonder how I can possibly confine my concern to privileged America when most of the world is beset by extreme poverty, hunger, and unemployment. In defense, I can only argue that the best I can think to do to address the problems of the Third World is to start by creating a more just society with sustainable growth at home.[4]

THE MISMATCH OF WORK AND ECONOMICS

The greatest political economists of the last two centuries have all addressed themselves to a common question. Adam Smith, Karl Marx, J. S. Mill, Lord Keynes, and such contemporaries as J. K. Galbraith and Milton Friedman have all asked: How can a just society be created? Although they have real and obvious differences of opinion about what constitutes justice, these economists agree that the ultimate goal of the "dismal science" is to find the best way for society to satisfy the basic needs of all its members. And not merely to satisfy their physical needs. Recall that Smith cared about restoring to consumers the sovereignty that had been lost to state-granted monopolies; Marx was concerned with ending work place alienation; Mill sought to create equality of opportunity for women and for people of all social classes; and Friedman's highest aim is to secure political liberty. The great economists have, thus, all placed non-economic goals higher than purely physical or monetary goals.

Although most contemporary economists have forsaken this great tradition of political economy in favor of econometrics and other scientific pursuits, a few still continue to ask the ultimate question. But, during the last decade, a twist has been added to that question. Today we ask: How can a just society be created *in a world of limited natural resources*?

This little addendum seems almost insignificant given the immensity of the basic question. But, in fact, the added clause has had the effect of making traditional economic thinking obsolete. For all traditional economists, regardless of their ideological proclivities, believe that *societies must grow their way to justice.* If Marxists can't have economic growth, they cannot achieve their cherished egalitarian dreams. If libertarians can't have economic growth, they can't have their cherished freedom. And so on down the line, the achievement of the primary goals of every traditional political/economic philosophy is predicated on the efficient optimization of the factors of production in order to maximize growth. Growth, in sum, is seen by all economists as the means to the higher ends they pursue.

While the ideas of economists are seldom right, they are, alas, invariably influential. We recall Lord Keynes' famous (and only slightly hyperbolic) statement that the notions of economists,

> both when they are right and when they are wrong, are more powerful than is commonly understood. Indeed the world is ruled by little else. Practical men, who believe themselves to be quite exempt from any intellectual influences, are usually the slaves of some defunct economist. Madmen in authority, who hear voices in the air, are distilling their frenzy from some academic scribbler of a few years back.[5]

And, Keynes might have added, the content and organization of work created by industrial engineers, corporate managers, trade unionists, and psychologists also are ruled by the basic assumptions of economists. In particular, every major theory about work place organization, worker motivation, and job satisfaction is predicated on the unstated assumption of economic growth. Indeed, these theories *depend* on growth for their successful application. I challenge the skeptical reader to take any leading philosophy of work and to ask if the goals of that philosophy could be achieved without high rates of economic growth. To illustrate my point, let me take the most difficult case: Maslow's humanistic hierarchy of needs. To Maslow, scarcity is a thing of the past, assumed away as America enters a new era of affluence and abundance. And the trade unionist psychology, expressed eloquently and precisely by Samuel Gompers in one word, "More," requires no elaboration on its attitude toward growth. Then, consider the clearest cast: F. W. Taylor's Scientific Management, the sum and substance of which is the maximization of output. Taylorism is merely the translation of economics into psychology.

With the notable exception of E. F. Schumacher, every major writer who has addressed the issues of the structure and functions of work has assumed economic growth as the basis for funding the alternative systems of rewards and incentives they have proposed. Look closely, and I suggest the reader will find the unstated assumption of continued economic growth underlying even the writings of the last two generations of organizational theorists, including Peter Drucker, Douglas MacGregor, Warren Bennis, Frederick Herzberg, Rensis Likert, and all the young authors of today who are concerned with sociotechnical systems and other "quality of working life" approaches. Like political economists, these scholars, too, are interested in creating greater social justice. Their specific concern, of course, is justice in the work place. And, like economists, growth is the means to the higher goals they pursue. Whether their goal is self-actualization, job satisfaction, individual development, greater equality, or whatever, economic growth is unconsciously factored into their thinking as the means to pay for the greater justice they envision. For example: One simply cannot imagine workers achieving self–actualization in a stagnating economy, or in a firm that is re-trenching. At any rate, no student of work has ever advanced a theory of organizational development or behavior on the assumption of declining growth, hard times, or the end of abundance.

While it is a matter of considerable controversy whether America's economy can or should grow again at the rates that were common before the 1970s, it, nevertheless, is a fact that few economists *expect* America to return to these high rates of growth. The U.S. economy is not now experiencing much real growth, and economists argue that it is highly unlikely that the country will soon return to the boom conditions of the past, primarily because these good times were fueled by cheap and abundant energy. And no expert is forecasting cheap and abundant energy to flow again in the next two decades.

We are now getting close to defining the central work place issue that arises in America as the result of slowing economic growth: *There is a mismatch between the philosophy and organization of work and economic reality.* The symptoms of the mismatch abound: high inflation, low productivity, declining innovation, a drop in the relative value of the dollar, and increasing export of jobs to foreign competitors. While this mismatch is not the simple, proximate cause of these symptoms, it is a major *contributing* factor to them all. There would, thus, seem to be a need for a philosophy and organization of work that do not depend on rapid and constant economic growth.

My aim in this chapter is to delineate the broad contours of such a philosophy, and to suggest some characteristics of work place practices and policies that would be consistent with it.

THE CURRENT STATE OF WORK

American workers have been conditioned by experience constantly to expect more from work—more money, more fringe benefits, better working conditions—while putting in less time and less effort on the job. Wanting more for less is not a problem if the engine of economic growth is running in fine tune. But the engine has started to knock. Productivity has faltered. Now, the nation is no longer able to sustain the level of output to match the ever-rising levels of worker expectations. In sum, the search for an appropriate philosophy and organization for work is immediately frustrated by the fact that *worker expectations* are predicated on growth.

A simple solution suggests itself: change worker expectations. Unfortunately, there is nothing simple about changing social attitudes and expectations. Of course, such values do change, but only slowly and only when they are confronted with a reality that consistently demonstrates their inappropriateness. Because American workers are currently receiving mixed signals—particularly from employers, unionists, and government, who all deny the need to switch to policies consistent with slower growth—workers see no convincing reason to change their values. And change is particularly difficult in this case because worker expectations have been transformed into *rights*; that is, during the last two decades a host of personal needs have been translated into social *entitlements*. And while it is possible to alter a privilege, rights are by definition guaranteed, inalienable, and unalterable forever.

Paradoxically, this new "rights consciousness" among American workers waxed just at the time America's economic fortunes began to wane. Once the wealthiest nation in the world, America now stands sixth in per capita income behind Switzerland, Sweden, Denmark, Norway, and Iceland, and is soon to be surpassed by West Germany and Belgium.[6] Impervious to this relative slippage, Americans demand not only new "rights," but "zero risk," and increased "selfishness." Let us quickly review these three complementary components of the peculiar value system of the 1970s, the better to understand how work has come to be thought of as it is, and the better to understand how it is likely to be viewed in the future.

The Entitlements Consciousness

Arthur Okun[7] divides the "things" provided by a society into two categories: entitlements, or things due to all individuals as rights by virtue of their membership in the society; and market goods and services, or things that are distributed unequally by virtue of individual preferences, luck, hard work, merit, or intelligence. In general, then, entitlements promote equality, while the marketplace promotes inequality. Significantly, several changes are occurring in the composition and characteristics of both these categories:

- The dimensions of rights have expanded *horizontally* to include larger and larger portions of the total population. Consider how the notion of what defines a citizen with full and equal social and political rights has expanded in the United States over the last 200 years: at the time of the Revolutionary War, the only people with full rights were white, male property-owners over the age of twenty-one. By the Civil War, property qualifications for the right to vote were gradually disappearing, and black males started to become full citizens shortly thereafter. By the middle of this century, women started to achieve full rights. Now, one hears talk about the rights of children, of the handicapped, of aliens, of prisoners, of homosexuals, and even of fetuses.

- The dimensions of rights have also expanded *vertically*, so that *every* individual is now entitled to a greater number of things that were formerly in the realm of individual obligations. As this occurred, the focus of rights shifted from the traditional domain of *personal liberties* (the right to vote, to marry, to procreate, to emigrate, to equal justice before the law, and to freedom of speech and religion), to the broader area of *social entitlement* (the right to food, clothing, shelter, education, housing, health care, and an income).

- The concept of what constitutes a right underwent a radical transformation: in the Roosevelt era, rights were tied to responsibilities; by the Lyndon Johnson era, the tie between these concepts was broken. Okun suggests that it is instructive to look at changes in the old-age benefit laws to understand this transformation in the notion of rights. Under Roosevelt, the basic philosophy of Social Security was *contributory*, "stressing the obligation of people to provide for themselves."[8] And so on down the line, contribution has been removed as a requirement from most national social welfare programs. Contribution—that is, individual responsibility—thus stands as a quaint anachronism, a mere vestige of our former values. And the new values of entitlement—freed now completely from the notion of individual responsibility—have spread from Social Security to almost every other aspect of working life.

- Whereas traditional rights were relatively cheap for society to provide (it costs the taxpayer little if his neighbor votes, goes to church, or decides to get married), the new social entitlements can be extremely costly. For example, if health care came to be defined as a right (as it is in England), its provision could cost the American taxpayer in excess of $200 billion (roughly, $2,000 per each employed person). Admittedly, the size of the current national bill for entitlements is problematic and controversial. But, if one includes every service and benefit provided by government, the bill runs as high as one-third the gross national product. While this figure clearly overstates costs that most people consider social entitlements (it includes, for example, the entitlement to security against foreign aggression), it also *understates* the total bill because it excludes entitlements provided directly by employers to employees.

● There has been a trend toward employer-provided entitlements. These are mandated by the government (e.g., the minimum wage), voluntarily-provided (e.g., private pensions), or contractually-provided as the result of collective bargaining (e.g., dental plans). The total cost of fringe benefits (the most easily measured entitlements) went from 17 percent of the total compensation of workers in the private sector in 1966, to 21.8 percent in 1974.[9] The figure for organized workers in 1974 was 24 percent.[10] In large corporations in the chemicals and primary metals industries, the figure was 43.3 percent in 1977.[11] The U.S. Chamber of Commerce reports that employee benefits cost a total of $310 billion in 1977, up from $100 billion in 1967.[12]

● The size of the entitlements domain has grown drastically at the expense of the market domain. Indeed, according to Burnham Beckwith (a California sociologist), if current trends continue, over 50 percent of all *consumer* goods will be entitlements within the next century. Beckwith believes that entitlements "depend far more upon the degree of industrialization than upon the ideology of the government."[13] Because he sees growth in entitlements as an inevitable by-product of economic growth, Beckwith argues that the United States and the Soviet Union are both regressing toward the same mean of entitlements, and both societies will eventually provide the same kinds of "free" goods and services, with the same kinds of goods and services for sale in the market.

● There are new demands for purely *social* employee entitlements, including:
—The right to "blow the whistle" on illegal and unethical practices of employers
—The right to privacy (e.g., confidentiality of personnel files)
—The right to conscientious objection to unethical orders
—The right to freedom in outside activities (e.g., political activities)
—The right to sexual freedom (e.g., homosexual rights)
—The right to freedom from sexual harrassment by superiors
—The right to individual choice of appearance (e.g., for men to wear long hair and hippy beads in a bank teller's cage)
—The right to vote on plant relocations
—The right to participate in all decisions directly affecting one's job

—The right to self-actualization on the job (i.e., the right to develop one's full productive potential)

—The right to adequate leisure time (e.g., adequate time to spend with one's family)

—The right to reject a cross-country transfer

—The right of all employees to full access to information about corporate activities.

These expansions of, and changes in, the nature of worker and social entitlements do not, by themselves, constitute a problem. In fact, the manifest benefits of entitlements in *reducing* poverty, ill-health, discrimination, and inequality, and in *increasing* personal freedom and equality of opportunity have been so great that few Americans bothered to question the costs involved when entitlements were being expanded prodigiously in the 1950s and 1960s. But recently, a growing number of observers have been calling attention to the following kinds of problems or costs associated with the expansion of entitlements:

● Entitlements are often in conflict with each other. Some firms are simultaneously faced with antidiscrimination *and* reverse discrimination suits. Other companies experience conflict between affirmative action and seniority rights.

● Entitlements, particularly such new rights as national health insurance, come at high economic costs. These are ultimately borne by taxpayers and consumers when entitlements reduce the nation's productivity and international economic competitiveness. For example, the National Planning Association has claimed that entitlements in Europe have pushed up production costs in some countries to the point where governments must subsidize certain domestic industries and impose import barriers.[14]

● Entitlements often require government interference with personal liberty and managerial prerogatives.

● Entitlements are inefficient, as is anything that by definition is given to all outside of market constraints and that cannot be bought, sold, or traded. Because entitlements come "without charge," people don't "economize" in their use.[15]

● Entitlements don't allow for comparative advantage. Since they go to everyone, they are used (or misused) by incompetents as well as by competents.[16]

● Entitlements do not serve as incentives because they are not distributed as rewards for performance.[17]

- Entitlements often reduce employee loyalty, commitment, and discipline.[18] The National Planning Association study concluded that the high rate of entitlements in Europe increases absenteeism, turnover, and idleness, and adversely affects productivity.[19]

- Entitlements add administrative burdens and overhead costs.

- Entitlements are a fixed or uncontrollable cost that can get out-of-hand rapidly, both in the public sector (e.g., Medicare) and in the private sector (e.g., pensions).

Whether such costs outweigh the benefits of a particular entitlement is a subjective value choice. Okun offers a neat little test, ''The Leaky–Bucket Experiment,'' for helping to decide how one feels about the cost of a particular entitlement.[20] Suppose a tax and income transfer program required that the money be carried from the rich to the poor in a leaky bucket (the leaks being, of course, all the inefficiencies described in the above list). The test is this: Would you favor the program if 10 percent of the money leaked out? Would you still favor it if 50 percent leaked? As Okun points out, Milton Friedman probably wouldn't tolerate much more than a 10 percent leakage, while John Rawls would keep filling the bucket even at a rate of 100 percent leakage. There is clearly no objective answer as to how much leakage should be tolerated. But it is probably still within the realm of objectivity to claim that the total leakage from public and private entitlements is increasing in the United States. The reason for this increase in leakage (or social overhead, if you will) is simply that demands for new rights have increased dramatically in the last decade.

Daniel Bell was the first to identify, and Daniel Yankelovich has spent the past decade measuring, the new ''rights consciousness'' that transforms personal needs and privileges into social entitlements. The two Daniels entered a political lion's den when they took up the issue. For they argued that what was once viewed as a valuable benefit to be earned is today seen as an indispensable right, not given but *owed*. This is not an argument that is particularly appreciated by the New (or Old) Left, for it implies a kind of welfare-state paternalism and a corresponding erosion of moral fiber.

Yankelovich has found a distinct difference in attitudes about entitlements between the Baby Boom generation and their parents (whom I will call Depression Kids).[21] The under 35s, as distinct from their parents, are most concerned about purely social and psychological rights. For example, young workers feel they are entitled to self-actualization on the job and to a voice in the decisions that affect their jobs. Older workers, in contrast, tend to stress purely economic rights, such as pensions, job security, and the like.

For purposes of clarity, Yankelovich divides the workforce into two broad but distinct categories:

1. *Traditionalists*, the shrinking 56 percent majority of workers who are still motivated by money, status, and security. Included in this category are the older, blue-collar workers for whom work is a habit, the older, "silent majority" white-collar workers with conservative values and life-styles, and a small number of young "go-getters" who have traditional goals and life-styles (Jaycees, for example).

2. *Nontraditionalists*, the growing 44 percent of the work force, almost all of whom are under 35. This category includes both the "turned-off," hedonistic, poorly-educated, unenculturated working-class and poverty youths, and the highly-educated, bright, creative, underemployed young people who have turned to leisure and nonwork avocations for the challenge and fulfillment they can't find at work. The nontraditionalist's strongest demand for change in working conditions is more time away from the job. In fact, University of Michigan researchers have recently found that half of employed Americans report that they have problems with the inflexibility of their working schedules, and that the demands of work leave them with inadequate time for leisure activities.[22] One-third of these workers report that inconvenient or excessive working hours interfere with their family life. In a separate study, Fred Best of the Department of Labor found that 70 percent of American workers would be willing to give up 2 percent *or more* of their income for less time at work.[23] In particular, the workers said they would like to take this additional free time in the form of a sabbatical or a longer vacation.

Employers find that motivating this growing population of nontraditionalists is difficult. The threat of being fired doesn't work because welfare, unemployment compensation, food stamps, and other sources of income provide a cushion that turns brief spells of joblessness into vacations. Moreover, because many young people have two-career marriages (and no children), one partner is always free to quit his or her job without a consequent severe drop in the family standard of living. The carrot of money is often not much of a motivator in these two-career marriages, either; for example, a raise means only half as much as in a one-career marriage. All of this adds to a further erosion of loyalty, motivation, and commitment.

For predictive purposes, Yankelovich's findings add more to complexity than to clarification, for there are at least two reasonable interpretations. The first is that young people can be expected to follow in their parents' footsteps as they "mature," leaving behind youthful concerns for the quality of life and embracing "adult" concerns for security. The second interpretation is that values are determined more by experience than by age. Many who were at an impressionable age during the Depression (including most corporate executives aged 55-65) had their values formed by the fact that they had trouble finding first jobs, that they had to postpone their educations and forego the purchase not only of luxuries but, in many instances, of necessities. Worse, they saw their adult relatives suffer the terror and humiliation of economic insecurity. It is not surprising that people who had thus seen and suffered would place security as the Sirius in their firmament of values.

In contrast, the postwar Boom Babies (the 18-35 year-olds who are now on the bottom rungs of the organizations headed by the Depression Kids), were at impressionable ages during an era of incredible affluence. They saw blacks and women arguing to restructure centuries-old social patterns and, most excruciatingly, they were on the front line during the Vietnam protests. The brightest stars in their heaven of values are flexibility, choice, options, variety, and diversity. This is not to argue that they don't like money—human nature hasn't changed, just work values—but they see plenty of different avenues to material security.

If this second explanation is accurate, it could mean that the values of youth will come to dominate in the work place of tomorrow for, by 1985, the postwar Boom Babies (the largest age cohort in the nation's history) will represent about one-third of the entire U.S. work force.

Indeed, there is some evidence—although it is far from conclusive—that instead of young people growing out of their values, older Americans are growing into the Boom Babies' values. David Ewing of the *Harvard Business Review* has discovered a growing concern for the social rights of workers among the nation's business leaders.[24] Comparing the 1977 attitudes of corporate executives about employee rights to their attitudes in 1971, Ewing found a marked tendency among executives to be more tolerant of such rights as whistle-blowing, privacy of personnel files, conscientious objection to unethical orders, "nonbusiness-like" dress and appearance on the job, employee voting on plant relocation, and the need for corporate ombudsmen. Clearly, some values can be contagious.

Zero-Risk

At the same time that this rights consciousness-raising was going on, there was also a trend toward what Aaron Wildavsky calls the "chicken little" mentality—or the avoidance of all risk.[25] The clearest example of this is the famous Delaney Clause that requires the banning of any substance that is found to produce cancer in humans or in animals. While zero-risk has been applied most frequently to environmental and consumer issues, the notion has recently been extended to the work place, most notably and clearly in the area of occupational health and safety. For example, it used to be that a steeplejack assumed all the risks of his dangerous job. Today, he is protected by OSHA, worker's compensation, health insurance, and disability insurance, all of which seem only fair to most people. But workers can't take risks even if they want to: witness the safety lines and nets that circus high wire acrobats are now forced to use. Most controversially, the reduction of risk goes beyond physical safety. Through unemployment insurance and so forth, individuals today bear very little economic risk or job insecurity at all. For example, recent court rulings have made it increasingly difficult to fire even the most undeserving of employees. (Academic tenure is the clearest application of the notion of zero-risk to an employment issue.)

And government regulations dealing with affirmative action also reduce the risks of workers. Indeed, affirmative action can be viewed as the ultimate in risk reduction: it provides insurance against the bad luck of having been born black or female. Add to these trends the power of labor unions, unemployment benefits, and cost-of-living adjustments and it is clear that workers are more secure today than ever before in America's history.

Thus, a major reason why young people are so unconcerned with material security may well be that they already have all the security they could ever possibly need. Indeed, it is the issue of security that unites the concepts of the rights consciousness and zero-risk. Moreover, under both concepts, individual risk is socialized. As Wildavsky points out, this does not mean that risks are reduced; they are merely displaced. Health insurance doesn't make people any healthier, or reduce the overall costs of health care. In fact, what is controversial about such programs of social insurance is that they put by-standers, third parties, or society in general at risk. For example, there is evidence that individuals covered by worker's compensation will behave more irresponsibly on the job than if they didn't have this accident insurance (or if there were a penalty for fault). And recent studies show that individuals with unemployment insurance are less likely to look for work than those without it.[26]

What is significant about zero-risk is its cost. Economists demonstrate rather convincingly that the marginal costs of reducing risk increase exponentially as the zero-state is approached. For example, the costs of providing due process for employees is manageable, acceptable, and justifiable on a cost/benefit basis. But experience in Italy, where *every* employee has tenure, shows that the costs of zero-risk of termination are unacceptably high in terms of lost labor force mobility and productivity. In the U.S., ironically, demands for zero-risk are increasing just as the economic growth that might have paid for it is giving out.

The "Culture of Narcissism"

Christopher Lasch has recently described "American Life in an age of diminishing expectations."[27] He notes that, for the first time in recent American history, there is pessimism about the future. And when the future looks bleak, one tries "to grab all the gusto one can." After all, "If I've only one life, let me live it as a blonde." In such an era, citizens become narcissistic. The signs of such selfishness abound in the "self-awareness" movement: est, gestalt therapy, rolfing, massage, jogging, health foods, and meditation are all manifestations of a new self-consciousness.

The most fully-documented aspect of the behavior of the "Me Generation" is the contemporary unwillingness to have children. For example, the percentage of single person households doubled in California over the last twenty years, and the number of childless households increased by over 6 percent between 1970 and 1977.[28] Since one cannot have a movement in America without having a national organization to legitimate it, childless couples have formed anti-kid clubs across the nation. And there is some evidence that Americans are spending less time with the children they have; employed women spent 20 percent less time on child rearing and housework in 1978 than in 1965, and even full-time housewives spent 12 percent less time on these domestic activities.[29] And child abuse among the working class rose at about the same time it became faddish for upper–middle-class women to "get some space of their own." These women liberated themselves from their children to the surprise of their ex-husbands (these men then found themselves part of the "single parent" statistic).

What is important about all this is that the "Me Generation" may be missing out on an essential stage of human development. What has occurred to nearly everyone who has thought seriously about the child rearing process is that it has unique benefits *for parents*. In particular, parenting helps to make one into an adult, because it makes infantile selfishness rather difficult to maintain. One simply has to make some sacrifices in terms of time and money for one's children.

But the "Looking Out For No. 1" philosophy eschews self-denial in favor of self-indulgence. This may produce prodigious consumers, but it makes for lousy workers. The new narcissism, like zero-risk and the rights consciousness, stresses entitlements without concomitant responsibilities. While the rights of the narcissistic generation have expanded, in general, their responsibility for the quantity and quality of their work has diminished, their responsibility for taking the initiative on the job has decreased, and their attitudes toward fellow workers, supervisors, and customers has grown less considerate—in a word, they are more *selfish*.

This is a strong assertion. Especially so because it is impossible to offer scientific evidence to support the existence of greater irresponsibility and selfishness among the youngest half of the population. Nevertheless, the impression that young workers are unreliable is broadly shared among American employers in particular, and the older half of the population in general. Witness the following story from the Associated Press:

> Boulder, Colorado, July 18, 1979—Employers in this college town have horror tales to tell about today's young workers, a survey by the *Denver Post* found.
> Home builder Bob White says he is quitting the construction business because of his frustration with the work habits of area residents. Restaurant owner Pete Brophy says

he loses $700 a month in dishes and silverware because of careless employees.
Absenteeism is another general complaint of Boulder employers, the *Post* reported. . . .
Boulder's average employee is between 20 and 35, well-educated and too often not
well-trained to work, office managers said. Several employers said 10% of those hired
for jobs fail to show up for the first day of work.
White said the construction pace in Boulder is about half that in the East, where he
grew up in the building trades of New York City.
"There were crews of Lithuanians, Latvians, Russians . . . all kinds of people who
took pride in their work," White said of New York. But he said that Old World work
ethic is missing in Boulder. . . .
These young guys are saying 'Look at me. I'm a carpenter. I have a pickup truck,'
but they don't really want to know how to do the work. I can't deal with it. I can't
accept it anymore," White said.
Brophy said he uses a crew of 70 in his restaurant. Last year he hired 192 employees.
Like other employers, he blamed problems with the work force on reluctance to settle
down, the ease of group living in a university town, unemployment-pay and food
stamps.
"They're single. It isn't like being 25 and married with a baby and a house payment
and a car payment. They just don't have any responsibilities," he said.

Coincidentally, the same week this article appeared, I received a phone call
from the chief executive officer of one of the largest corporations in the country.
I can't say if his tone was more of exasperation or desperation, but my notes
confirm that he said roughly the following:

My managers—even the top ones—don't seem to care about their work anymore.
There is no real dedication, no commitment, no sense of obligation. I've tried giving
them everything—money, long vacations, big bonuses—but nothing seems to motivate
them like it used to. . . . How long can the nation keep going on like this?

Permit me one more anecdote: While waiting for my wife in a large department
store, I recently witnessed an encounter that seems to sum up the present problems of work in America. I noticed three clerks—all under 30, I would
guess—standing idly chatting while about a half dozen customers waited for
service. Finally, a customer approached one of the clerks and politely asked for
help. The clerk reluctantly and sullenly agreed. As it turned out, the customer
wanted an article that had to be ordered from the stockroom. In a desultory
manner, the clerk sent the order to the stockroom and went back to chatting
with her fellow clerks. The customer waited patiently for about twenty minutes
(punctuality is not my wife's strong trait, so I was able to watch this little drama
unfold in its entirety), then the customer asked the clerk to please check if
anything was wrong with her order. The reply came back with the sting of a
razor strap: "Are you kidding, lady? If you don't want to wait, that's cool with
me. But I'm sure as hell not going to run around finding out what's holding up
your order."

Of course, the story by itself "proves" nothing. But such stories add up: A secretary takes a day of "sick leave" to go to the beach during the busiest week of the year. Managers buy luxury limousines for themselves with company money, while failing to provide clean restrooms for their workers. A worker refuses to walk fifteen yards to pick up an essential piece of material that has not been delivered to his work station; he assembles the product without the key piece. A worker in a factory, or office, or store—it doesn't matter which—is temporarily overloaded with work, but his underutilized co-workers will not help by picking up a larger share of his load. The new American credo seems to be: "Who gives a damn?"

All of the behavior I have described could be dismissed as trivial if it were not for the mounting evidence that the world will no longer "buy" the behavior of American workers and managers. Because of increasing balance of payments deficits and the declining dollar, American managers may no longer be able to afford to play golf on company time, and American workers may no longer be able to afford to goof-off. The nation can only afford such behavior if it doesn't want Arabian oil, Jamaican bauxite, Brazilian coffee, French wines, and Japanese radios. Workers and managers can only behave this way if they don't want zero-risk, and if they don't want rising entitlements.

The mismatch between the behavior of workers and the new economic realities threatens the long-term viability of the American economy. As the American economy becomes more labor-intensive as a result of the shift toward service, clerical, and knowledge work, the *attitudes* of workers become central factors in national productivity. In our mature, post-industrial economy, the success or failure of the national enterprise rests on the willingness of individual workers to take responsibility for the quality and quantity of their work, to take initiative in those increasingly frequent work situations that cannot be routinely handled, to show a real interest in the welfare of customers, suppliers, and fellow workers—in short, to *care* about one's work.

Trade unionists deny the existence of this problem: it is, after all, "illiberal" to suggest that workers need more self-discipline. And managers wrongly assume that the problem can be corrected by the workers: it is, after all, dangerously "leftish" to suggest that it is the responsibility of *managers* to organize work in a way that elicits greater worker self-discipline. Of all those who have studied work, only E. F. Schumacher saw what was happening and, more important, recognized that it is *the result of the existing philosophy and organization of work*, ". . . the result [of which], not surprisingly, is a spirit of sullen irresponsibility which refuses to be modified by higher wage.[30]

The problem with the current state of work in America is *not*, then, simply the rising costs of entitlements. Rather, the problem is the social costs of entitlements that are demanded and granted without the productive wherewithal to afford them. The problem, then, is *not* simply workers who have too little commitment to their jobs. Rather, the problem is the current organization and philosophy of work that does not tie responsibilities to rights.

Echoing the corporate executive quoted above, we may well ask: How long can America go on with the expectations of an affluent nation and with an organization of work that prevents it from affording these entitlements? Put another way, will the young "grow out" of their expectations and values?

Knowing that "a good depression will shape them right up" is an interesting but nonoperable insight. (At any rate, with the exception of the obnoxious traits of selfishness and irresponsibility, the values of the Boom Babies may be *socially preferable* to those of the Depression Kids. For a future of declining growth, diversity and flexibility may be more appropriate values than security and rigidity. Thus, there may be good reason to try to preserve the values of the young while finding ways to make them considerate and cooperative workers.) But the insight about a depression *does* at least have the virtue of calling attention to the relationship between the broader social and economic environment and worker values. Without understanding this relationship, one cannot hope to understand the way the future may unfold. Hence, as a first step toward looking at the future of work, let us examine the social context that produced these new values.

ALTERNATIVE FUTURES

Historical Disjunctions

The new values all surfaced in the early 1970s, suggesting that some kind of historical watershed may have been reached at that time. Indeed, a strong case can be made that the 1968-1975 period may have witnessed the most significant historical disjunction in the United States since the Great Depression. Clearly, the postwar 1945-1968 period was relatively free of revolutionary value change. Although the period included such momentous events as the Korean War, the Joseph McCarthy era, the Kennedy assassination, and several severe but short recessions, in general it was characterized by economic progress and social stability. Even the traumas of the McCarthy period and the Civil Rights protests could be seen, in hindsight, as affirmations of the nation's ultimate commitment to do what was right in the long run.

It may well be that the nation was able to absorb the potentially devastating blows of such events because they were played out against the backdrop of high rates of economic growth, low rates of unemployment and inflation, cheap and abundant energy, and rising individual affluence and educational attainment. During this period, the United States was master of the world both politically and economically. With unshaken faith that the country faced an ever-rosier future, Americans believed that their personal fortunes would rise along with the nation's. Not surprisingly, it was during this era of ever-rising affluence that corporations greatly increased the benefits they offered workers, and government entitlement programs grew like weeds on a suburban lawn. By the end of this period, security was taken for granted by young, educated workers, according to the first national Survey of Working Conditions undertaken in 1969.[31] It was thus no accident of history that it was during this era that Abraham Maslow's "hierarchy of needs"—with all the basic security needs fulfilled for the typical worker—enjoyed such popularity. Maslow's theory can be viewed as an historically-valid description of the rising affluence, educational levels, and social expectations of the post–World War II era. This era ended appropriately with an event that symbolized the faith in the progress that had characterized it: the landing of two Americans on the moon in July 1969.

The next year—when the first of the large Baby Boom Generation was getting ready to graduate from college—America bombed Cambodia. Then things fell apart, the nation's center failed to hold, and anarchy was loosed upon the land. In quick succession, the country experienced a string of morally-debilitating events: campus unrest, Watergate, the oil embargo, corporate bribery, and political contribution scandals.

In the economic sphere, the value of the dollar declined, rates of unemployment and inflation were simultaneously higher than ever before in recent history, national productivity and technological innovation drooped, and international economic competition became intense for the first time in over thirty years. In short, the five-year period beginning in 1970 was characterized by the marked relative decline of the United States as a world economic and political power.

Domestically, the social arena was strewn with the bodies of bloodied contestants as the nation rejected two decades of centrist consensus politics to engage in interest-group mayhem. The participants in the last half decade of discord can be listed as litigants, for often it was in the courts that they fought their oversimplified, two-sided battles: Men v. Women; Blacks v. Whites; Public Sector Employees v. Local Government Employers; Students v. Teachers; Enlisted Men v. Officers; Guards v. Prisoners; Homosexuals v. Heterosexuals; Environmentalists v. Corporations; Landlords v. Tenants; Consumers v. Producers; and Workers v. Managers. As a result of this considerable wrenching of the social cerebral cortex, social expectations fell. Faith in institutions declined. The collective view of the nation's future darkened over with black clouds of cynicism and pessimism. Is it any wonder that Maslow's hierarchy has now come under attack? When the blood-dimmed tide is loosed, and the ceremony of innocence drowned, people are unlikely to put much faith in a universal prescription based on economic growth, collective progress, and individual self-betterment.

The validity of social theories lasts for as long as the times in which they are promulgated. As good as Freud and Keynes were at explaining the behavior of the prewar era, their brilliant insights did not help much in the 1960s. And the social theories of the 1960s have been made obsolete by the fundamental changes that recently have occurred in the traditional relationships between categories of people. The superior/subordinate relationship of men to women became one of relative equality. The relationship of producer and consumer went from *caveat emptor* to *caveat vendor*. The relationship of landlord and tenant was turned completely on its ear—the landlord, who once had almost all the rights, now is often left unable to evict a tenant who refuses to pay rent. The relationship of worker to manager that was characterized by managerial prerogatives is now characterized by worker rights. In essence, the rules of the game have changed; there are new social relationships, new values, and a new economic reality. What does this historical disjunction portend for the future?

A Nominal Scenario

In order to construct some concrete underpinnings for our analysis, let us extrapolate a few current developments into the future. For instance, we might assume that the next decade will not see any real increase in the GNP. No decrease, either, for this is not a doomsday scenario, rather it is one with the same characteristics the nation has grown accustomed to over the last few years. Things just get a little worse with time. The next decade thus would witness slightly higher inflation and interest rates, much higher energy prices, lower corporate profits, more debt, less saving, less investment—all for the same interrelated reasons that such conditions exist today. But since these trends feed on each other, with time they naturally will grow worse, not better.

Mind you, disasters will not occur in this scenario. For example, OPEC will keep selling oil. Gas will be available at about $3.50 a gallon toward the end of the decade—just under the price of synfuel, which, like all government programs, turns out to have a slight cost-overrun problem. International trade competition will, of course, continue to stiffen. Domestically, the main effect of this particular congeries of developments would be unemployment—not depression-type unemployment, but a decade of stagflation.

I hasten to add that this is not a future that I personally advocate. I, too, hanker for a return to the conditions of the good old days of the 1950s and 1960s. But I am, nevertheless, unwilling to place all my bets about the future on a return to levels of economic performance experienced in the United States only twice in the last hundred years. Nostalgic I may be, but a long shot bettor, no. I'm hedging my bets in this scenario, playing the short odds that America is in for hard times but not bad times, prolonged recession but not depression. Of course, if Herman Kahn is correct that prosperity is right around the corner, you'll find no one happier than I. If he is right, we will be able to go on just as before, working less and getting more, and everything will be copacetic. But in the event he *might* be wrong (he bats only about .200, after all), I suggest that there is more than theoretical value in exploring how American workers might respond to a prolonged period of high inflation, high unemployment, a declining standard of living, and a diminished national ability to cope with the international conditions that exacerbate these domestic problems. While such an examination of the future can be no more than speculation, I suggest that it would be reasonable to expect the following kinds of developments:

- Workers would "hunker down," and their primary goal would become economic security.
- Unionization would increase.
- Industrial unions would negotiate a guaranteed number of work days per year; nonindustrial unions would demand Japanese–style lifetime job security.
- Unions would demand legal restrictions on the freedom of corporations to close uneconomical plants, divisions, and facilities. These regulations might include:

    —a required one year advance notice before closing a plant;
    —a required filing of a statement of economic justification and economic impact of a closure;
    —sufficient opportunity for workers and local communities to purchase a plant;
    —indemnities to affected workers and communities; and
    —requirement for the corporation to find alternative employment for affected workers.

- Federal restrictions would be placed on the introduction of labor-saving technology.
- National manpower planning would be introduced.
- Inflation would wipe out the value of most pensions funds, leading to the federalization of all pensions in the one fully–indexed system, Social Security.
- A nationalized health system would be established.
- Permanent wage/price controls would be enacted (or, alternatively, wages would be indexed).
- Unemployment benefits would be greatly increased, as would worker compensation benefits, and both programs would be federalized.
- A U.S. Labor Party would be formed; like European labor movements, it would look to government to provide entitlements, rather than seeking these through collective bargaining.

Although this future is quite probable and may even be desirable to a minority of Americans, from the point of view of the majority it would be repugnant. It would lead to British-style labor unrest, ideologically–based class conflict and low productivity. It would satisfy neither the economic needs for greater efficiency and higher productivity, nor the humanistic new values of young workers.

It would cause government to become more bureaucratically centralized and force unions to abdicate collective bargaining for a political role (a role, to their credit, they have repeatedly rejected).

Another Scenario

Ironically, these outcomes all run counter to the expressed wishes of the key stakeholder groups in the society. But it is a paradoxical fact that societies often end up inheriting futures that no one had consciously chosen. There are many reasons for this unfortunate historical pattern, not the least of which is that key individuals or groups who might have changed the course of events for the better acted unwisely or, more to the point, failed to act at all. Often, these reluctant players—political parties, newspapers, unions, corporations, the intelligensia, the church, the military—assumed that they were *helpless* to influence events positively. They concluded that they could only *react* to the moves of other players.

In the scenario I have just described, corporations play a passive, reactive role. This is probably unrealistic. Historically, corporations have been activists, not ineffectual bystanders along the parade route of American history. While corporations can do little directly to increase national economic growth, create full employment in the economy, reduce inflation, or increase the value of the dollar, they *can* greatly influence the future of the work place. (Indeed, even if they choose a course of reaction over the more prudent course of proaction, this policy, too, will inadvertently shape the future.) Of course, this does not mean that corporations are free to create any kind of future they desire. The American labor market is pluralistic; therefore, workers, unions, and the government will all have a share of influence as well. Nonetheless, corporations have the *opportunity* to shape the future of work in a way that meets the needs of the society and the economy, *if* they are able to link these needs to the self-interest of the other stakeholder groups. In short, it seems to me that corporations have the greatest leverage to produce an alternative future to the one just described, an alternative in which:

- Workers change their entitlement expectations to be consistent with their productivity and the availability of resources to afford the entitlements.
- The government changes its posture to that of a goal-setter with regard to entitlements, leaving industry with flexibility as to how to meet those goals.
- Unions change from their adversarial posture to one of cooperation.
- Corporations adopt *pro*active policies that encourage workers, government and unions to act in these ways.

By any reckoning, this is an extremely improbable future. While few Americans have any confidence that this scenario would occur in the next score years, few influential Americans (with the possible exception of some radical labor unionists) seem opposed to it. At least theoretically, most Americans seem to favor a future of industrial cooperation that resembles certain aspects of the current situation found in such countries as Sweden, Germany, and Japan. It would seem that a future that so many people find desirable is worthwhile exploring, no matter how improbable it may seem at the present. For the analysis might reveal some things that could be done to enhance the odds of its occurrence.

The kicker, of course, is that corporations do not know how to create such a future. The current philosophy and organization of work that pervades American industry is, after all, a cause of our current and, most likely, future problems; therefore, that with which we are familiar will be of little of no help. Thus, to create working conditions that are compatible with economic reality, it first will be necessary to get back to basics. It will be necessary to begin with a reassessment of the economic assumptions on which the current philosophy and organization of work are based.

TOWARD A NEW LABOR ECONOMICS

Economists are right at least about one thing: the trick is to "optimize" a set of conflicting factors. Unfortunately, the set of factors traditional economists have been optimizing for two centuries is, today, incomplete, insufficient, and inappropriate. While traditional economists could content themselves with optimizing production functions that include only three variables—land, labor, and capital—the new economist of the 1980s will have a more complex task.

In the emerging resource-scarce, highly-industrialized, highly-polluted, highly-populated, highly-interdependent, and highly-contentious world, it is increasingly anachronistic to make policy choices solely and simply by trading-off more expensive factors of production for cheaper ones (or, *apparently* cheaper ones, for what is cheap today may be dear tomorrow). A nation could succeed in creating a just society following the guidelines of the old economics only if:

- it were isolated from all other societies and economies (e.g., if it were entirely self-sufficient; if it didn't have to worry about foreign competition);
- its population shared consensus on social goals and values (e.g., if it didn't have pesky ecologists, consumerists, trade unionists, libertarians, and the like who focused on aims other than growth and efficiency); and
- it had solved the problems of energy availability and pollution (e.g., if it didn't have to depend on foreign sources of energy or didn't have to rely on dirty coal or unsafe nuclear energy).

In other words, for traditional economics to serve as an appropriate guide to national policy, the nation would have to possess the equivalent of Saudi Arabia's energy, America's agriculture, Albania's isolation, and Switzerland's social and political values. Since I know of no nation so blessed (or cursed), I conclude that traditional economics will not solve the problems of work in an era of slow growth.

Traditional economics (as it is practiced in America today) is concerned primarily with *efficiency*—that is, with maximizing the three traditional factors of production. Unfortunately, as Arthur Okun has demonstrated, efficiency is gained by trading off equality;[32] as E. F. Schumacher has shown, efficiency is gained by trading off the quality of life;[33] and as J. K. Galbraith has shown, efficiency is gained by trading off certain freedoms—particularly consumer sovereignty and competitive markets.[34] In effect, the scope of traditional economics is simply too narrow. For, by excluding liberty, equality, and the quality of life from the scope of its concerns, it excludes the highest values of the majority of the population. Optimizing land, labor, and capital the way economists do with their production functions leads to unemployment, inequality, pollution, and social alienation—outcomes that may be acceptable to some economists, but are unacceptable to most other Americans. There is, thus, a misfit between the operating philosophy of the economy and the new goals of the pluralistic polity.

While it is not useful here to offer a thorough critique of the discipline of economics, it is, nevertheless, important to at least briefly call attention to its manifest shortcomings. For it is this discipline that informs—even dictates—the prevailing philosophy and organization of work. Following willy-nilly the dictates of economists, people are replaced by machines; interesting craftwork is replaced by dull machine–tending; organizations are structured hierarchically; workers are abruptly laid off when the economy turns down; and maximization, optimization, least-cost, and all the other notions of industrial efficiency are applied indiscriminately to the organization of work and to the treatment of workers. This is not to argue that machines should never replace workers or that workers should never be laid off. But the problem with economics is that it does not indicate *when* it is appropriate to do such things. Economic assumptions, because they are misconstrued as laws, dictate that machines *always* should replace workers when this leads to greater efficiency.

Sometimes it is *inappropriate* to replace people with machines—particularly if one is trying to achieve liberty and equality, and to improve the quality of life, *in addition to* increasing efficiency. But the old economics is single-minded. It can maximize the factors of production *only* to achieve efficiency. Thus, the new economics would have to be more complex than the old variety. The new economics would have to pursue all four of the prime goals of political/economics (liberty, equality, efficiency, and the quality of life) *simultaneously*.[35] Such a new economics does not yet exist (I will shortly argue that Schumacher's Buddhist economics does not fill the bill). But whatever form it ultimately takes, the concerns of the new economics undoubtedly must transcend the simple maximization of the traditional factors of production. What is needed is a new, complex calculus for satisfying many extremely broad factors of production *and consumption*. These disparate, nonfungible factors, which can be neither measured nor equated, might include: employment, energy, ecology, efficiency, equality, excellence, enfranchisement, education, (and enjoyment?).

My argument is that in choosing an appropriate organization of work the new labor economist will have to simultaneously satisfy all these E-Factors. These E-Factors are, in effect, performance criteria for work in a new slow-growth economy. Because of such new conditions as resource–scarcity, new values, world interdependence, and industrial complexity, America will *not* be considered a just society *unless* it satisfies nearly all the following performance criteria:

Employment: A job for every citizen who wants one.

Energy: An adequate, long-term supply of energy.

Environment: A safe environment with clean air and water.

Ecology:Husband its natural resources so these will be available to future generations.

Efficiency: An adequate standard of living, on the average no lower than what is enjoyed today. This requires improved productivity and technological innovation so the country can remain competitive in the world economy.

Equality: Greater economic security for the poor, reducing the gap between the poor and the middle class by bringing the poor up to a decent standard of living, *without* reducing the standard of the middle class.

Excellence: Encourage and recognize individual merit, rewarding those who contribute to the satisfaction of the other factors on this list.

Enfranchisement: Every citizen must be provided with a voice in the affairs that determine his or her daily life. This will mean increasing individual freedom of choice in all aspects of life, particularly work.

Education: Equal access to a form of education that is adequate preparation for the world conditions that are emerging. (An explanation of this appears in the coda to this chapter).

(Enjoyment: Here I am uncertain. Must America provide conditions for its citizens in which they are happy, satisfied, and where they find transcendent rewards? Or will satisfying the other factors on this list bring all the enjoyment that a society is expected to bring to individuals?)

This is quite some list. I wish to make clear that it is *not* my personal wish list. Rather, it is designed to reflect the goals of the major interest groups in America: unions, corporations, minorities, the professions, environmentalists, journalists, educators, the "new class" of bureaucrats—even economists. Because I assume that America will continue to be a pluralistic democracy, I suggest that any new philosophy and organization of work will have to score high across-the-board on the above performance criteria. Failure to do so is likely to lead to rejection by one or more of the powerful groups which have the power of nullification in the delicately balanced American political system. This list thus provides a "reality test" for alternatives to the current organization of work.

I am sad to say that it is this test that Schumacher's Buddhist economics flunks. (Although I single out Schumacher, I mean, of course, to include in my criticism all the minor variations on the voluntary simplicity theme by such composers as Amory Lovins, Willis Harmon, and others.) While I personally share Schumacher's values, I have no illusion that they are shared by the groups that have power in America. While Buddhist economics satisfies a different set of performance criteria appropriate for Asia and Africa, it will be rejected in the United States of America because it does not satisfy the requirements of Efficiency, Enfranchisement, or Excellence.

When applied to the American work place, Schumacher's "small is beautiful" philosophy is nostalgically appealing, but is not a realistic guide to changing the organization of work. For the philosophy would require the dismantling of at least the 5,000 largest U.S. corporations—an impractical idea, to say the least. The nation could obviously not afford to write off what surely amounts to over 75 percent of its total invested capital. Moreover, these very same corporations provide many good, secure jobs because they are the most effective competitors in world markets. For example, if the United States tried to compete with Japan with an economy composed of the equivalent of lemonade stands, its markets would be swamped with foreign goods, and millions of jobs would be exported. As Galbraith points out, the giant corporations are the show cases of the modern economy, providing most of the productivity and wealth to pay for the social entitlements of the welfare state.

The challenge, then, is *not* to dismantle the large corporations, but to *reform* them so they will meet the humanistic and environmental goals that Schumacher wishes to achieve—without giving up the required efficiency that is the bulwark of the nation's standard of living. On this task, Schumacher and his followers are silent. Schumacher's only reply was that Americans should reduce their standard of living. Such arguments strike a discordant note with perhaps 90 percent of the American public. In addition, his notion that technology is the demon does not sit well with the American citizenry. For example, his damning of the 1969 moon landing might have gone down well with the folks at the British Soil Association, but to most Americans the lunar program was the best thing the country did collectively since it saved Europe from the Nazis. Schumacher talked about appropriate technology. But, more important if one *really* wants to change things, is *appropriate ideology*. When it comes to convincing Americans to mend their ways, Schumacher's inappropriate Buddhist connection mars his otherwise brilliant contribution.

I have argued that the traditional *discipline* of economics has considerable shortcomings as a guide to a more appropriate organization of work. But, as this review of Schumacher suggests, economics is not just a discipline, it is also a collection of *ideologies*. And each of the major economic ideologies addresses itself directly or indirectly to the issues of work. Moreover, like Schumacher's Buddhist economics, each of the other standard economic ideologies is an inappropriate guide because each also requires a consensus of values, that is, a national willingness to coalesce around a single goal.[36] This flies in the face of political reality. As Schumacher was willing to foresake liberty, equality, and efficiency to maximize the quality of life, Milton Friedman and other Libertarians will sacrifice equality, efficiency, and the quality of life to maximize liberty. Marxists and other egalitarians will sacrifice all other values to maximize equality, and Corporatists (an inelegant term that describes the prevalent ideology among American economists and businessmen) will sacrifice all other values to achieve efficiency. While of the four ideologies Schumacher's has the saving grace of being the only one not dependent on growth, all four are, nonetheless, equally out of phase with the exigencies of a complex, modern democracy. In effect, the four major ideologies are equally inappropriate in a society where policies must deliver on liberty *and* equality *and* efficiency *and* the quality of life. None of these ideologies, then, can cope with the complexity of the E-Factors.

The search for a just philosophy and organization of work must therefore proceed without the discipline or the standard ideologies of economics as guides, but with the E-Factors as tentative criteria against which to evaluate alternative work policies.

As an illustration of what this would mean in practice, consider the major public policy issue of providing "full employment." The classic economic answer to this problem is "growth." The assumption of the ascendant economic ideology is that full employment flows as naturally from growth as pure water flows from mountain springs. *If* this assumption were valid, it would take something like nine years of uninterrupted 5 percent real growth in GNP to reduce unemployment from 7 to 4 percent.[37] Realistically, the odds against the nation achieving that kind of growth are about a thousand to one. But even if it were possible, the Environment E-Factor would not be satisfied because all the stops on pollution would have to be removed, and the Equality factor would not be satisfied because growth does not provide jobs for the "structurally" unemployed at the foot of the employment hierarchy.

Egalitarian economists offer a different answer to the problem: "government jobs." But when the government acts as the employer of last resort, bureaucratic social overhead is enormous. Thus, the factor of Efficiency is sacrificed.

A new economist would look at the problem still differently. He would look to investments that simultaneously satisfy the demands for equality *and* efficiency. In fact, one can easily come up with a long list of employment programs that would satisfy not only these two criteria, but most of the other E-Factors as well.[38] For example, one could turn the egalitarian argument on its head and make the private sector the employer of last resort. One way to accomplish this would be to borrow the insurance notion of "assigned risk," and make every corporation responsible for providing minimum-wage jobs for people from a pool of chronically unemployable workers (the number of slots assigned would depend on the size of the corporation). The advantages of this approach to the society as a whole would be tremendous. Social overhead would be greatly reduced by getting the government out of the entitlements business almost completely (welfare, unemployment insurance, public service jobs, public housing, food stamps, and the like could be cut back to the bone). The costs of crime, ill-health, mistreated children, and other by-products of poverty would be brought down to the levels found in Northern Europe. Although corporations would object at the outset, they would have a tremendous incentive to find productive ways to use the workers assigned to their payrolls. And, since corporations are far better at manpower training than the government, the chances are that they would find creative ways of making use of this currently untapped human resource.

This would be especially true if Ivar Berg is right that the characteristics of employed and unemployed ghetto residents are identical.[39] He posits that, since some ghetto residents are demonstrably productive, employers might find that many more can be employed productively than managers currently assume. Certainly, employers would benefit from an economy in which everyone was working. Because the costs of social entitlements would be decreased (as I indicate above) everyone would be paying less in taxes. This would create demand for goods and services. Eventually, the less productive workers would pay for themselves through their demands for goods and services. America would have "full employment" in the broadest sense.

While this approach has certain practical problems, I suggest that these are fewer and less significant than the problems involved in the Schumacher approach to full employment (i.e., a return to small-scale, labor-intensive, cottage industries) or the classical economic approach (i.e., grow at full tilt and wait for the jobs to trickle down).

A NEW PHILOSOPHY AND ORGANIZATION OF WORK

The E-Factors are *macro*-criteria by which to judge how alternatives to the current philosophy of work would contribute to social justice in the hard times ahead. Basically, they are performance criteria for national policy; they are not as useful at the level of a plant or an office. Thus, the next logical step in this analysis is to identify *micro*-criteria for work at the level of the single organization. Alternatives to the status quo will have to be not only politically acceptable in general, but also acceptable to managers, workers, unionists and others directly concerned with workaday conditions.

Workers Are Diverse

One's first inclination is to "go to the literature" in search of such guidelines; some social scientists have spent the last decade measuring what workers want from their jobs, what motivates them, what makes them productive. Alas, just as we find a loss of faith in economists, we discover that the findings of work place sociologists, psychologists, and survey researchers are being greatly discounted. We had more confidence in these folk in the early 1970s. Recall that social scientists at that time claimed they had data to *prove* that American workers were bored and alienated. It was authoritatively claimed that the desire of workers for "self-actualization" on the job was frustrated by the inhuman ways work tasks were designed. The crux of this argument was the supposed existence of a universal "hierarchy of human needs."

As postulated by Maslow, all human beings were said to have the same order of needs, and as each level of need was fulfilled the subsequent level became salient. Not unreasonably, social scientists in the 1970s argued that, since American workers in general were safe, secure, and affluent, they must then be searching for self-fulfillment through creative and challenging work. Once Maslow's assumptions about human nature were accepted, the following litany of conclusions about work flowed as naturally as marbles off an uneven table:

Because work is organized in an assembly-line manner, job satisfaction is thwarted.
There is wide-scale discontent with working conditions, particularly among blue-collar workers.
America's affluent workers are more likely to be motivated by interesting jobs than by additional money or other benefits extrinsic to the work itself.
Workers are demanding jobs that offer such intrinsic benefits as challenge and the opportunity to learn and to grow.
Jobs can be readily redesigned (or "enriched") to reduce worker alienation and to increase productivity.

During the 1970s, survey researchers at the University of Michigan spent about $1 million of federal money in three monumental efforts to measure the quality of working conditions in the United States and the levels of satisfaction of workers with those conditions. Then these researchers tried to correlate this information with individual worker characteristics and behavior. On first reading, it appeared that the Michigan data supported the above list of conclusions. But recent analyses by social scientists in the 1970s now indicate that the few statistically-valid correlations in the surveys do *not* lend credence to the Maslovian hypothesis.[40] It isn't that the Michigan studies prove the opposite of what social scientists were claiming (that workers *like* boring work, etc.); rather, the studies don't identify much of anything at all.

Through these and other similar operations, the entire job dissatisfaction argument, so popular and so persuasive in the 1970s, has had its theoretical and empirical underpinnings chopped from under it by skillful academic amputators. These surgeons have rather successfully demonstrated that no quantitative evidence can be adduced to prove the existence of a universal hierarchy of needs; widescale or growing discontent with work; a blue-collar revolt against the assembly-line; low productivity among "overeducated" youths; that workers are demanding redesigned or enriched jobs; or that job redesigners know how to organize work in a way that would elicit sustained, high levels of human satisfaction and productivity.

Today, then, no analyst of sound mind would frame an argument for work reform based on the social science findings about job satisfaction or on the quantitative findings about the changes in worker attitudes and productivity that are claimed to result from enriched jobs. These "hard" data too often have turned out to be no more than social science fiction.

Nevertheless, one strong conclusion can be drawn from the studies of work that have been conducted during the last decade, that is: workers are not all alike; they have different needs, interests, and motivations. Moreover, these characteristics constantly change over the career of each worker.

Criteria of Performance

The only generalization that holds today is that workers and their values are *diverse*. The implications of this scientifically verifiable conclusion are enormous. It probably means that no single set of working conditions is appropriate for all workers, or even for one worker during the course of his or her career. The one valid finding of the social scientists thus suggests the appropriateness of the following micro-performance criteria for the organization of work:

Diversity: A just work place would offer a wide range of tasks, each requiring different levels of skills, abilities, effort, degree of commitment, supervision, creativity, and length of time for completion. There would also be a spectrum of rewards and incentives.
Choice: A just work place would make it possible for workers themselves to select jobs that meet their individual needs (since it is impossible for an organization to fairly identify and match individual needs with task requirements).
Flexibility: A just work place would accommodate differences in the off-the-job commitments and interests of workers. To meet various levels of family responsibilities, working conditions and career paths would be individually tailored as far as practically possible.
Mobility: A just work place would provide the opportunity for workers to move to other organizations, within limits, without penalty. That is, because worker needs, wants, and interests are constantly changing, and because few organizations can provide the full spectrum of diversity, choice, and flexibility necessary to accommodate these, there is need for unencumbered opportunity to

move when a worker cannot find the conditions he or she seeks. Within organizations, there would also be ample opportunity for lateral as well as noninvidious downward mobility.

Assuming that workers are paid a fair salary and retain all their current entitlements, it is hard to believe that they could ask from their work much more than diversity, flexibility, choice, and mobility. But the purpose of work is not just to meet the needs of workers. (Unfortunately, for too long work was organized without any concern for their needs, but that is no excuse to compensate now by overreaction.) In addition to satisfying the needs of workers, the other purposes of work in the future must continue to be to provide the goods and services society needs and wants. Since the desires and requirements of American society are high, this means that productivity at work, too, must be high. And since worker productivity cannot be commanded, it must be elicited through self–interest. This suggests the need for two additional performance criteria:

Participation: A just work place would provide workers with full information and authority to manage their own jobs. Workers would design their own jobs, set their own work schedules, decide their own salaries, indeed, be fully self–managing as far as their *own* work went, and to the extent that it did not affect anyone else's work. Workers would also participate in profits and ownership of stock in the companies in which they worked.

Rights Tied to Responsibilities: As a necessary corollary to the participation criterion, a just work place would be one in which every individual is held responsible for all the decisions he makes or actions she takes.

Many workers, of course, are unwilling to assume such responsibility; this calls for one final criterion:

Security: A just work place would be one in which managers planned for full employment. This is not to say that workers would have academic-style tenure. Sadly, it is not recognized by either unions or management that workers can receive a real sense of job security through stock ownership, elaborate due process and grievance procedures, Lincoln Electric-type annual hour guarantees,[41] and so forth, all without offering Japanese-style life-long sinecures.

Stated generally, these criteria may appear quite radical. It may indeed be difficult to imagine a work place that would satisfy these factors and still be profitable and productive. But we needn't confine ourselves to imagining. Nor need we "dream of a system in which no one inside has to be good," to use Schumacher's paraphrase of Gandhi. In fact, we don't have to dream or imagine at all. Instead, we could simply apply the performance criteria to the various policies and practices of American corporations to see what satisfied the criteria and what didn't. If working conditions are found that satisfy the criteria, then obviously they are not too radical, for they have passed the ultimate pragmatic test: they work.

E-Factors in Real Life

What would an organization look like that satisfies these criteria? Because imperfection is the rule of life, no one organization ever could be found that perfectly satisfies all the criteria. Moreover, a message of this chapter is that there are no monolithic solutions to the problems of work. There is no way to "solve" a production function using the E-Factors or the micro-criteria. Not only is the work force diverse, but places of employment vary by size, location, technology, the products they make, and the services they provide, and many other significant variables. And, finally, for any organization there is a spectrum of appropriate conditions that would meet the established criteria.

Although no "ideal" organization can be identified, for the sake of illustration, let us consider a work place that at least approximates what we have in mind. For the sheer challenge of it, let us identify a work place in an industry where there are few exceptions to archaic working conditions. After all, anyone can change for the better the organization of a plant that uses "continuous process" technology, but it takes real talent to devise a just alternative to the monolithic model to which, for example, almost all large law firms conform. Although few work places are less progressive, more inhumane—in fact, more medieval—than the typical corporate law office, the existing model is accepted without question "because that is the way things are done." But there are just and productive alternatives even to the tradition-bound organization of law firms, as the following case illustrates.

Typically, large law firms hire young lawyers (associates) and condemn them to dog work for seven years. Young lawyers willingly submit in the hope of one day becoming partners and making "megabucks," as they say. The name of the game (it is literally a game) is "billable hours," in the service of which associates forego vacations, weekends, and a good many lunches in order to impress partners with their 12-hour-a-day dedication. In contrast, the Los Angeles firm of Munger, Tolles and Rickerhauser was founded by a group of young lawyers (aged 25-35) who decided to trade-off the long-term possibility of hitting the megabuck jackpot in favor of current satisfaction. Some of the policies they have initiated are purely hedonistic: first year lawyers get a month's vacation. But most of their radical policies affect the work itself. For example, new associates deal with real clients, not just footnotes on partner briefs. Right from the start, associates are given full responsibility for entire cases. Promotions to partnership are made at the end of three years. This means that the firm is top-heavy (two partners for every associate, instead of the typical two associates for every partner). Consequently, the difference in compensation between partners and associates is much smaller than in typical firms. But in lieu of Beverly Hills mansions, attorneys have the opportunity to take sabbaticals and to do extensive *pro bono* work on firm time. They also treat each other in civil, unlawyerly ways (for example, all the lawyers lunch together twice a week. At one lunch they hold seminars with outside experts, often on nonlegal topics). Significantly, the firm has the reputation of having the finest lawyers in Southern California.

Flexibility, choice, participation, responsibility, and security—all of the essential ingredients that are missing in a traditional corporate law firm are present at Munger, et al. Yet, in some ways, it is unrealistic to compare this firm with the typical law firm. Since Munger and friends recruit only the top graduates from top law schools, it could be claimed that it is relatively easy to shove these young people out of the nest on day one, or that these top graduates are the only ones who would want to be given so much responsibility so soon. After all, the typical lawyer or the typical professional worker is, by definition, *average*.

But one ought to be able to get the same results with *average* professionals that they get at Munger, Tolles and Rickerhauser. For example, one could reinstitute the old apprentice notion that somehow got lost in the misguided rush to vocationalize the nation's schools and universities. For almost all professional tasks (and most other white-collar work in the fast-growing information industries), an excellent way to organize work is to put a young person under the wing of a much older one (so neither will be threatened by the other). The young person would be free to stay in the protege status until he or she is ready to fly solo. But when he finally does go it alone, he would be held responsible *only for his final product*. That is, there would be no mid-course snooping or second–guessing by "management." If the young person wanted to return to his or her old mentor for additional mid-flight tutelage, that would be his business.

I'm not too proud to admit that I'm average, at least in that I am reluctant to take full responsibility the first day on a new job. But I've been lucky; I've found myself in a protected situation three times in my checkered working career. Unhappily, the practice is quite rare at work today, as simple and as logical as it seems. (It is coming back, I'm happy to say, under the with-it name of "mentoring.") And this fact highlights how far the nation must go to get appropriate performance criteria disseminated throughout American work places. I know of one of the nation's 100 largest firms that makes even its executives punch in on a time clock. Talk about encouraging responsibility!

I have, of course, just committed heresy. I have discussed working conditions without reference to blue-collar workers or to the assembly line. While blue-collar workers probably deserve the highest priority in reforming work, it should be remembered that they are a *minority* of the work force. In fact, job redesigners have spent 95 percent of their time on the problems of the less than 5 percent of the work force who work on assembly lines and with continuous process technology. But priority should not mean exclusivity.

At any rate, to prove myself true-blue, let me now bridge the spectrum by examining a company with a semi-skilled and unskilled work force. Again, no "ideal" exists; Donnelly Mirrors only should be considered illustrative of a firm that seems to be developing an organization of work that will be appropriate for the future.

All of Donnelly's 500 employees participate in managerial decisions, and all participate in the firm's profits. The entire work force is organized into teams of eight to twelve workers. Each team has access to all the information it needs to be self-managing. Using this information, the teams have full rights and responsibility for making all decisions directly affecting their work, e.g., how fast to run the production line, when to shut it down for maintenance, when and how to implement a schedule increase, a product variation, or a method change. Decisions are made democratically, based on a consensus of the members of a team.

Each team elects a "linking pin" who then becomes a member of a team at the next level. In this way, communication and participation flow in all directions throughout the firm. For example, ideas bubble up to an elected Suggestion Committee whose function it is to consider all suggestions that reach it, and to do something constructive about every suggestion. If the suggestion is unfeasible, for example, the Committee ensures that the person who made the suggestion is promptly told why his idea is not being implemented. The Committee has sweeping powers: for example, based on the suggestions of workers, the Committee eliminated hourly wages and time clocks in favor of salaries.

There is also a company-wide representative committee that deals with grievances, pay policies, fringe benefits, promotions, and all other personnel policies and problems. Employees elect fellow-workers to this committee who each represent six to eight teams. The representatives serve two-year terms, subject to recall by their constituents.

Every month a bonus is paid to the workers, based on all increases resulting from innovations, productivity gains, and savings in labor, materials, and operating supplies. Workers get special bonuses (and promotions) for finding ways to eliminate their own jobs. To provide the security needed to encourage this dedication to productivity, the company offers a guaranteed annual wage and a guarantee against technological layoffs. There is also widespread employee stock ownership in the company.

Although this description may sound futuristic and utopian, it is, in fact, based on a 1977 description from the *Harvard Business Review*.[42] As a result of adopting the system described above, Donnelly increased its sales from $3 million in 1965 to $18 million in 1975, and increased the size of its work force from under 200 employees to over 500 during the same ten-year period. While labor and material costs escalated rapidly during this decade, the company succeeded in holding the line on the prices of most of its products and *reduced* the prices of all the others. Donnelly now has one-tenth the U.S. national average rate of absenteeism and one-third the average turnover rate.

The company's policies are acceptable to managers and shareholders because it is efficient and profitable, acceptable to traditional workers because there is plenty of security, acceptable to nontraditionalists because there is unlimited opportunity to participate in decisions, and acceptable to unionists because the system is, after all, just a variation on their very own baby, the Scanlon Plan.[43] In a nutshell, the company satisfies the needs of its major constituents, scores high on the various performance criteria, and offers a range of motivations that satisfies traditionalist and nontraditionalist workers alike.

Significantly, entitlements are *not* a problem at Donnelly. The Donnelly experience illustrates the point underscored earlier: entitlements, per se, do not constitute a problem for American corporations or American society. The problem is entitlements or wages that are demanded and awarded without the wherewithal to afford them. That is, a highly–productive company or highly-productive nation can afford a high level of compensation, while nonproductive companies and countries simply cannot. In this regard, it is instructive to recall that the level of social and worker entitlements is higher in productive Sweden than in nonproductive England, higher at productive Donnelly Mirrors than at hundreds of thousands of traditionally managed American firms. The crucial difference is that in countries such as Sweden, Germany, and Japan, and in companies like Donnelly Mirrors, workers don't demand or get entitlements or wages that exceed the productive capacity to pay for them.

None of this is news to America's corporate leaders. They have recognized this fact of life for years. They've preached it to unions, workers, to the government, and to the press. But while they've said all the right things, few of these corporate leaders have acted to make their corporate policies conform to their beliefs. In real life, there is no way of reversing unsupportable levels of entitlements by bemoaning them. Preaching never got rid of sin.

Industry analysts have been warning for years that American steel was being priced out of world markets because of low productivity, outmoded technology, and high labor costs. Recently, it has been estimated that labor costs at the Youngstown Sheet and Tube plant in Ohio could have been reduced by 21 percent by changing union seniority rules and by reducing plant manning levels that had been bargained into contracts.[44] Failure on the part of managers and the union to make such changes led to the permanent closing of the plant. Now, when it is too late, Youngstown workers are saying that they would gladly behave more responsibly if they were given the chance to try again. But they won't get that chance. Nor is there to be a second chance in the hundreds of other plants that have recently closed or will soon close because their low productivity makes foreign goods cheaper for American consumers. It is not the case—as trade unionists are wont to claim—that the reason for the export of jobs is that America's competitors have lower labor costs. In fact, total labor costs in Sweden and Germany are *higher* than in the United States, and not much lower in Japan. A more relevant figure would be some kind of productivity/labor cost ratio, for it is here where America's recent record is abysmal.

Today, the best way to change a company's or a nation's productivity/labor cost ratio is to alter the entire congeries of fringe benefits/incentive schemes/working conditions. These conditions have to be such that workers can clearly see that it is in their *self-interest* to improve the productivity/labor cost ratio. Jaw boning won't do it. Systems like the one developed at Donnelly Mirrors will.

Rights with Responsibilities

It was my earlier conclusion that the inevitable outcome of current entitlement trends is undesirable. Here I am arguing that the only positive and realistic way of avoiding such a future is for companies to infuse concomitant *responsibilities* into the existing and probably inviolable arena of *rights*.

Recently, there have been some interesting experiments in which workers have been encouraged to take responsibility for their behavior. At places like Donnelly, workers have been given financial and technical information, and given both the rights and the responsibilities for organizing their own work accordingly.[45] In several instances, such information has been given to workers and they have been free to set their own salaries.[46] In other instances, responsibility has been encouraged among both workers and managers by adopting a system of worker capitalism.[47] That is, employees have become the owners of the firms in which they work. In these various experiments, workers have *not* given up their rights, but they have assumed concomitant responsibilities. The risks of self-management, of setting one's own salary, and of assuming ownership seem to provide a necessary balance to the idea that rights come without responsibility, that zero-risk comes without cost, and that entitlements need not be tied to productivity.

The managerial challenge before the nation is to take the best from the experience of small, innovative firms like Donnelly, Lincoln Electric, Harman International, and a few dozen others with at least some conditions that satisfy the micro-performance criteria and to translate these for application in large corporations. A partial list of fringe benefits, incentive systems, and working conditions that have been tried in American work places and that would seem to score high on the micro–performance criteria would include: The Scanlon Plan; cooperative stock ownership; productivity sharing; incentives for teaching other workers; peer-set salaries and raises; salaries instead of hourly wages; well-pay, safety-pay; pay for task completed, not hours worked (e.g., finish job, go home); flextime; autonomous work teams; peer-established work/plant rules; self-managed quality control; worker-designed jobs; job posting; job sharing; and permanent part-time employment.

To repeat: Not all these conditions are appropriate in all work places at all times. Nor is the list rigorously all-inclusive; it is merely suggestive of the kinds of conditions that have the characteristics of diversity, choice, flexibility, mobility, participation, rights tied to responsibilities, and security. Significantly, few traditional fringe benefits appear on the list, nor do such new benefits as group legal and auto insurance. Because workers in large corporations are "benefited-up," fringes do not seem to offer much in the way of motivation, productivity, or enhanced responsibility. Contrarily, several incentive systems are on the list. This runs counter to the Maslovian wisdom of the 1970s, the crux of which was that money no longer mattered much to workers. Finally, conspicuously absent from the list are job enrichment, which is merely the replacement of one monolithic job design with another, and traditional profit-sharing and employee stock option plans which, because they typically offer *inauthentic* levels of participation, fail also to tie rights with responsibilities.

While one can cite all the examples on one hand, a few progressive, large corporations are abandoning monolithic approaches to the problems of work and workers and are trying to find ways to provide individual workers with the choice among, and mobility between, a spectrum of different kinds of jobs. There is now some experimentation in firms such as Polaroid with methods to assess the various rewards, demands, and challenges of jobs; to assess the interests, needs, and desires of workers; and to let workers self-select into the jobs that meet their needs. The company facilitates the process by providing training, placement, and other career services. Other innovative approaches in large corporations include Xerox's sabbatical program for workers who wish to spend a year doing public service; and General Motors' half-dozen new accessory plants with self-managing, autonomous work teams. Unfortunately, most large corporations have been willing to alter their philosophy and organization of work only when there is no alternative to doing so. Some, like Youngstown Steel, wait until it is too late. Others, like the example below, change just in the knick of time.

In 1973, the managers of Kaiser Steel decided to permanently close their mill in Fontana, California because they were convinced that the mill could not compete with imported Japanese steel. Faced with the loss of their jobs, the workers at Fontana persuaded Kaiser to keep the plant open by agreeing to change their behavior. The hitherto adversary labor relations at the plant were quickly turned around, and workers started accepting the necessity for such things as limited lay-offs and overtime work. The managers changed their behavior, too. Previously unwilling to listen to the advice of "dumb" workers, they now started listening. For example, they followed the suggestion of the workers to improve the maintenance of machines, a simple act they had long resisted, but one which reduced the rate of rejects by 39% in the first instance it was tried. The managers had never believed the plant could run profitably; but it is still in operation in 1979. A union official identified the secret: "Nobody really knew what the workers could do if they cooperated."[48]

While Kaiser Steel is hardly a paradigm of progressive management, its case is nonetheless instructive, for steel was one of the earliest industries to experience the conditions that all of American industry is likely to experience during the 1980s. The case also shows that it is possible, even in an industry with outmoded technology, to increase productivity through encouraging worker responsibilities. And, significantly, the Fontana workers gave up no rights of any consequence. If something positive can be done about the "worst case," then there is hope for the future.

In summary, there appear to be two alternatives open to American corporations in an era of declining growth. Either: they continue their current philosophy and organization of work, to which workers are likely to respond with demands for more entitlements and less responsibility in which the Northern American economy will slip further and further behind the economies of Northern Europe and Japan; or: they can create the conditions in which it is clearly in the self-interest of workers to tie entitlements to productivity by assuming greater responsibility for the quantity and quality of their work.

The Bottom Line

It might be argued that the new philosophy and organization of work outlined here is merely "another growth model." That is, if work comes to be organized along the principles suggested, it is likely that economic growth will ensue. To this charge, I plead guilty in one respect: I think America needs a certain amount of *appropriate* growth. For example, there is an unqualified need for growth in employment opportunities for poor black men and women. To the extent that my proposals would lead to fuller employment then, yes, the E-Factors would produce some economic growth. But I would suggest that full employment based on these new criteria would be qualitatively different than full employment achieved through policies designed solely to maximize economic efficiency.

At the level of the individual firm, the companies which adopted the micro-criteria might also grow. But why would this be bad, per se? More to the point, the macro-criteria would lead to worker motivation, loyalty, commitment, and responsibility *even when there is no growth* to finance new fringe benefits, new entitlements, or higher salaries. The philosophy and organization of work proposed in these pages is appropriate to an era of slow growth or no growth because it offers meaningful rewards to workers in place of bribing them with meaningless and expensive fringe benefits that increase the overall cost of labor without increasing either productivity or the quality of working life.

It would seem that using the macro E-Factors and the micro–performance criteria as guides to the future would help to produce sustainable growth in America. For the new philosophy of work would stress the use of America's greatest asset: the skills, intelligence, education, abilities, and ingenuity of the work force. Growth in the use of these resources is, appropriately, nonpolluting, energy-conserving, and ecologically sound. Fully tapping these resources might even give the United States a new, competitive advantage in the world economy in the long run. Then we would be able to sustain our standard of living without sacrificing the quality of our lives. We would be able to increase equality without sacrificing liberty. Indeed, the closer we came to satisfying the pluralistic E-Factors, the closer we would come to achieving the just society that has eluded the single-minded economists who have dominated Western thought for too long.

CODA: EDUCATION FOR THE FUTURE OF WORK

The prevailing philosophy and organization of work is buttressed by a complementary philosophy and organization of education. Two conspicuous characteristics of this educational system are *specialization* (the departmentalization of learning) and *vocationalism* (including career education in the schools and undergraduate professional training in the universities).

The workers produced by American high schools and junior colleges are trained to perform unquestioningly a single task; but too many of them are unable to read, write, compute, and think analytically, critically, or ethically. Sadly for them, the routine, grunt work they are trained to do will all be done by machines in the future. This means that "people work" will mostly be nonroutine and characterized by participative self-management. What America will then require is workers who are humane individuals, with analytical and entrepreneurial skills, who know how to work in groups, and who know how to solve problems. Unhappily, vocationally-trained workers are manifestly unprepared to design their own work, to pacticipate in decision making, to assume control over their own working conditions, to work as members of a community of equals, or to take responsibility for the quantity or quality of their own work. In fact, these vocationally–trained workers are easily threatened by change. They too often act defensively, inflexibly, and in ways society deems *irresponsible* when circumstances force them to adapt (as the panicked behavior of the workers at Three Mile Island illustrates).

And the managers trained by the higher educational components of this system cannot cope with the competitive crises that America faces because they have been narrowly and ill-liberally trained. When the world changes, the specialist—whether an engineer or an accountant—has a difficult time adapting to the change. Such individuals are, like vocationally–trained workers, threatened by change—which is exactly the wrong mentality when what is demanded by the moment is innovation and initiative. Moreover, threatened managers often behave in ways that society deems irresponsible. Too often it was narrowly trained engineers, scientists, and people with undergraduate degrees in business who panicked at places like Lockheed and paid bribes to foreign officials, who panicked at places like Allied Chemical and dumped Kepone into the James River, and who panicked at places like General Motors and spied on Ralph Nader when he was challenging the safety of the Corvair. Of course, there were also liberally–educated people involved in these and other instances of corporate irresponsibility. The point is not that liberally–educated people are more moral or ethical than narrowly–trained specialists. Rather, corporate executives are now finding that over 50 percent of their time is devoted to social issues—complex, ambiguous, sensitive, human issues that technicians are less prepared than generalists to handle.

It is clear that, in the future, American industry will face a challenging set of emerging issues, including a growing estrangement between corporate behavior and the needs and interests of society, new competitive challenges from abroad, and the need to restore the corporation as the nation's source of technological innovation and organizational creativity. For higher education to produce managers and professionals who could cope adequately with such problems, it would have to reverse the trends of which it is most proud: namely, departmentalization, specialization, and undergraduate professionalism.

Somehow, amidst all the trendy brouhaha about making education relevant for work, the message has not gotten through to educators that the problems people face at work are complex, interdependent, and ignorant of the departmental structures of universities. Most of the really tough problems people face at work are not technical—the computer can be made to solve these. Indeed, the tough problems are not really problems at all (if a problem is defined as having a single solution). There are *no* "solutions" to the tough policy and organizational problems of work—there is only a spectrum of alternative responses, some more appropriate than others, but few that are simply either right or wrong. Significantly, it is problems of this nature that have the deepest precedents in the history of social affairs. It is such problems that the broadly-educated, truly encultured person is best equipped to handle on the job.

In conclusion, it doesn't seem too far-fetched to suggest that it was the narrowly-trained specialists who got the world into the growth fix to start with. It was myopic, compartmentalized thinking that failed to recognize that all parts of the world system—economic, political, technological, and natural—are *complexly interrelated*. Not only does this suggest that to cope with declining growth, change must occur on many fronts simultaneously—in education, work, and political institutions—but that liberally-educated people who understand the interrelatedness of all things are the best equipped to lead those institutions.

NOTES

1. Jean François Revel, *Without Marx or Jesus: The New American Revolution Has Begun* (Garden City, N.Y.: Doubleday, 1971).

2. Herman Kahn, *World Economic Development* (Boulder, Colo.: Westview Press, 1979).

3. E. F. Schumacher, *Small is Beautiful: Economics as if People Mattered* (New York: Harper and Row, 1973).

4. I should like to acknowledge the contributions of August Ralston and Kenneth Brousseau to a study of the future of worker entitlements that was undertaken at the USC Center for Futures Research in parallel with this paper.

5. Quoted on p. 2 of Robert Heilbroner, *The Worldly Philosophers* (New York: Simon and Schuster, 1961).

6. O.E.C.D., Figures quoted in *International Economic Indicators*, U.S. Department of Commerce, Vol. IV, No. 4, December 1978.

7. Arthur Okun, *Equality and Efficiency: The Big Tradeoff* (Washington, D.C.: The Brookings Institution, 1975).

8. Ibid., pp. 18-19.

9. *Statistical Abstract of The United States*, Washington, D.C., U.S. Department of Commerce, 1978.

10. "Employee Benefits 1977," Washington, D.C., Chamber of Commerce of The United States, 1979.

11. Ibid.

12. Ibid.

13. Burnham P. Beckwith, "Free! Free! Free! The Priceless World of Tomorrow," *The Futurist*, October 1978, pp. 307-12.

14. Study undertaken by Theodore Geiger, Director of International Studies, National Planning Association, Washington, D.C., 1978.

15. Okun, *Equality and Efficiency*, p. 6.

16. Ibid., p. 7.

17. Ibid., p. 8.

18. See James O'Toole, "The Irresponsible Society," in *Working in the 21st Century* (New York: Wiley Interscience, 1980).

19. Geiger study.

20. Okun, *Equality and Efficiency*, pp. 91-95.

21. Daniel Yankelovich, "New Approaches to Worker Productivity," Address before National Conference on Human Resource Systems, Dallas, Texas, October 25, 1978.

22. *Quality of Employment Survey*, Survey Research Center, Institute of Social Research, University of Michigan, Ann Arbor, 1977.

23. Fred Best, "Exchanging Earnings for Leisure," Washington, D.C., National Commission for Employment Policy, January 15, 1979.

24. David W. Ewing, "What Business Thinks About Employee Rights," *Harvard Business Review*, September-October 1977, pp. 81-96.

25. Aaron Wildavsky, "The Guiding Force in America Today? Chicken Little," *Los Angeles Times*, Sunday, March 4, 1979, Business Outlook Section, pp. 3, 9.

26. "Labor Letter," Report on Research by Purdue University's John Barron and Labor Department economist Wesley Mellow, *Wall Street Journal*, January 16, 1979, p. 1.

27. Christopher Lasch, *The Culture of Narcissism* (New York: W. W. Norton, 1978).

28. Peter Steinhart, "Children: Victims in the Culture of Narcissism," *Los Angeles Times*, Sunday, May 20, 1979, Opinion Section p. 1.

29. Ibid.

30. E. F. Schumacher, *Good Work* (New York: Harper and Row, 1979), p. 28.

31. *Quality of Employment Survey*.

32. Okun, *Equality and Efficiency*.

33. Schumacher, *Small is Beautiful*.

34. John Kenneth Galbraith, *The New Industrial State* (New York: Mentor Books, 1971).

35. James O'Toole, "What's Ahead for the Business-Government Relationship," *Harvard Business Review*, March-April 1979.

36. Ibid.

37. For an explanation see James O'Toole, *Work, Learning and The American Future* (San Francisco, Jossey-Bass, 1977), p. 211.

38. Ibid., pp. 86-88.

39. Idea presented at a Workshop on Work, Education and Leisure, Preconference Woodlands Workshop, 1979 at the University of Houston, from Berg's current research on "structural unemployment."

40. See James O'Toole, Making America Work: *Productivity and Responsibility* (New York: Continuum, 1981).

41. Robert Zager, "Managing Guaranteed Employment," *Harvard Business Review*, May-June 1978, pp. 103-15.

42. David W. Ewing, "Participative Management at Work," *Harvard Business Review*, January-February 1977, pp. 117-27.

43. A Scanlon plan is a contractual agreement between a union and management to share with workers a bargained percentage of increases in profits that result from worker suggestions or increases in productivity that result from their unusual efforts.

44. Nick Kotz, "Worker-Community Group Sees Dream of Saving Steel Mill Fade," *Los Angeles Times*, July 9, 1979, Part III, p. 14.

45. James O'Toole, "Thank God, It's Monday," *The Wilson Quarterly*, Winter 1980.

46. "Firm Finds Pay Scheme That Works: Employees Vote on Each Other's Raises," *Los Angeles Times*, January 10, 1979, Part III, p. 1a; and Edward Lawler, "Workers Can Set Their Own Wages—Responsibly," *Psychology Today*, February 1977, pp. 109-12.

47. James O'Toole, "The Uneven Record of Employee Ownership," *Harvard Business Review*, November-December 1979, pp. 185-97.

48. Daniel Zwerdling, "Workers Seize Plant!" *Working Papers*, Fall 1974, p. 17.

Part Three

GROWTH POLICY IN GLOBAL PERSPECTIVE

Chapter 8

THE INTERACTION OF FOOD, CLIMATE, AND POPULATION

Walter Orr Roberts and Lloyd E. Slater

INTRODUCTION

We were asked in the Third Woodlands Conference on Growth Policy to seek creative solutions to problems that imperil our planet, to generate new perceptions of how to attain sustainable growth within the finite resources of our good Earth, and to visualize how human needs can be met in every corner of the globe. But we will first approach the state of man from a cosmic perspective.

Harlow Shapley wrote a book in the 1930's called *Flights from Chaos*. "It is fortunate," he wrote, "that we are customarily unaware of our twisting, slipping, whirling, flying motion through space; we might otherwise lack the courage to explore and analyze the surrounding world."[1]

We earthlings are fortunate, he wrote, that we do not directly sense the spin of our globe on its axis, propelling Houston ever eastward at almost 800 miles per hour. Fortunate that we fail to feel our planet's monthly oscillation toward and away from the Sun as we pivot, at 30 miles per hour, about the center of gravity of the Earth and Moon system. Lucky that we neither sense Earth's 600 million mile circuit of the Sun at 20 miles a second, nor the synchronized tilting of its rotational axis. Favored that we have no sensation of our 150 mile per second headlong plunge, Earth and Sun together, toward the constellation Cygnus, which we in the Northern Hemisphere see high overhead on late summer evenings. We are even spared everyday consciousness of participating in the vast inexorable cosmic explosion: that most fascinating mystery of our expanding universe, its hundred billion sun-like stars, and its speed-of-light borders.

We are blissfully unaware of these gyrations. Had we been hourly forced to be conscious of the plunges and twists of our planet in its headlong flight through the cosmos, we might never have had the serenity to compose the music of Palestrina, to write the plays of Sophocles, to paint as Rembrandt or Picasso. Perhaps if we had been conscious, from moment to moment, of our breathtaking speed in space we would have been distracted from mathematics, science, and engineering, or even from the possibilities of applying the products of human imagination to achieving peace, justice, and the equitable access of all humanity to the richness of our earthly abode.

Human occupancy of planet Earth has been but a mere moment in the life of the planet itself—perhaps a million years or so of its many billions. Yet in this moment, due to evolution of high order intelligence and adaptability, *Homo sapiens* not only dominate the planet but expand their numbers and extend their exploitation of earthly resources at a cataclysmically accelerating pace.

Human perspectives have also altered drastically. Our world view changed forever with that first human footprint on the Moon and those first close-up views of our unimaginable sister planets, Venus, Mars, and Jupiter. These events shocked us into an awareness of the speed and fundamental nature of the revolutionary changes that characterize these latter years of the twentieth century.

Not only in space research, but in every domain of science and technology we see ever more rapid advances in the process of control over nature. We are characters in the now-expanding chronicle of man's governance of Earth. The challenge we face is to be not passive but active characters in this drama, and by our efforts to bring rationality, wisdom, and compassion to the governance of human affairs within the plentiful but finite resources of our biosphere. Through mindful use of our goods, human and material, we can, indeed, bring about the growth of limits to the quality of life and to the numbers sharing in this enlargement.

THE PROSPECT WE FACE—A PESSIMISTIC VIEW

Many analysts of today's world visualize a gloomy future prospect. They see us beset by a world population growth rate that leads toward a doubling within the next 50 or 60 years and, ultimately, a possible tripling of today's numbers before nature or man exerts control.

Coupled with this soaring growth in people is rising energy demand as the poor nations seek a share in world energy usage at, they deem only just, a favorable low price. This is in the face of dwindling cheap oil or gas and mounting environmental ravages from nonrenewable coal extraction to meet mankind's rocketing demands.

We have chosen the years 2030 to 2040 as an arbitrary reference time because it seems that by then not only will the world's population be doubled, but its food demands will be tripled and its energy appetite quadrupled. Moreover, by that date the atmospheric burden of carbon dioxide will be doubled as a result of consumed fossil fuel and reduced world forestation. Most climate experts anticipate from this a climate warming without precedent since the dawn of human history. Furthermore, within fifty to sixty years it is expected that little if any new land will remain to be opened for agriculture, and that much of today's more marginal land will have been ravaged by salinization and loss of topsoil due to overcultivation, inadequate and inefficient irrigation water, and high fertilizer costs.

Let us now briefly consider some of today's prevailing outlooks for a year 2040 scenario:

Climate variations and bad weather extremes will continue to hurt agricultural food production. Figure 8.1 shows how grain production varied in the three major growing nations in 1960-77. Together the United States, the USSR, and China today produce almost half of the world's yearly harvest of about 1,300 million metric tons of grain.

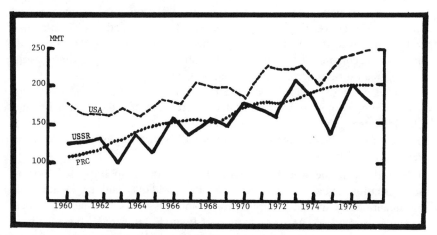

Fig. 8.1 Grain Production 1960-1977
Source: Adapted from Schnittker Associates, 1978.

Weather had a big impact in the Soviet Union where shortfalls of 30 to 50 million metric tons happened during recent bad years. Notice especially the dip in 1975. United States production fluctuated as much due to changing policies (land withheld from production under federal cash incentives) as to weather, although the big crop loss in 1974 was clearly weather-related. China's productive stability is generally credited to good weather, intensive irrigation, and intensive care in cultivation, though some experts suggest the figures may have been smoothed out before they were released. In any case, variations of only a few percent in the yearly harvests of any one of "the big three" will profoundly affect the availability and price of the grain needed by the import nations. In 1972-73, for example, a downturn of only about 2 percent was followed by large reductions in world reserves and a tripling of prices.

It is a widely held capitalist-nation perception that American or Western agricultural practices yield vastly higher production rates than those of the Soviet

Union. We often say that this comes about because of the alleged low incentives and the bureaucratic inefficiencies of communism. It is the view, however, of a number of our own Western experts who have traveled extensively in Soviet agricultural lands that the demeaning yields are mostly the result of a harsh and unfriendly climate. Russia, if it had the relatively benign climate of America's vast Midwest breadbasket, could probably flood the world with export grain, even under communist management. Moreover, the People's Republic of China, despite limited mechanization, is a food-producing marvel, furnishing modest but adequate and steadily improving nutrition to nearly four times as many people as the United States on about half the amount of cultivated land.

But to return to the main theme, history tells us that weather and climate fluctuations can bring on, directly or indirectly, serious foodcrop shortfalls, regional famines, and untold human suffering. Millions died in the grim drought of 1899 in India. In modern times, we have seen the food disasters in Ethiopia and the Sahel region of Africa as overstress of land, economic exploitation, and drought conspired to bring vast human misery. Even in an advanced nation like the United States, the severe drought of the 1930s in the Great Plains coupled with a deep economic recession resulted in abandoned farms, blowing topsoil, and the greatest out-migration in our nation's history.

There is every expectation such disasters will repeat and, *if* human societies continue to grow more and more unstable, future weather and climate anomalies will undoubtedly wreak as great or greater future human havoc. The big question, however, for us is: Must these disasters occur? Is there any reasonable way to avert or control them?

Most nations of the world will depend increasingly on imported food to meet nutritional goals. In recent decades, the world distributive system moved vast quantities of available surplus food—mainly grain—to ameliorate local shortfalls and famines. At the same time, as figure 8.2 illustrates, there has been a dramatic change in world grain trade since World War II, with most former exporters relying now on imports mainly from the North American breadbasket. This trend away from food self-sufficiency is expected to intensify, with the poorer nations excluded from necessary and emergency food imports by the disappearance of surplus grain as prices rise and supplies dwindle. And the disturbing question might well be asked: Will there still be important emergency food sources available in the world in the next century?

Debilitating malnutrition will engulf more and more people and further widen the gap between affluent and impoverished. While localized climate-induced famines are sharp reminders that we face a world food crisis, the chronic malnutrition of the poor of the developing nations is a less dramatic yet more fundamental expression of the problem. Simply producing more food is clearly not the answer. During the past decade, while food production figures everywhere moved steadily upward, the amount of calories and protein reaching poor people in developing countries steadily declined. Figure 8.3 reveals the situation today, where a large fraction of the world's people—an estimated half of the

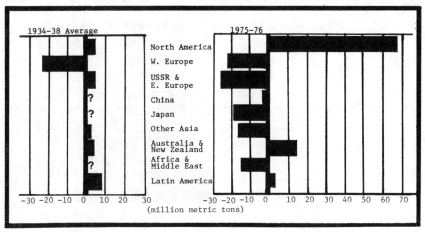

Fig. 8.2   Changes in World Net Grain Trade 1934-1976
Source: Adapted from *Scientific American*, September 1976.

population in most Third World countries—get less than the average minimum 2,700 calories and 75 grams of protein deemed necessary for vigorous human growth, mental development, and continuing body strength. Some of the more obvious side-effects of this tragic situation give us serious worry about any future world scenario. One is increasing migration of rural poor, misguidedly to seek urban "security" through more readily available food or increased income to buy food. Usually neither happens. Less obvious and infinitely more complex is the relationship between poverty, malnutrition, and total population growth. It raises perplexing questions such as: Is rapid population growth an effect rather than a cause of malnutrition? Or, conversely: Will better food supplies and distribution furnish the incentive or otherwise help encourage family planning?

Our list of trends and conditioners for a Year 2040 World Food Scenario could obviously be a long one. We have chosen only three—climate, food productivity, and the malnutrition issue—because they are so fundamental to the theme of this chapter. Other equally powerful shapers of the scenario should have high priority in our thinking. One is our energy habits. Another is the increasing tendency of all nations to invest vast resources, talent, and capital in military systems. Still another is our tendency, so to speak, to foul the human nest by polluting the air, the lakes and rivers, and the oceans of the world.

WHERE WILL IT ALL LEAD?—THE 2040 SCENARIO

Where will all these ongoing trends and conditioners—these widely held perceptions of hungry mankind constrained by ravaged environment, limited resources, and the innate foibles of human nature—take us? What picture should we paint of the climate/food/population interaction in the Year 2040?

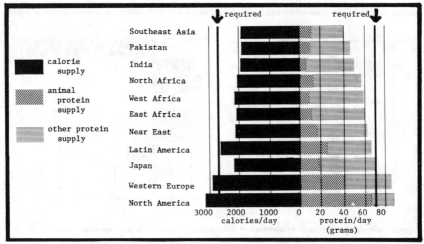

Fig. 8.3 Protein and Calorie Intake
Source: FAO, 1970.

The pessimistic view that many hold paints a grim picture of the environment that world strife for food will produce. It is an appalling vision of degradation and depleted resources. Oil and gas will be exhausted; nuclear power too dangerous to use; energy from coal either highly expensive, environmentally destructive, or both. Vast amounts of woodlands removed for food production and fuel will, in this ominous scenario, encourage the washing and blowing away of topsoil and a general worldwide decline in farm productivity as clean, sweet water becomes too expensive for agricultural use. Stressed use of marginal arid lands will add untold millions of hectares to the world's deserts. World climate, perturbed by man's effects, will be less favorable for increasingly necessary food production. And man's air pollution and waste carbon dioxide, along with widespread forest degradation, will begin to alter world climate in a way that impairs agriculture. All of this will contribute to a faltering world food system.

What about people in this pessimistic scenario? Will war be a greater threat, a grimmer reality? Will we face food riots and major conflicts for access to the sources of food production? Will the population growth still be rampant? How many of those 8 billion or more living in 2040 A.D. will have adequate nutrition, economic and political freedom, comfortable shelter, compassionate health care, full educational opportunity? How many will remain as today, surviving pitifully, desperately malnourished, and dying of the many diseases that breed on hunger, devoid of joy for the present or hope for the future?

If 2040 brings but an extrapolation and heightening of the problems of 1980, what can we think of ourselves? Is this the world for creatures that can today walk on the moon, live for months in deep space, send their artifacts to other planets, and transmit messages among the stars? Is this the world for men and women who will, in 2040, harness the atom and the Sun, carry vast electronic libraries in their pockets, write incredible new symphonies, perhaps even converse regularly with intelligent cosmic neighbors?

CHALLENGING THE FUTURE

Just how valid is this fear of a hungry, strife-filled, resource poor, people-polluted world? This is a hard matter to judge, for it depends strongly on the character of human nature. It relies on the strength of common resolve among the world's diverse people to cooperate in preserving values that have created a possibly unique experiment of intelligent life on an astronomical body. Some, like Robert Heilbroner, take a gloomy view of man's ability to overcome his present myopic, selfish, greedy, and narrow nature. Others are far more sanguine. We see the goal of a far different and happier world as attainable, and the challenge as one that both demands and merits the best that is in the nature of humankind. We shall elaborate on this in the pages to follow.

What the future brings in 50 to 60 years will probably be least of all what one derives from simple linear or quadratic extrapolation. Rather, it will be the result of forces strongly affected by human intervention, brilliant or mindless, compassionate or ruthless. It will be the work of leaders and of followers, good or bad, strong or weak. It will gain its impact from the still exponentially growing control of nature that comes, whether we like it or not, mostly from the products of science and technology wielded by the hands of men and women in power, be they of noble purpose or of evil design or of bumbling incompetence.

If there is an imperative for today, however, it is that the new conditions of this planet—with all of us drawn closely together, and every day still more closely—demand active participation in the drama of history that we are writing. We must bring rationality, wisdom, and compassion to the governance of human affairs; to the husbanding of our planets' goods, human and material.

The question is what is rationale, what wisdom, what compassion? With our world today segmented into many value systems, the choices are wide and varied. One's outlook is conditioned, strongly or subtly, by the society where one resides as well as the culture and knowledge one has absorbed. The dynamic competing forces of communism, socialism, and capitalism all pose alternatives in human thought and response. Yet people in all societies, and especially those most prone to be active characters in the drama of history, usually share the same pride in the accomplishments of science, the same joy in the beauties of art and music, the same gratifications for work well done, and the same desire for the pleasure of leisure occupations. Our hope, given these shared joys and hopes of humanity, is an eventual harmonious coexistence of the diverse perceptions that define our societies and shape our futures where differences between ideologies will resolve in a stimulating contest in the world of ideas and good works rather than in war or use of force to validate human preference or loyalty.

Let us turn again to some of the problems and limits perceived in such dire fashion by those more pessimistic than we about future human prospects, for they do indeed describe our obstacles. Briefly they suggest that

- climatic variability, impacting on food growing, is an incontestable adverse force;

- the limits of arable land are just about here, those for quality fresh water will soon follow;
- cheap energy is running out, hence high productivity, mechanized farming is doomed;
- producing food by alternative technologies will never successfully compete with conventional agriculture;
- livestock, because of their inefficient conversion of feed grains, will eventually disappear in the world food system;
- farming in the tropics, where most food problems are, is inherently marginal, hence food self-sufficiency in many tropical nations is an unattainable goal.

These issues and many more need careful analysis. For each there is a constructive counterprospect. For each an answer can be given that brings hope for a better life in 2040 A.D. But the costs of some of the answers will be high, and the social innovations required for others may strain our style and values.

A recently completed major work of the International Federation of Institutes for Advanced Study (IFIAS), co-sponsored by our Aspen Institute program, illustrates our view. For the past three years, working with a worldwide team of investigators, Dr. Rolando Garcia, a distinguished Argentinian meteorologist and scholar, assessed the many facets behind the so-called world food crisis that climaxed in 1972 with a series of severe droughts in the Soviet Union, African Sahel, India, and elsewhere. A telling commentary on the findings of this ambitious project is the title of its book: *Nature Pleads Not Guilty*.[2]

The Garcia study strongly challenges many conventional perceptions, particularly those in our Western society, on the forces that shape any future world food scenario. In essence, the study proposes that the 1973 depletion in world grain resources, the subsequent escalation in grain and food prices, and increasing hunger and malnutrition among the poor were inevitable sooner or later, with or without the bad weather that surely triggered them off. They are, says the study, much more a result of misdirected human policies and undampened human greed than the product of natural calamities. Bad weather merely hastened the process of collapse.

UNDERSTANDING THE SYSTEM—A DYNAMIC VIEWPOINT

As the work by Rolando Garcia and his team demonstrates, enlightenment in our complex world rarely comes through simple analysis of obvious cause and effect relationships. Most problems of society result from intricate, subtle, and often hidden interactions. Professor Mesarovic, in a study for the Club of Rome,[3] puts it well when he describes societal problems as almost always multilevel, multidisciplinary, multidimensional; and, we might add, multimindboggling in their complexity. Above all, today's societal problems represent *systems* that are dynamic in behavior and, thanks to man's constant intervention, always in a state of change.

So if we are to form and assess new perceptions on climate, food, water, land use, energy, or include any of the problems of the human prospect, it is essential to understand how all these elements integrate or interact as a "system." The need for this holistic approach is a compelling one. True, it is easy to give lip service to the "systems approach," but it is fiendishly difficult to carry it out well; and, when one finishes up, the results are usually controversial. We suspect that some of today's misleading pessimism, particularly in our Judeo-Christian society, is rooted in a righteous and simplistic cause-and-effect viewpoint or ethic; there are good guys and bad guys, and one wins and one loses.

Behavior and achievement in a complex problem area, such as that embracing climate, food, and population, defy prediction through such simple linear analysis. Rather, the many elements involved interact to form a dynamic feedback system where failures and successes in any one component, as well as in the system itself, are greatly influenced by the performance of all other components.

When a person employs simple cause-and-effect or intuitive analysis as the basis for actions to ameliorate a complex problem, the results, as Jay Forrester so aptly puts it, are often "counterproductive."[4] A classic example is what happened in the African Sahel after an extended plentiful rainfall period when well-intentioned funds from international aid organizations were poured into drilling deep tubewells to water the flourishing livestock herds. This caused formerly nomadic herdmen to settle around the new water sources. A subsequent concentrated overgrazing and trampling of vegetation paved the way for severe soil erosion and desert when the dry climate returned. The net result was a series of desolate wells surrounded by desert scattered with the carcasses of dead animals.

Careful and thorough systems analysis reveals how components and guiding forces within a complex problem area interact. Quite often, the analysis uncovers the phenomenon of synergism, where cooperative action between system elements leads to better results than the sum of each taken independently. It is through this understanding of system behavior and the potential for improving system performance that appropriate and useful intervention, or feedback, can be identified and applied. The result can be a revelation of strategies and policies in human governance that reflect a "growth of limits" or a dynamic for sustainable growth for the system and, we might add, hope for the future.

Figure 8.4 is a schematic and highly simplified view of the climate/food/population interaction as a dynamic system, displaying the most important elements and feedback paths that work together to furnish food to a nation. What the diagram underscores, more vividly than reams of words, is the futility of simple cause-and-effect or linear analysis. It also reveals, most powerfully, the central influence of government and its policies in the system—the potentially useful or destructive feedback force we alluded to earlier. Climate and weather are as yet a generally uncontrolled input, but fortunately they are of restricted variation in amplitude. Unless we face unprecedented weather and climate disasters, it is how capital and resources, both human and material, are allocated that will determine the success of food production and distribution in most nations.

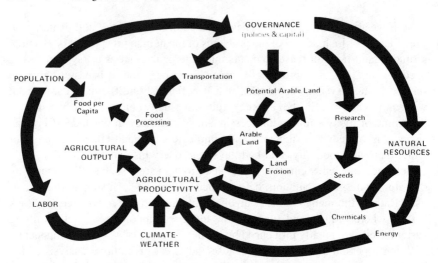

Fig. 8.4. Dynamics in a nation's food production system.

FORMING NEW PERCEPTIONS—OR, RESHAPING THE OLD

While the systems approach is a cogent way to expose outmoded perceptions, it also stimulates alternatives for their replacement. Old ideas and viewpoints can be weighed against promising new technologies and alternate policies. Using computer-held models of the system, such as the world food system model prepared for the Club of Rome by Mesarovic and Pestel,[5] these alternatives can be previewed in scenarios of the future before their widespread use in the real world by enlightened governments.

A recent conference held by the Aspen Institute Program on Food, Climate and the World's Future, to consider the effects of a hypothetical global warming due to an increased carbon dioxide burden in the atmosphere, illustrates application of the global modeling and scenario technique. The methodology suggested by conference participants entails a variety of possible future climate scenarios, some based on unusually warm periods in the past. These scenarios will be used to preview the economic, social, and political consequences that are liable to result from a global warming. While not construed as a forecasting tool, the results may help identify vulnerable regions on the Earth and suggest adaptive political and technological changes.

In the remainder of this chapter we will question many currently held viewpoints, particularly those half-dozen referred to earlier that suggest a hopelessly underfed world 50 years from now, by posing some promising possible alternatives.

The Climate/Environment Impact

Must we consider continuing bad weather and deteriorating environment as inevitable shapers and limiters in our future world system? Certainly not, for there are far too many interesting developments under way right now that could diminish these influences.

The prospects for anticipating and reacting defensively against bad weather in food production are most encouraging. While forecasts which can be furnished to farmers are, at present, considered reliable only up to five days, there is concerted world effort to extend this to a useful "reactive" period, which might be one month or more. It's not certain how far this will succeed. Much hope rests on new understanding of how weather systems work. A good deal of this is expected to materialize from the massive amount of information being generated by the Global Atmospheric Research Program (GARP). GARP, with 5,000 specialists and 72 nations participating, started collecting data on weather and its influencing environment using ships, aircraft, balloons, satellites, and instrument platforms in 1974. In 1979, an intensive GARP effort, assessing weather readings from all over the world for a 12 month period, furnished forecast modelers an opportunity to compare what was predicted with what actually happened. This will be especially useful to general circulation model builders like Leith and his global weather forecasting group at the National Center for Atmospheric Research (NCAR), Smagorinsky and colleagues at the Geophysical Fluids Dynamics Laboratory of NOAA at Princeton, and to the team working with Gates at Oregon State University.

But even if a forecasting breakthrough is achieved at one month or one season lead time, we face a real problem in getting localized bad weather predictions out to farmers in a form that will permit real agricultural strategy decisions to be beneficial. Some help may be on the way here in a growing field of technology called crop/weather modeling. One example, developed by Control Data Corporation is shown in figure 8.5. By adding agronomic, crop physiology, and

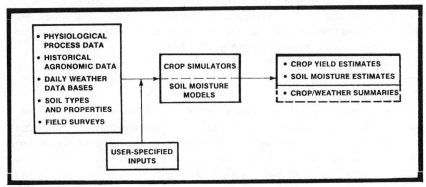

Fig. 8.5 "AGSERV" - Crop/Climate Model
Source: Control Data Corporation, 1978.

historical data to real-time, incoming weather information, it becomes possible to forecast an actual shortfall or crop failure and carry out crop protective measures or strategic replanting. For the past few years, this system has been experimentally monitoring the complete soybean and corn crops of Midwest America. Control Data is also collaborating with the Aspen Institute Food and Climate Forum in a project in Venezuela to adapt this technology to the needs of a developing country and its agricultural and food distribution system.

Many institutions and individuals are presently at work perfecting crop/weather models. A group at Purdue University, for example, has concentrated on developing quantitative relationships between weather, soils, and crop growth. They have come up with an Energy-Crop growth index which follows weather effects on a day-to-day basis.[6] Another major activity is the LACIE, or Large Area Crop Inventory Experiment, which was launched under the sponsorship of NASA, USDA, and NOAA in 1974. The LACIE effort developed an experimental system to monitor and estimate yield projections of wheat production throughout the world using remote sensing imagery from space, ground station weather, and other environmental data. It is anticipated that LACIE technology will be adapted to inventory production of other food and fiber crops, including corn, rice, soybeans, and forestry products, as well as to monitor the world's important rangelands. This increased capability could conceivably be developed and completed in the mid-to-late 1980s.

Defensive measures by farmers against predicted or predictable bad weather or unfavorable climate are not limited to emergency actions such as early harvesting, replanting, or massive irrigation. They also include agronomic improvements which essentially reduce the vulnerability of their crops and fields to nature's vicissitudes. Some are newly revived, ancient water conserving techniques that were swept away by the high-efficiency monoculture farming now in practice. They include windbreaks, strip-cropping, subsoil irrigation, intercropping, crop rotation, evaporation control, and mulching. Most of these were arrived at over the centuries to conserve water, prevent erosion, and increase the possibility of a successful crop despite unexpected weather calamities. Modern agricultural science is also adding to the climate-defensive arsenal by studying the relationship between crops and their environment to find ways, via cultivation, fertilization, or genetics, to reduce water needs and improve photosynthetic efficiency, or to stabilize water supplies to crops through improved irrigation practices.

An even more hopeful expectation that more food can be grown in the world despite climatic limitations is suggested in figure 8.6, an analysis of the climate regimes and their food-growing possibilities in the developing countries of Africa and Asia. Far too often, what is grown and how the land is used does not match well the local climate conditions or local food needs. In predominantly dry Mali, West Africa, for example, cotton and rice (which have high water requirements) are grown extensively for export. Yet, during the severe Sahelian drought of 1968-71 when cotton, rice, and groundnut production for export reached record highs, corn production for local consumption fell by one-third and many starved.

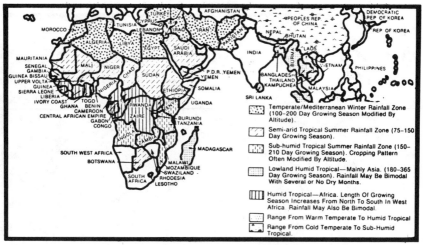

Fig. 8.6 Climate Regimes in Africa and Asia
Source: International Food Policy Research Institute, 1975.

This pattern, which prevails through much of the tropical developing world, could be changed through enlightened government policies and incentives to vastly improve the stability and availability of food supplies for local, at present malnourished, people. Some quite responsible experts, like Peter Buringh and colleagues at the Agricultural University in Wageningen, the Netherlands, estimate that world agricultural production could be doubled and doubled again with no increase—indeed with a decrease—in the land under cultivation.[7]

The Limits of Resources

Land, fresh water, and energy are the primary resources required for agriculture, and the widely held view is they are fast reaching their limits. Can this perception be challenged?

If one looks at the hard numbers in figure 8.7, a map showing present cultivated area as a percentage of all potentially arable land, the situation does not look too hopeful. Only Africa and South America seem to offer some hope for significantly increased food production through expanded acreage, but both have serious regional problems which make that promise difficult to achieve. Some 700 million hectares of sub-Saharan Africa are closed to cultivation and foraging because of the tsetse fly and onchocerciasis, the river-blindness disease. Heavy tropical rains and leached, generally low fertility, soil hinder further agricultural development of the vast Amazon region.

The issue, however, is agricultural production and not how many more acres can be squeezed from the Earth for farming. Only a small fraction of the world's arable land now in use is being farmed productively, if crop yields in Japan, China, North America, and Western Europe are paragons of efficiency. India, with its 623 million people greatly dependent on locally grown food and one-third less arable land than the United States, offers a good example of the

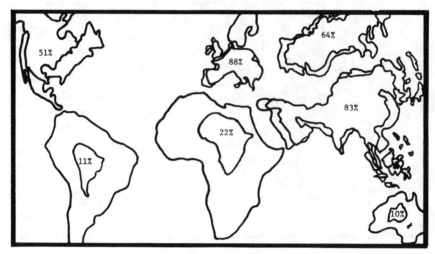

Fig. 8.7. Present Cultivated Area as percentage of Total Arable Land
Source: Adapted from *Scientific American*, 1976.

possibilities. In 1950, food grain yields in India averaged less than one ton per hectare, about one-fourth of the equivalent crop yield in the United States. Concerted national effort to increase productivity has raised yield to a present average of almost two tons per hectare. Now the fourth largest grain producer in the world, India could become the leader were it to obtain maximum crop yields and fulfill its tropical potential for growing two grain crops per year on the same land.

India's food production success story, as yet only partly fulfilled, could be duplicated in many other developing countries where land is being used at marginal efficiency. Nigeria, most populous country in Africa, is a good example. About a third of its presently cultivated land is in cereals, with an average yield of 0.6 tons per hectare. Responding to a drive to improve productivity, Nigeria could easily eliminate its present need to import about a half million metric tons of cereal each year and become a significant net exporter.

Before leaving the limits of arable land, why not consider briefly the limits of unarable land? It has been estimated that only about 3.2 billion hectares of the total 13 billion on the Earth's ice-free land surface can be cultivated. Now only one-tenth of this area grows crops. Desert soils, considered unarable because of insufficient rain or nearby irrigation water to support plant growth, amount to about 1.7 billion hectares. Imagine if water could be brought or rain induced to fall onto these enormous tracts of desert wasteland. There is potential for actually doubling today's farming area.

The prospects for irrigating the deserts and otherwise using them productively keep improving. Those desert areas near the sea are being made to bloom through exciting work on saline irrigation and salt-tolerant species by the Israelis and others. Vast untapped sources of underground water are known to be available, many below the deserts; only an estimated 1 percent of the world aquifer has been drawn up thus far for irrigation. With almost uninterrupted, intense sunlight and cheap land, the deserts are also highly eligible for greenhouse and hydroponic production of food crops. Fast developing technology in controlled environment and soilless farming, producing yields many times that of conventional agriculture, are rapidly improving cost-effectiveness in this essentially climate-independent form of food production.

Some futurists now claim that water, not land, could become the principal restraint on world food production. Although water covers 70 percent of the Earth's surface, only 1 percent is the fresh water required for conventional farm irrigation and, of that, 99 percent is underground. Nevertheless, large quantities of easily available fresh surface water remain untapped; over two-thirds still flow unused to the oceans.

Because of lack of rainfall, about two-thirds of the Earth's land would require irrigation for successful crop production. Yet, only 12 percent of all land now being farmed, somewhat over 200 million hectares, is irrigated. The mounting costs and energy requirements of new irrigation schemes, not the availability of fresh water, have been the prohibiting factors. As the world food crunch intensifies, vast new acreage under irrigation will have to take the highest priority in national planning and development.

An impressive example of a vast scheme to better utilize fresh water resources is being planned in the USSR. The Soviets hope to divert one-tenth of the northward flowing River Ob—about 25 cubic km. per year of water—southward to Kazakhstan. Later they plan even more. This giant source of diverted water will be used to irrigate the new breadbasket of the Soviet Union.

Enormous investments, however, are not necessarily required to extend the limits of water for agriculture. Estimates have less than 50 percent of all water now deployed in irrigation systems being used efficiently. Mismanagement and poor technology often furnish crops with much more water than they need for successful growth. Techniques to correct this are well known and often require only modest capital outlay.

Figure 8.8, showing the number of irrigated hectares as well as percentages of cultivated land irrigated in eight nations, offers a glimpse of the possibilities. What it does not show, however, is the relative efficiency in managing irrigation. China, with only 11 percent of its total land area able to be cultivated, manages to meet almost all food needs of its 900 million people through highly efficient irrigation over about two-thirds of the farmed area. Egypt, on the other hand, less successful in feeding its people without large food imports, irrigates 100 percent of its cultivated land. In Egypt, water is free to anyone with power to lift it to the fields; hence, with animal power plentiful, overwatering of crops is widespread. Besides water waste, poor drainage and subsequent soil salinization have steadily decreased fertility in what was once the richest food producing area in the world. The control of the flooding, which previously brought new soil and nutrients, may also contribute to decreased soil fertility.

Now for some brief comments about the prospects for furnishing energy to stoke what many regard as the insatiable needs of our future world food system.

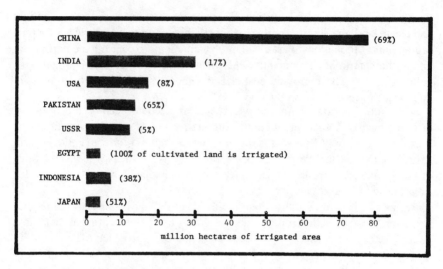

Fig. 8.8. Percent of Cultivated Land Irrigated
Source: FAO, 1971.

As figure 8.9 reveals, modern, high-production agriculture as practiced in the United States and proposed by many as a model for the world requires large expenditures of energy. At present, almost all nonhuman energy in the system is derived from fossil fuels, mainly petroleum. Hence, farm energy costs are expected to rise severely as oil disappears as a plentiful resource. While it is probable that other energy sources (biomass-alcohol, gasified coal, solar, wind,

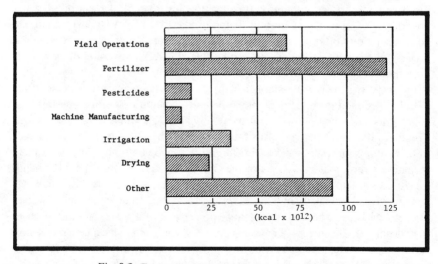

Fig. 8.9 Energy Required for Modern U.S. Farming
Source: American Association of Agricultural Engineers, 1975.

even mini-nuclear power units) will make their way to replace gasoline and diesel oil for driving farm machines, there appears a more serious problem when it comes to alternatives for today's petro-based fertilizers and pesticides. After all, massive fertilizer application is generally considered the key to future dramatic improvements in food productivity in the Third World. Fertilizer consumption has doubled every ten years since World War II. It is held mainly responsible for an almost 40 percent gain in major crop production in the United States in the past three decades.

It seems to us that highly mechanized farming will always have a place in the future of man, despite the often-advanced argument for labor-intensive and "appropriate" technology in those countries now teeming with the underemployed and underfed. Motor driven devices such as pumps, conveyors, dryers, coolers, and lighting generators are so essential in today's farm enterprise that their replacement by human or animal labor seems inconceivable. As suggested, the fuel for these devices will surely materialize. Motor-driven vehicles and tilling units will probably also survive, although some real competition may come if animals stage a comeback as an important limited farm power source, even in developed countries.

Alternatives for high energy fertilizers and pesticides, however, are another matter. Increasing attention has been given lately to the virtues of pre-World War II farm technology, when much food was grown without need for giant application of inorganic fertilizers and massive doses of insecticides. Manure was applied, fields were rotated, mulches were developed, crops were intermixed; all practices which we now call "organic" (or biological) farming, a term which still provokes derision among agri-bureaucrats. But arguments for a large-scale return to organic farming cannot be dismissed lightly. The water, soil, and energy conservation it offers are important imperatives. But the most interesting prospect of all is that, in some recent comparative tests, well-run organic farms have proven equal to equivalent crop monoculture in average year-to-year productivity. One important reason is the defense against bad weather inherent in mixed crop, organic soil farming.

New technology to support low-energy, or energy-efficient farming is also on its way, at a cost. Crop refuse, animal wastes, and even human garbage can be turned back into the production system through fuel or feed producing fermentation units located right on the farm. Soil additive bacteria, organic foliar sprays, and rich fertilizers made from seaweed and municipal sludge are now on the market competing with chemical fertilizers. Research and development in food crop genetics to add nitrogen-fixing capabilities is widely under way. Biological controls, where selected inoffensive insects or diseases prey on economically important crop pests, are displacing and are potential replacements for costly, environmentally-dangerous insecticides.

Climate-Invulnerable Food Production

In briefly surveying the various models or scenarios that project future world food production, we were struck by an interesting omission. All hypothesized that tomorrow's food will be produced pretty much as it is today; that is, by conventional soil, sun, and rain farming, augmented by some harvest from the sea. True, most at least mentioned in passing the notion of growing food from biomass, or synthesizing food in factories. But the general impression was that these developments, even though they are proving technically feasible, stand small chance of competing with traditional farming, hence will contribute insignificantly to tomorrow's food supply.

If we are to accept the prevailing view that we will soon reach the limits of land, water, and perhaps even energy for world food production, then we must be obliged, also, to take the position that other ways of producing food are bound to eventually become more attractive and cost-effective. Since it is implied these alternatives will be somewhat if not completely independent of natural environment, it makes sense to call them techniques for "climate-invulnerable food production."

Solar greenhouses, of course, are a well proven way to produce food free of rain and soil, although still dependent on a relatively sunny climate for efficiency. However, as figure 8.10 suggests, greenhouses are evolving toward completely controlled environment agriculture, where even sunlight can be eliminated as a variable. Programmed use of high intensity lighting, along with controlled microclimate, nutrients, and water, yields three or four times the productivity

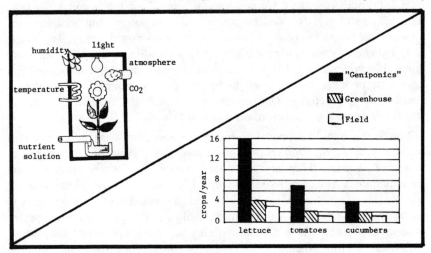

Fig. 8.10 A Fully Programmed Plant Environment "Geniponics"
Source: General Electric Company, 1978.

possible in equivalent conventional greenhouse culture, says General Electric. The approach appears to have great promise in cold locations where energy is in good supply and there is a lively market for fresh produce.

Controlled environment food growing which still takes advantage of free solar energy has also advanced significantly in recent years. Better greenhouse thermal design and new techniques in hydroponic (soilless) culture are extending the range of food crops successfully grown and marketed. While a sophisticated installation is high in capital costs—anywhere from $50,000 to $200,000 per hectare according to its complexity—it is expected that engineering improvements, economies of scale, conservation of water and nutrients, and alternative demands on the land will make its widespread use highly probable during the next century. Controlled environment greenhouses, with water requirements as low as one-tenth of field crops, can be located on relatively inexpensive arid land or desert where sunshine is maximum.

Closed environment fish culture is another long-established way to produce food essentially independent of climate. In 1975, an estimated 6 million metric tons of fish were cultured worldwide, with most growing in small ponds cultivated over the centuries, mainly in Eastern Asia. Recent progress in closed-cycle aquaculture (where fish are raised in high-density, highly mechanized growing enclosures) and in polyculture (where satellite fish and food crops are grown in the system) suggest we may be on the threshold of a major production breakthrough. But these technologies demand major capital investment. Large-scale industry backing is clearly required if this source of food is to attain the 50 million metric tons per annum level by the year 2000 estimated by the National Research Council in its 1978 report.[8]

Climate-independent, highly mechanized aquaculture in the United States is, as yet, confined to relatively small-scale production of luxury species such as oysters, crayfish, and trout. A good example of things to come, however, is the shrimp-growing venture developed by Professor Carl Hodges and his group at the University of Sonora and the University of Arizona.[9] Their pilot plant at Puerto Peñasco on the Gulf of California in Mexico's Sonora state is an exciting venture with high prospect of economic success. At present, in closely-controlled shrimp "feedlots," yields are approaching 3,500 kgs. (7,700 lbs.) of whole animal per hectare of surface water per year. Initial marketing of the product furnished the exciting news that it returns a higher profit per hectare than any other cultivated foodstuff. With Coca Cola as the project's sponsoring firm, its expansion into a major source for prime frozen shrimp is clearly in the cards.

Large-scale biosynthesis, or the conversion of organic bulk material into food by bacteria or enzymes, is another emerging technology with great promise for the future. Several years ago, when petroleum seemed abundant and cheap, a number of big (100,000 metric tons/year) ventures to grow single cell protein on this substrate were planned. Most of these schemes were abandoned when oil prices went up. Today, the main hope for producing a significant amount of the world's protein by this method lies in the use of carbohydrate or cellulosic (biomass) materials as the carbon sources feeding the proteinaceous microorganisms. While the costly hydrocarbons are a richer and more efficient substrate, their fermentation requires more stirred-in oxygen than carbohydrate materials, hence significantly more energy input.

Having a major source of future food protein through biosynthesis very much depends on efficient collection of normally unutilized or cheap field and crop wastes as substrate material. The quantities available today worldwide are impressive: there are 58 million metric tons of wheat chaff, 30 million metric tons of corn cob material, 83 million metric tons of sugar bagasse, 9 million metric tons of molasses, and vast quantities of unused waste from the big food processing industries, such as starchwater and peelings from potatoes, organic solids in corn and soy waste streams, and trim and peelings from fruit and vegetable canneries.

A word should be said at this point about the prospects for cultivating forests as a substrate for biosynthetic food production. During two world wars, large factories converted wood by hydrolysis into edible sugars which, in turn, fed yeast to produce protein-rich food and animal feed. Some natural forests of the tropics have been found to contain biomass of over 1,700 metric tons per hectare, with growth rates exceeding 50 metric tons per hectare per year. It has been demonstrated that wisely managed forest areas will produce two to three times more biomass than unmanaged land produces. For example, one species which grows widely in the tropics, *Leucaena*, can be cultivated to yield, through periodic harvest of its leaves, 90 metric tons per hectare per year of palatable green feed, equivalent to 23 metric tons of hay containing 26 percent protein. Without question, agroforestry has an important future in the developing world, particularly in the hot, humid tropics where rain and sun are ideal for abundant growth.

The promise inherent in biosynthesis technology is obvious in its production efficiency. A 453 kg. (1,000 lb.) steer grows less than 2 kg. (4 lbs.) of protein a day; 453 kg. of soybeans, cultivated well, will yield over 40 kg. (92 lbs.) of protein every 24 hours; but 453 kg. of microorganisms, given enough raw materials to grow on, will turn out over 45,000 kg. (100,000 lbs.) of protein in a single day.

The most dramatic alternative to conventional agriculture is to produce food by chemical synthesis on a giant scale in factories. The chemical process shown in figure 8.11 proves the notion is not a wild one. It is the way Germany manufactured 100 million kilograms of edible fat during World War II, starting from coal; a vivid example of how, when conventional food is scarce, a synthetic food or nutrient can be made by a chemical or biochemical process from nonliving raw material.

All of the basic nutrients of food—protein, fats, carbohydrates, vitamins—can be produced by synthesis; and balanced, attractive, edible products can be fab-

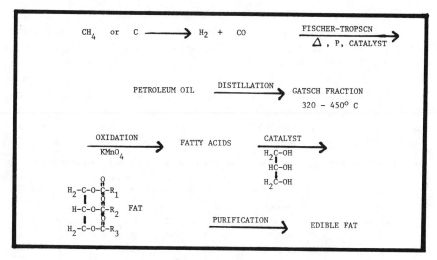

Fig. 8.11. Food Produced by Chemical Synthesis
Source: I. G. Farben Company, 1943.

ricated from these "building blocks" by modern food engineering. Synthetic vitamin production is already a worldwide industry. Some of the important micronutrients in protein, the amino acids, are also being made commercially by both synthesis and fermentation. Pilot methods to build whole proteins from these constituents have been developed but are not as yet practical outside the laboratory. Techniques for synthesizing edible sugar, the basic food carbohydrate, also have been perfected, and its economic manufacture is predicted (if desired) by 1985. Meanwhile, a number of high energy substitutes for sugar have been synthesized and only await FDA-type approval before a predicted widespread use.

With significant synthetic food production already commercially under way, many believe the present low-volume production of essentially high-unit-value items will expand naturally to have some impact on the world food supply by the year 2000. However, if a conscious world effort in research and development were to be mounted to achieve economic production technology and necessary social acceptance, food synthesis could furnish an important part of the world food budget in the century ahead.

More Livestock as a Food Panacea?

The notion of climate-invulnerable food production brings up the thought that farm animals, especially ruminant livestock, probably face a strong revival in food systems of the future. This is because they serve as a food source when crops fail, and as a way to store grain during harvest gluts. Of course, this view is directly counter to the often stated prediction that livestock will disappear because they compete inefficiently for grain that could be fed directly to humans.

Figure 8.12 shows the central role that ruminants play in an integrated, energy-conserving food producing system. Besides being a capital reserve in lean years, the animals furnish meat, milk, fertilizer, and other useful by-products. With their multicompartmental stomachs, ruminants are able to prosper on marginal vegetation in areas too dry or unfit for cultivation. Some ruminants such as zebu cattle, goats, reindeer, and camels have adapted to difficult climatic regimes and can survive where other livestock fail. At present about 12 percent of the world's population relies almost entirely on ruminants for their livelihood and food.

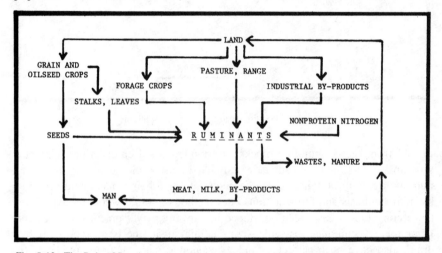

Fig. 8.12. The Role of Ruminants in an Integrated, Energy-Conserving Food Processing System
Source: Agricultural Council for Science and Technology, 1975.

Should world agriculture, because of the escalating costs of petroleum–derived fertilizer and pest controls, be forced toward more conservative organic practices, livestock/crop rotation will surely reemerge as a key element in the farm production system. The combination contributes to soil fertility and erosion control, helps check soil-borne diseases of both crops and animals, and, when rangeland is properly managed, grazing can be a stimulus to growth and productivity of herbage. Further, the way ruminants are fed in the farm scheme can be usefully flexible. Avoiding costly feed grains, which today furnish about 30 percent of livestock diet in the developed countries, ruminants can thrive on a wide range of agri-industrial wastes and by-products which normally go unused in many parts of the world.

A Post-Harvest Bonanza

In our studies of food losses in crop shortfalls, due directly or indirectly to unfavorable weather, we became aware of a startling fact. It appears that in many parts of the world the amount of food lost in various ways after harvest is often of equal and greater magnitude than most shortfalls. If these post-harvest losses could be reduced or eliminated, it would furnish an unexpected bonanza to many countries now deficient in food.

Figure 8.13, which tabulates losses of important food crops after harvest in various parts of the world, hints at the possibilities. In many countries, particularly those least developed and with the most troublesome food storage climates, losses of harvested food may run as high as 50 to 60 percent by weight. Lack of proper storage facilities, fungi, pests, and poor handling all combine to create the problem.

Crop	Country	Weight Loss (%)	Storage (# of months)
Legumes	Upper Volta	50 – 100	12
	Tanzania	50	12
	Ghana	9.3	12
Maize	Zambia	90 – 100	12
	Benin	30 – 50	5
	USA	0.5	12
Rice	Malaysia	17	8 – 9
	Japan	5	12
	United Arab Republic	0.5	12
Sorghum (unthreshed)	Nigeria	2 – 62	14
(threshed)	USA	3.4	12
Wheat	Nigeria	34	24
	USA	3.0	12
	India	8.3	12

Fig. 8.13 Post-Harvest Losses by Weight
Source: FAO, 1970.

To get some sense of the monumental amount of food lost in this manner, consider the conservative figure of 25 percent post-harvest attrition in cereal and pulse crops in semi-arid Africa. In 1971, a typical year, this amounted to about 7 million metric tons. At $7/hundredweight, this approximates a $1 billion loss, year-after-year, on two important food crops in one poor region of the world. Since one ton of cereal will feed six people for a year, the food saved could sustain literally millions of people.

A successful worldwide drive to minimize post-harvest food losses would eliminate dependency on hugh grain imports and improve local food supplies dramatically in most Third World countries. Much of the technology required for difficult hot and humid locations is well known, since effective storage systems have been well developed for tropical export commodities. While the cost of applying this technology on a broad scale would be high, the investment should amortize rapidly. Furthermore, the smaller-scale, village–level food crop storage required so desperately in rural Africa and Asia need not be sophisticated and high-priced. In Benin, Nigeria, for example, effective low-cost storage facilities were designed by the nation's agricultural extension service assisted by volunteer workers and aid agencies. Built with inexpensive local materials and by local labor, the new food storage units have almost eliminated rodent foraging in that area.

Extending the Food System

Proper food storage after harvest is an essential first step in evolving toward the modern food system which is diagrammed in figure 8.14, typical of the food industries of North America and Western Europe. Most developing countries, particularly those in the tropics, have relatively primitive food systems, with most activities at the agrarian level. Facilities to store, preserve, package, and distribute food to local and urban consumers, and the infrastructure required to support these functions, are largely missing. Hence, native food supplies come in seasonal bursts, depend on marginal and fragile distribution paths, and are especially vulnerable to losses due to bad weather and climate as well as to pests and predators after crops leave the field.

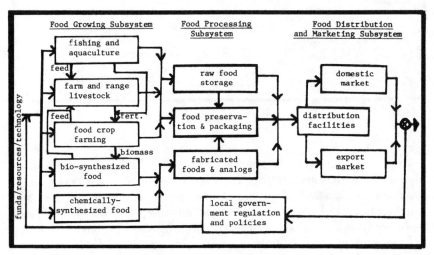

Fig. 8.14 Elements in a Developed Nation Food System
Source: L. E. Slater, 1978.

Food processing—its preparation and packaging in preserved form—not only adds value and convenience to the product but extends its life for several years of safe storage. Furthermore, investment in processing technology, which for efficiency should be near the raw food source, fosters the rural development so desperately needed in most Third World nations. Processing plants create jobs, require roads and utilities development, and bring many usually forgotten people into the nation's cash economy. Small farmers, assured of income from processing plants for their crops, are finally able to secure credit for the agricultural inputs needed to maximize production and combat the impact of a difficult climate or environment. The process of industrializing food also opens the way to use of the abundant biomass and farm wastes of the tropics in schemes for making fuels and feeds in centrally located fermenters.

In a way, those Third World nations with only primitive agrarian systems are in an enviable position. Should their leaders decide to give food production and distribution a high priority in national planning, they face a fine range of choices to achieve what amounts to an optimum system. Through the modern techniques of information systems, modeling, and systems analysis they can determine what can best be grown, processed, and distributed under local climate, agronomic, environmental, and socioeconomic conditions. For example, present use of best lands and resources to sustain export crops can be traded-off against the benefits from growing more food for internal consumption. The latest developments in food technology, such as biomass to protein conversion and aseptic container-izing to avoid costly refrigeration, can be evaluated. These efforts to optimize a nation's food producing and distributing capabilities, if widespread, could add greatly to net world food productivity and reduce malnutrition in the forthcoming century.

What about Population?

Finally, we must come to the most predicted, yet most elusive element in our future world food equation: Population. It is difficult for us to take a position on the population question or to consider it as a manageable dynamic variable in our climate, food, and population system. Its laws are simply not that well known. Rather, we are inclined to accept the widely held prediction that world population will essentially double by the year 2030 or 2040 and let it go at that.

On the other hand, responding to the analytical bent of the systems approach, it is possible to put forth some modest conjecture on how increasing success in food production and distribution worldwide may interact with population dynamics. It seems reasonable to us, for example, that family size in most developing societies must integrate closely with the availability of food and the quality of nutrition. This suggests a few assumptions for a possibly more hopeful outlook than a Malthusian view of population and its limits.

Can we not assume that when infant and child mortality is high because of malnutrition and the diseases which follow, the incentive is to produce more children? Conversely, when well fed children can be counted on to survive, the incentive to have large families surely must decrease. We can also readily assume that improving family purchasing power in the poor usually leads to fuller stomachs, provided the needed food is available. Might not this also further result in a level of security and awareness in the family where decisions on its size can be consciously made and implemented?

A final assumption is that better fed individuals within a developing society will add energy to the development process. This, in turn, will foster the economic progress among the poor which is suggested as a helpful condition for facilitating birth control.

There is, of course, great circularity in the logic supporting these few assumptions. This sort of analysis is one of the beguiling pitfalls as well as hopeful benefits of the systems approach. What it suggests, however, is that there may be important synergistic possibilities in the food/population interaction: that improved food supplies may lead, ultimately, to lessening demand for food and to a possible steady-state in world food needs, perhaps by the year 2050.

THE REAL WORLD ALWAYS INTRUDES

To conclude this somewhat optimistic inventory of new perceptions which can possibly shape a well fed world by the year 2040, we must not avoid some comments on the most critical element in our climate/food/population interaction: the dynamic force in our system diagram that is labeled *Governance*. Another name for this force might be the Real World.

We are well aware that all of the arguments we have made for ways to reduce the hazards in our future are simply whistling in the wind if no one takes them seriously. Even the finest shopping list of promising technology and policies for its application has small relevance if chances for its use are minimal. In the final analysis, it will be the governments of nations, subject to major world economic forces as well as the tugs and pulls of ideologies and power, that will decide. Improvements in the values, arts, and capabilities of governance over the next several decades are unquestionably the key to realistic, hopeful scenarios in all aspects of the future world condition.

The driving force for governments to assume the frame of mind and movement toward a new concept of growth based on human goals, integrative thought, and selective, imaginative use of technology and supporting policies can only come from an enlightened body politic. At this juncture in history, our most valuable effort toward this end probably lies in the interchange resulting in this and similar conferences and in the widespread public airing of our findings. We have no clues, however, on how to carry out the massive worldwide unbrainwashing that will prepare the mind and spirit of men and women of every region of the Earth for the large human commitment that will alone bring peace, prosperity, and justice to us all.

CONCLUSION

How are we to overcome the ultimate paradox of our time? Never before in history has mankind had the technological power to feed, clothe, educate, and provide health care abundantly for all. Yet, we continue to give the effort second place in our goals. It is easier to marshal support for guns and rockets for military security. Larger sums are spent for mind-modifiers in alcohol and drugs. How far short of the human potential we fall.

In the perspective of deep space, what happens here on Earth may be of small consequence. Yet we sometimes imagine the great thrill we would feel to stand on the Moon's surface and look back from that harsh lunar landscape to the softness of the luminous, hospitable Earth. From that vantage point, we could view our planet in its oneness. Then we could fully understand that not just a few nations but all mankind shares in the fantastic achievement of a lunar landing or flights to Mars, Jupiter, Saturn, and beyond. After all, science, engineering, and management skills, the common heritage of mankind, are what make possible such feats. All nations and races have contributed to the unbroken threads of knowledge that are man's highest cultural treasures.

From space, our astronaut and cosmonaut friends tell us, the Earth appears as a rare and beautiful place in the vastness of the universe. From this perspective, it seems unimaginable that the people of the Earth will not turn their vast new skills and powers toward making a better world for all mankind. If we choose to pay the price, it is clearly within our capabilities to construct world systems that achieve sound harmony between man and nature on this magnificent, and possibly unique, planet.

NOTES

1. H. Shapley, *Flights from Chaos* (New York: McGraw Hill, 1930), p. 3.

2. Rolondo, Garcia, *Nature Pleads Not Guilty* (Oxford: Pergamon Press, 1980).

3. M. Mesarovic and E. Pestel, *Mankind at the Turning Point*, The Second Report to the Club of Rome. (New York: E. P. Dutton, 1974).

4. J. Forrester, "New Perspective in Economic Growth," in *Alternatives to Growth*, edited by D. L. Meadows. (Cambridge: Ballinger, 1977).

5. Mesarovic and Pestel, *Mankind at the Turning Point*.

6. H. F. Reetz, *Physiology-Based Computer Simulation Models in Large Area Crop Forecasting*. Report of the Agronomy Department, Purdue University, 1978.

7. P. Buringh, private communication, 1979.

8. Commission on International Relations, National Research Council, *World Food and Nutrition Study* (Washington, D.C.: National Academy of Sciences, 1977).

9. C. Hodges, "New Options in Climate-Defensive Food Production," in *Climate's Impact on Food Supplies*, ed. by L. E. Slater and S. K. Levin (Boulder: Westview Press, 1980).

Chapter 9

CHANGING GROWTH PATTERNS AND WORLD ORDER

Lincoln Gordon

INTRODUCTION

The decade of the 1970s was marked by profound upheavals in the fabric of international economic relationships. In 1971, the Bretton Woods monetary system was brusquely terminated by the unlinking of the dollar from gold. In 1973-74, the sharp rise in oil prices brought about a far-reaching international redistribution of income, strengthened the forces of simultaneous recession and inflation, and helped to create an international debt structure of colossal magnitude and uncertain stability. At the same time, the developing countries, organized in the United Nations under the banner of the "Group of 77," put forth their claims for a new international economic order, leading to a frustrating and inconclusive process of "North-South dialogue."

Within the "North," there were conflicting tendencies. Increased unemployment led to the dismissal and return home of many imported "guest workers" and to a resurgence of protectionism, especially for labor-intensive industries. Yet the European Communities were enlarged from six to nine members, with the early prospect of adding three more. Mutual criticism and rudimentary harmonization of macroeconomic policies became institutionalized, both in the Organization for Economic Co-operation and Development (OECD) and at periodic summit meetings of the seven principal market economy countries. As the decade ended, a new experiment in regional financial cooperation was launched under the rubric "European Monetary System"; and, on a broader front, the Tokyo Round of multinational trade negotiations in Geneva concluded with substantial additional liberalization of trade and pioneering agreements to limit non-tariff barriers—rescuing these prolonged negotiations from a series of near collapses. The communist countries were also showing increasing interest in participating in international trade and securing technology from the industrial democracies.

Macroeconomic trends showed an apparent break around 1973-74, when the deepest recession since the Great Depression of the 1930s was accompanied by a speed-up of inflation. The first signs of these tendencies antedated the oil price rise which, therefore, cannot be blamed as their sole cause. But the "oil shock" greatly aggravated both recession and inflation, while imposing severe pressures on the balance of payments of both industrial and developing oil-importing countries. A quarter century of sustained economic growth at the highest rates in recorded history, interrupted by only mild and short–lived cyclical recessions, seemed to have given way to a new phase of uncertainty and perplexity.

Figure 9.1A charts the basic facts on growth and inflation for the market-economy industrial countries—the members of the OECD. It helps to explain the variety of interpretations. On the side of growth, optimists can attribute the depth of the 1975 recession to "Special circumstances" and point to the ensuing recovery as a sign of basic economic health. Pessimists note that recent recoveries have peaked at similar levels while successive recessions have become steadily deeper, with ominous implications for the early 1980s. On the side of inflation, on the other hand, there is little room for optimism. The net effect has been a general loss of confidence in Keynesian demand management through fiscal and monetary policies as an adequate means for keeping modern economies on the narrow track of high employment without undue inflation. Inflation control has come to displace high employment as the cardinal objective of economic policy management. Acceptable levels of unemployment and inflation have both been revised upwards, but at the cost of severe tension in all but a few fortunate countries, the most notable exceptions being Switzerland and Austria.

Figures 9.1B and 9.2 show a somewhat better record of recent growth in developing countries, but there is little confidence in its sustainability. The variety of experience is much greater in this group. The oil exporting countries are obviously in a class by themselves, with the special problems of applying windfall resources to a process of modernization in which noneconomic obstacles are much more formidable than economic constraints. Among the other developing countries, the majority also appear to have been fighting losing battles against inflation while facing new balance-of-payments constraints, even with some lowering of sights on desired growth rates. At the same time, there are signs of growing disparities within the group, with growth rates in the more industrialized developing countries steadily outpacing the poorest group, mainly in tropical Africa and South Asia—the "Fourth World."

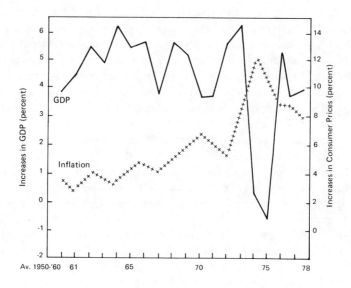

Fig. 9.1A. Industrial Countries, Rates of GDP Growth and Inflation, 1950-78.
Sources: United Nations, *Yearbook of National Accounts Statistics, International Tables*, 1968, 1975, 1978 (New York); International Monetary Fund, *International Financial Statistics*, May 1973, May 1976, November 1980 (Washington); International Monetary Fund, *Annual Report 1980* (Washington).

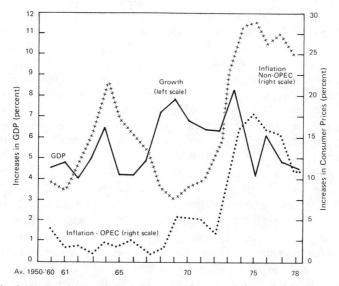

Fig. 9.1B. Developing Countries. Rates of GDP growth and inflation, 1950-78.
Sources: Same as 9.1A.

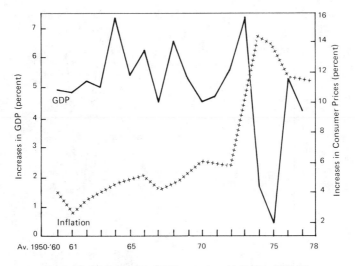

Fig. 9.1C. World Rates of GDP growth and inflation, 1950-78.
Sources: Same as 9.1A.

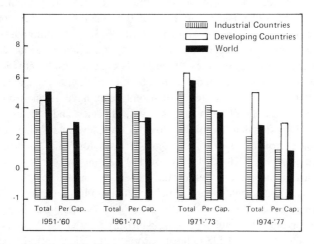

Fig. 9.2. Growth Rates of GDP, Totals and per capita, 1951-77.
Sources: Same as 9.1A and estimated by author or projected from data in International Bank for
Reconstruction and Development (World Bank), 1978.

Perhaps the most surprising phenomenon of these recent years has been the
maintenance of far-reaching economic interdependence in the face of powerful
forces working for domestic protectionism and insulation from external influence. The simplest measure of interdependence is the volume of goods and
services traded in relation to the volume produced (or, for a given country, the

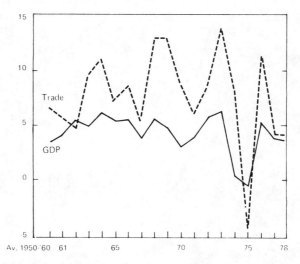

Fig. 9.3A. Industrial Countries. Growth rates of GDP and exports, 1950-78.

proportion of its total output involved in imports and exports). Financial interdependence is also of great importance but is too complex to be appraised by
any similar, single measure. As shown in figure 9.3(A-C), although the growth
rates of world trade no longer exceed those of production by as large a margin
as in the earlier postwar decades, trade levels continue high, notwithstanding
the malaise of "stagflation" and the uncertainties of fluctuating exchange rates.

Whatever their causes, the trends of recent years have given strong support
to more profound questions of growth objectives, strategies, and policies—questions
whose roots long antedate the mid-1970s. As the Cleveland-Wilson essay puts
it, issues of "Growth for Whom?" and "Growth for What?" have come to
prominence alongside "How Much Growth?"[1] In the industrial countries, the
classic issues of income distribution have been exacerbated by the slowing of
overall growth; and a major share of public attention has gone to new issues of
environmental protection, the scope of public services, trade-offs between work
and leisure, and other elements of what is broadly termed the "quality of life."
A highly articulate and visible minority professes to reject entirely any interest
in material goods beyond a bare minimum. In some developing countries, and
in the "development community" of professionals and intellectuals concerned
with developmental planning and assistance, increasing doubts are expressed
about whether the conventional path of modernization and industrialization is
either feasible or desirable for many of today's poorer countries. Alternative
paradigms of developmental strategy are struggling for recognition. Their implications for relative growth rates and patterns in various groups of countries,
and their impact on international interdependence are only beginning to be
seriously explored.[2]

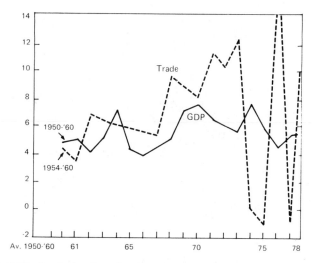

Fig. 9.3B. Developing Countries. Growth rates of GDP and exports, 1950-78.

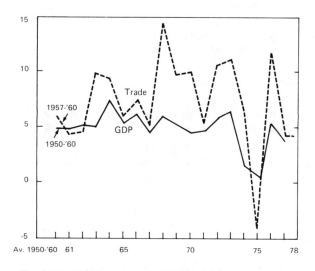

Fig. 9.3C. World. Growth rates of GDP and exports, 1950-78.

The thesis of this chapter can be summarized in six propositions:

1. Domestic tranquility as well as international harmony will require a more affirmative approach to long-term growth policies and problems of structural change than has heretofore been the accepted doctrine in the United States.
2. The international dimension of such policies is of cardinal importance.

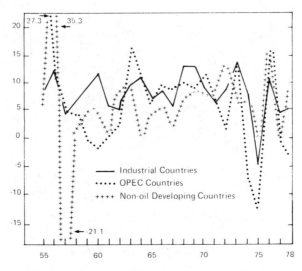

Fig. 9.4. Industrial and developing countries. Growth rate of exports, 1955-78.

3. The United States can neither insulate itself from the international environment nor alone control the shaping of that environment.

4. New forms of international cooperation are essential to the adoption of growth policies which maximize the shared interests and minimize the zero or negative-sum games among nations.

5. Collective leadership of differing groups of nations wil be required on specific issues, but the active participation of the United States will be indispensable to successful management of all the more substantial ones.

6. Institutional and policy development will be mainly incremental rather than revolutionary, but incremental means are capable of producing revolutionary results.

Before examining these concepts in more detail, however, it will be useful to clear away some potential misunderstandings on the nature of the "growth transition."

ALTERNATIVE VIEWS ON THE GROWTH TRANSITION

The Cleveland-Wilson essay on *Humangrowth*, taken as a starting point for the discussions at the Third Woodlands Conference on Growth Policy, presents a stimulating array of premises and interpretations concerning the nature of the contemporary transition in three respects: the objective character of economic growth; changing values and attitudes toward the goals and modes of growth in various groups of nations; and changes in the structures of governance and in growth policies at both national and international levels. While not disguising the many ills and discontents of today's world, the essay leaves the reader with a highly optimistic sense of growing harmony: harmony between mankind and the natural environment; harmony among aspirations, possibilities, and accomplishments; the universal satisfaction of basic human needs; and a benign international regime for the management of interdependence.

Such harmonious outcomes of the contemporary transitional period are certainly not to be taken for granted. Their realization can flow only from widespread consensus on the diagnosis and effective leadership in implementing the prescriptions. In fact, however, there is substantial disagreement on both diagnosis and prescription, even within the limited circles actively concerned with long-term growth issues, while active concern of any kind extends to only a small fraction of leadership groups. A long process of public education and political mobilization on these issues lies ahead.

This author, although sympathetic to the broad spirit of the essay, does not accept some of the premises as stated. In particular, the repeated characterization of historial experience as "indiscriminate economic growth" looking toward GNP expansion as an end in itself, regardless of the content of production, is a caricature of the economic and political history of today's industrialized democracies.[3] Since their economies are predominantly of the open market type, consumer demand has been the main force in determining patterns of production. Their effectiveness in meeting the needs and desires of the great majority of their populations—remaining pockets of poverty notwithstanding—has been anything but "indiscriminate."

Alongside high productivity and high rates of economic growth as conventionally measured, these countries have achieved the highest ranking on all the scales of social indicators: health, education, shelter, recreational opportunities, innovations in both high and popular culture, civil liberties, and political freedoms. Whether equality of income distribution is inherently a social good is disputed among both philosophers and politicians; but, on this scale also, the industrial democracies generally rank near the top, while all of them have made enormous strides toward equalizing opportunity for all classes, regions, and ethnic groups in their populations.

Of course, it must be recognized that GNP alone is not a satisfactory measure of social welfare and that market forces have their limitations. In particular, market forces do not reflect "externalities" (costs or benefits not accruing to the firms or individuals responsible for them); they are responsive to "effective demand" based on existing income shares; they are often distorted by "market imperfections" (inadequate knowledge and monopoly power); and they cannot be applied to the provision of "public goods." In this author's view, they are much less effective in guiding capital investments for production with long technological lead times than they are in promoting efficiency in the current allocation of resources. But all of the industrial democracies have developed some kind of "mixed economy" in which market forces are supplemented or supplanted by political decisions—in the opinion of many, too much so.

Rather than contrasting "human" or "organic" growth with a mythical straw man condemned as "indiscriminate growth," this author prefers to regard the transition as one of ongoing structural change, in which new constraints, new technological and institutional opportunities, and, above all, changing attitudes and preferences (perhaps changing "values") interact to alter the rates and directions of growth and to raise new issues for national and international governance. In open and dynamic societies, it is to be expected that the solution to every old problem will create its own new problems. The speed of technological and social change in today's world often seems to outrun our institutional capacities. But that is no reason to denounce yesterday's adaptations as narrow and undiscriminating, any more than to romanticize the past as a lost golden age.

A second caution concerns the profundity of today's transition, which *Humangrowth* suggests may be comparable to the introduction of settled agriculture, the evolution of the nation-state, or the Copernican and Industrial Revolutions which initiated the modern era. If that turns out to be so, it will become evident only to our grandchildren. In the politico-military sphere, nuclear weaponry has certainly constituted a basic revolutionary change. By placing in jeopardy entire large societies and perhaps the entire human species, it has profoundly altered the terms of all equations risking overt hostilities between nuclear powers. But in the domain covered in this chapter, the author's analysis of long-term trends has identified only five clear cut discontinuities: the exceptional surge of growth in the three decades following World War II, which could not be sustained at those rates; the explosion of population growth in the developing world, accompanied by imminent population stabilization in the industrialized world; the drive for economic modernization and autonomy in the decolonized continents of Asia, Africa, and Latin America; the prospective shift of energy sources from low-cost fossil fuels to some mixture of higher-cost but essentially limitless alternatives; and the expansion of human activities to a scale placing severe stress on some local environmental capacities and potentially affecting the global environment.

These discontinuities are occurring alongside more normal processes of change: technological evolution; new kinds of private and public claims on the economy; growing international interdependence; and shifts in relative affluence and influence of individual nations—with good reason to focus special attention on the "newly industrializing countries." Taken together, these movements surely constitute a major transition. But in historical perspective, the changes of the next 50 years seem more likely in magnitude to resemble those of the first half of this century (without, one hopes, the punctuation of two World Wars and a Great Depression) than the earlier profound transformations which altered the entire fabric of human civilization.

With respect to the outlook for developing countries, this author questions the extent to which the satisfaction of "basic human needs" has, in fact, moved to a central role in the policymaking of most national governments, as contrasted with the international "development community." Nor is he satisfied that the commendable objective of placing a floor under some kind of poverty line appropriate to the circumstances of each nation has been matched by workable developmental strategies—even in intellectual concept, not to speak of political feasibility. The term "entitlements" in this connection has unfortunate connotations of costless privileges which can be secured without effort. A preferable approach would place the main stress on opportunity for full participation in development and its benefits by the hitherto marginalized rural and urban poor. Nor is it realistic to suppose that any nation will give to basic needs an absolute priority over its other legitimate aspirations.

For the North, the "new growth ethic" must be regarded more as an aspiration than as a confident prediction. As described by Cleveland and Wilson, it would make for more harmonious societies in which lower growth rates are readily accepted because of changing preferences and a greater variety of noneconomic satisfactions. Against that vision can be set the pessimism of Fred Hirsch, who saw the demand for "positional goods" as inherently incapable of saturation;[4] the astringent diagnosis of Mancur Olson, who sees the power of special interest groups and institutional sclerosis as brakes on productivity and growth ardently desired by publics-at-large;[5] or the mute testimony of two European elections in May 1979—the British overturning a social-democratic government with a record of low growth and high inflation, and the Austrians strengthening a social-democratic government presiding over high growth and low inflation. One underlying cause of inflation in industrial societies is rivalry for larger increments in goods and services than the economic machine is generating—a clear indication that the "new growth ethic" is far from reality.

On the global plane, the hope that nations will move to realize the full positive-sum potential inherent in international cooperation is a consummation devoutly to be wished. Yet, the whole concept of positive-sum games, especially in economic intercourse, is a relatively new idea in human history, and it must still contend with the ancient tradition of assumed hostility among separate races, tribes, clans, or nations.

If slower growth is in store for the industrial countries, that will not automatically redound to the advantage of the developing world; the contrary is more likely. Nor would a new growth ethic necessarily entail a more outward-looking and sympathetic concern for development in remote societies. Against the evidence pointing toward a "global fairness revolution" in the 1970s are the falling proportion of resources devoted to official development assistance, the prolonged failure to implement the food security plans unanimously agreed at the World Food Conference of 1974, the narrow vision of many coastal country negotiators in the Conference on the Law of the Seas, and the revival of protectionism as an influential force throughout the world. In contrast with the institutional creativeness of the mid-1940s, the present dispersion of leadership and weakening of all forms of authority and hierarchy portends an uphill struggle to secure even second-best solutions to many of the newer international issues of growth policy.

THE CONCEPT OF SUSTAINABLE GROWTH

The theme of the Third Woodlands Conference is stated as "The Management of Sustainable Growth." Management implies a series of deliberate actions by individuals, private organizations, governments at various levels, and international institutions. The meaning of the term "sustainable growth" is more ambiguous.

In common usage, "sustainable growth" refers to short-term cyclical fluctuations and is often described as the desirable objective of macroeconomic demand management policies. That implies that growth should be maximized within such constraints as the size of the labor force (and its rate of growth), the amount of available capital (and the rate of investment), and expected increases in productivity. Productivity improvement will be determined not only by capital investment and technological change but also by investment in human capital (education and health), job satisfaction, and attitudes toward mobility and adaptation. In that framework, growth rates may be unsustainably high during the phase of recovery from a recession, while unemployed labor and capital are being reabsorbed into productive uses. But if demand management tries to keep growth rates that high, bottlenecks will soon be encountered which set off inflationary pressures, distort the allocation of resources, induce wage increases beyond the gains in productivity, and lead to another and unnecessarily severe recession. A policy of "sustainable growth," on the other hand, would gradually contain overall demand as recovery takes place and, perhaps, encourage investment in potential bottleneck sectors. That kind of policy would minimize cyclical fluctuations and keep growth rates close to the potential maximum made possible by labor force growth, additions to the capital stock, and technological and institutional improvements.

Since our conference is concerned with longer-term growth issues, however, the term "sustainable growth" must have broader connotations. It implies compatibility with limitations of natural resources and environmental absorption capacities. It may involve issues of intergenerational equity. And insofar as constraints on growth are institutional and attitudinal, it raises fundamental questions of the differing valuations placed on economic growth by different cultures or by individuals and groups within a given culture.[6]

It is easy to point to examples of unsustainably high rates of growth in specific environments. The most obvious example involves drawing on renewable resources beyond their rates of potential regeneration. Destruction of forests, "mining the soil," and polluting rivers and lakes beyond redemption all come to mind. The practice of sustained yield forestry and agriculture has long been recognized as a hallmark of prudent civilizations. Rigorous analysis can demonstrate some marginal tension between the technological and economic definitions of "sustained yield," but these are second-order refinements of a valid basic concept.

When it comes to depletable resources, however, there are no easy guides to optimum rates of use. One can start by rejecting the simplistic notion—surprisingly widespread in some current writings on resources—that renewables are always preferable to depletables because "it is unfair to future generations" to "use up" the depletables. Leaving aside the point that non-fuel minerals are not "used up," but simply shifted about, with ample opportunity for recycling many of them (although others admittedly become too widely dispersed for recycling), the logical conclusion of that simplistic notion would be that mankind should never have burned a pound of coal or a drop of oil, and should never have entered the Stone Age, much less the ages of bronze and iron. Yet there is also substance in complaints of "profligate" and "wasteful" use by some societies and in the broad notion of "stewardship" for generations yet unborn. The idea of stewardship is most clear-cut in relation to literally irreplaceable and unique treasures—whether natural (Grand Canyon) or man-made (works of art and architecture). But when it comes to raw materials, the human record of ingenuity in technological means of mineral discovery, extraction, and use, and in substitution when particular items become scarce and costly, suggests that a general policy of self-denial by the present generation would be fruitless and perhaps even positively detrimental to the interests of the future.[7]

In any event, there is no way of relating resource depletion systematically to a quantified notion of rates of "sustainable growth" over the long term. Insofar as raw material costs affect growth rates—and their importance is exaggerated in much of the popular literature—the pertinent question is whether, at a given time, their relative costs are falling or rising. If falling, as has been the case for most minerals during the past 100 years, they permit slightly faster rates of overall growth. If rising, as costs of energy are now doing, they will somewhat retard overall growth. But that effect is not very large, provided that the changes occur gradually. Econometric estimates for the United States, for example, suggest that a doubling of energy costs, spread over the twelve years 1978 to 1990, would reduce growth rates by only about 0.4 percent per year.[8]

The issue of population growth is also conceptually puzzling. If welfare could be measured in fungible units of happiness, an argument might be made that humanity is "better off" with 10 billion people at 125 units each than with 5 billion at 200 units each. (In the former case, there would be a total of 1,250 billion "happiness units" in the latter, only 1,000 billion.) That argument would be rejected by this author, but with the recognition that it is a subjective judgment which gives higher value to the material conditions for dignity and freedom of choice for the living than it does to sheer numbers on the face of the Earth.

A further complication in the effort to define sustainable growth arises from long-term changes in the content of production. In the short term, fluctuations in GNP serve as a good measure of levels of economic activity, which can be related to employment, capacity utilization, inflationary pressures, and the balance of payments. Some kind of structural change is also taking place, but at a slow enough pace to leave most of the structure unchanged over the few years of the typical business cycle. In the longer term, however, structural change becomes dominant, with basic alterations in what is produced and consumed and in where and how it is produced. Those alterations have major effects on resource and environmental impacts. At the same time, innovations in resource use and environmental protection are altering the impact of any given level and pattern of production.

For most goods and services, consumption does not follow a linear or exponential path of continuous growth. The more common path is a logistic (or S-shaped) curve leading to saturation, stability, and perhaps decline. The overall growth rate, which may appear as a fairly smooth curve, is the summation of a large number of goods in different phases of consumption—some in rapid expansion, some in stability, and others in decline. It follows from these characteristics that there can be no specific, determinable, rate of overall growth which is the "maximum sustainable" over a long period. What is sustainable in terms of resource limitations (if any) or environmental capacity depends on the composition of the baskets of goods and services, and that composition keeps changing in response to changing consumption preferences, perceived resource or environmental constraints, and technological capacities.

In practice, it is very unlikely that physical constraints will be the limiting factor on overall growth rates, which are more likely to be influenced by the changing character of demand and by social constraints on supply. Therefore, we may be well advised to focus on the policy implications and the management problems of alternative rates and patterns of growth without pressing further for an unambiguous and elusive definition of "sustainability."

PROSPECTIVE CHANGES IN GROWTH PATTERNS

Other papers prepared for the Third Woodlands Conference explore intensively the factors likely to influence growth patterns in the United States during the next few decades: attitudinal, technological, environmental political, and economic. This section deals more broadly with the same set of questions as they may apply to all the more industrialized countries, including the important class of "newly industrializing countries." This group, led by Brazil, the Republic of Korea, and Mexico (and also Taiwan, if that island can still be regarded as a country) is rapidly becoming differentiated from the rest of the so–called developing world. Some of the features summarized below are predictable with confidence, while others are merely conjectural. It should be noted that all of them work in the direction of slower growth than was experienced in the quarter century up to 1974.

Demography

The surest of all the causes of slower growth in the future, especially from the late 1980s forward, lies in reduced birth rates, which have fallen to or below the replacement level in most of the industrial countries. That trend might be reversed by a change in fashion, as happened after World War II, but a reversal would be surprising. After a short phase in which lower fertility may help to increase growth rates by increasing the proportion of working women and reducing that of dependent children, the longer-run effect will retard the growth in the labor force and also lead to a higher proportion of dependent older people. Medical breakthroughs on prolongation of life, coupled with deferred retirement, could either intensify or mitigate those effects, so that even the demographic influences on future growth cannot be foreseen with complete certainty.

In most of the developing countries, even though peak rates of population growth may now have been passed, the proportion of young people is so high that rapid increases in labor force will continue for several decades. In most of their economies, a shortage of jobs is a far greater concern than a shortage of labor, even though skilled labor is often scarce. In tropical Africa, moreover, death rates are still very high and life expectancy correspondingly low, so that several decades of increasing population growth rates may be expected before the point of inflection is reached. Although the growth in per capita incomes will generally be seriously retarded by rapid population growth, the overall national economic growth rates in most cases may not differ greatly from the results achievable with slower population growth.

The Occupational Shift to Services

The best established characteristic of structural change in all the more affluent societies is the sectoral shift of occupations—first out of agriculture into manufacturing and then out of both agriculture and manufacturing into services. It is generally believed that productivity gains if services are slower than in primary production and industry, so that a shift toward services tends to reduce average productivity. In some types of modern services, such as bookkeeping, clerical work, communications, and data handling, technology is evidently permitting enormous increases in productivity; but that is not the case for personal services such as education, health care, legal advice, or equipment maintenance. And there is a large class of services—including many kinds of government activities—whose output is not directly measurable and is simply assumed to be equivalent to the salaries paid. Whatever might be the judgment of some arch-angel with a magic measuring rod, the expressed opinion in most democratic societies indicates that their publics believe that they are getting progressively less return for their expenditures on government services.

Affluence and Saturation

The most elementary theorems of welfare economics suggest that as individuals, families, communities, or nations become richer, the marginal value of further increments in the output of goods and services should diminish. The progressive reduction in work time over the last century seems to confirm the theorem. Not only have weekly working hours been cut by one third or more, but holidays and vacations have been increased and the average span of working life reduced by more schooling at one end and earlier retirement at the other. Yet, one cannot project this trend automatically into a continually growing preference for leisure over work and into growing productivity from the fewer hours worked to the point where the minimum of desired work satisfies all of a society's material demands. (That would amount to what Lord Keynes once called the solution to the economic problem.) One of the most striking paradoxes of contemporary social history in the Western world is the failure of the enormous improvement in objectively measured material welfare to be matched by corresponding im-provement in subjective satisfaction.

On the evidence so far, the human capacity to keep new material wants up to or beyond the growing production potential seems likely to prevail for at least another generation, even though demand for specific items or categories (motor vehicles, food, clothing, refrigerators, space heaters) may become saturated. Some important types of growing demand are inherently open-ended, such as adult education and health care, while others are inherently unsatisfiable—notably the "positional goods" whose acquisition is by definition a zero-sum game. (Everyone cannot have a larger house or a higher rank in a job hierarchy than everyone else.)

Collective Goods

A strongly marked trend in the industrial countries throughout the twentieth century has been the increasing proportion (and even greater increases in absolute amounts) of collectively supplied goods and services—especially "public goods" (such as national defense or environmental improvement) whose benefits are pooled, cannot be individually appropriated, and therefore cannot be assigned a market valuation. In these cases, unless some new method of social accounting can be devised, the increments to welfare may be either greater or less than the conventional measurements we apply. If individuals could "buy" varying amounts and grades of purified air, the amount of resources devoted to air quality improvement would theoretically be balanced against all competing uses of resources. Wage gains based on productivity improvements would be allocated by each individual or family between environmental improvement and conventional family expenditures.

In theory, the democratic society (or local community) establishing environmental standards is applying a collective political judgment as a substitute for the sum of individual decisions on resource allocation. If those standards are achieved through direct public expenditure, as in the case of municipal sewage treatment, the needed resources are secured through taxation and the control expenditures appear in the national accounts as additions to GNP. But if the standards are achieved through regulation of the private sector, the costs must be passed on to consumers in higher prices; there is no apparent addition to real output; GNP is increased in "current dollars" but not in "real terms" after correction for inflation. Thus, productivity appears to have fallen even though purer air or water are being secured. The worker, instead of voluntarily allocating some income to environmental improvement, either directly or through political participation and decisions to tax, feels cheated by inflation which he then wants to offset by wage increases.

If environmental quality improvement requires heavy investment in pollution control equipment, thereby competing for capital supplies against investment which improves measured productivity and also raising the cost of capital generally, the downward effects on productivity and growth will be real as well as apparent. The gains in social welfare may still be worth the costs, but neither our methods of accounting nor the politics of environmental decision making are sufficiently sophisticated to examine such trade-offs explicitly.

It should also be recognized that, in an ideal system of social accounting, some aspects of environmental protection would be regarded as costs for producing other goods and services, while others would be treated as benefits in the form of additional environmental amenities. An increasingly complex industrial society is bound to incur larger environmental costs—either as consequences of environmental deterioration or as expenditures for pollution control and environmental protection. At the same time, an increasingly affluent society is likely to desire a larger amount of environmental amenities in its basket of goods and services. In either event, the measured effects will involve a reduction in growth rates.

Capital Shortage

The observed experience of the industrial world regarding voluntary savings and levels of investment varies greatly among countries and over time. In this, as in other fields, it is common to mistake a short-term fluctuation for a long-term trend. As the comparative study of economic growth experience makes clear, variations in the productivity of investment are often more important than differences in the rates of investment (or proportions of GNP saved and invested).

In the early stages of economic modernization, investment normally claims an increasing share of GNP. Since capital is scarce and returns to capital correspondingly high (i.e., incremental capital-output ratios are low), this type of structural change promotes high growth rates. A highly capitalized economy should expect lower marginal returns, and a society with elaborate arrangements for social security should expect lower voluntary private savings. The postwar record in most industrial countries has not borne out those expectations, but some observers believe that they are now finally coming to pass, especially in North America and parts of Europe (although less so in Japan). Whether the reduction in industrial investment really portends a fundamental long-term trend, or, rather, reflects the uncertainties of anti-inflation measures and other public policies, will not become evident until more time has elapsed.

Even if savings are not reduced, a capital shortage could result from an enlargement of investment needs to maintain a constant output (at least as measured). Pollution control is one example already mentioned. Both minerals extraction and continuing improvements in agricultural productivity may become more capital intensive in the decades ahead. The largest projected source of such expanded capital requirements is the energy transition, in which the shift to higher-cost sources will involve huge investments, both for energy conservation and for the development of alternative supplies.

Technological Slowdown

Another frequently suggested cause of slower growth is a reduced rate of technological innovation. In the absence of any satisfactory measure of either the quantity or the quality of innovation, it is impossible to verify the hypothesis. Proportions of GNP going to basic research, applied research, or research and development are very gross approximations; and the major scientific or technological breakthrough that might create a new "leading sector" a decade or two hence is unforeseeable, almost by definition. Since technological advance has accounted for so large a fraction of modern economic growth, a society intent on growth must have the means to generate new technology or the means to secure it from others. But the refrain that "there is nothing left to discover" has been sung by one or another prominent scientist at least once a decade, thus far always to be proven wrong.

What appears much more plausible is the argument that technological innovation contributed to exceptionally high growth in the postwar quarter century for two reasons which no longer apply. In the entire industrial world, it is first suggested, the Great Depression stifled new investment and therefore delayed the technological application of advancing science. Then World War II gave a strong impetus to both science and technology, but confined their applications to war purposes. After the war, there was a surge of civilian applications, making up for 15 years of lost time.

The second reason concerns the postwar "technological gap" between the United States and the rest of the world. Although some European writings on this topic, focused first on a supposedly permanent "dollar shortage" and later on a supposedly invincible "American challenge," look ludicrous in the light of subsequent events, the basic point holds true. In the late 1940s and 1950s, industrial productivity in the United States was much higher than in Europe or Japan, mainly because of better managerial and technological methods, including economies of scale. As markets were enlarged by the formation of the European Communities and by the worldwide reduction in tariffs, and as science, technology, and managerial methods migrated through a variety of means, including notably the activities of multinational enterprises, the gap was closed by Japan and most of Europe. During this "catching-up," Japanese and European growth rates—and average worldwide growth rates—could be exceptionally high. From now on, the possibility of high growth through gap closing no longer exists, since technology and productivity levels are fairly uniform all over the industrial world. (For the developing countries, in contrast, that route to high growth rates still remains open.)

A similar "catch up" phenomenon pertains to structural shifts in basic occupations, notably from agriculture to industry. For Japan and parts of Europe, that shift occurred on a large scale during the first two postwar decades, promoting increased productivity in agriculture and high growth rates in industry. With agriculture now accounting for only 3 to 10 percent of the labor force, there simply is no longer much room left for movement in that direction.

Institutional Rigidities

Mention was made earlier of the intriguing hypothesis that narrowly-based special interest groups can constitute retardants to growth notwithstanding the broader desires of their surrounding societies. In Mancur Olson's formulation, this unfortunate condition is likely to become increasingly common in stable political democracies enjoying sustained freedom from revolution, civil war, or enemy occupation. In his words, "the 'English disease' is not a unique ailment of one country, but rather the most advanced manifestation of an evolutionary process that has naturally reached its furthest point in the country with the greatest maturity in both industrialization and democracy, but which is also going on (at possibly an increasing rate) in all of the developed democracies."[9] His paper makes some comforting references to possible countervailing forces (short of war or antidemocratic revolution), but recent politico-economic history in most of the industrial world seems more in line with the pessimistic hypothesis than with the optimistic qualification.

Inflationary pressures, the spread of kidnapping, assassination, and other forms of violence, the failure to cope with urban deterioration, and the decline in social civility and in accepted legitimacy of any form of authority are all associated with tension between the interests of narrow groups, classes, occupations, or regions and some wider concept of national or global interest. In relation to economic growth, it is striking that several European experiments in moderating inflation through voluntary wage and price restraints negotiated with labor and business confederations have given way to "stagflation," with only Austria still pursuing that kind of "incomes policy" with apparent success.

The increased rates of inflation experienced by most of the industrialized countries in the 1970s cannot be attributed solely to institutional rigidities and special interest pressures. These factors have probably played a smaller role than those more commonly recognized, such as increased energy costs, slower productivity growth, excessive monetary expansion in several countries (often related to electoral timetables), and the surge in international liquidity associated with large deficits in the United States' balance of payments. Nevertheless, the Olson thesis points to an important kind of interaction between social and political institutions and economic growth—an interaction capable of generating either vicious or virtuous circles.

The Antigrowth Counterculture

The last of the more conjectural factors affecting growth rates is the possible spread of those forms of counterculture which regard economic growth as positively evil. There is a long history of religious and mystical rebellions against any kind of concern with worldly goods or other worldly affairs. Since the Industrial Revolution began, there have been periodic anti–industry movements, some popular (like the Luddites) and others intellectual (William Blake, William Morris, and many others). Conditions in the mines and factories of Victorian England make those reactions easy to understand, even though the peasantry in preindustrial England's "pastures green" also suffered on the average a short and deprived life with little opportunity to escape the privation.

In the more affluent societies of today, there is evidently far greater latitude for contrasting life-styles to coexist: the antimaterialists in their organic self-help communes, the consumerists in their well-tended suburbs, and many others in between. Generalized antigrowth sentiment seems thus far to have had little impact on public policies or "mainstream" life-styles in the industrial world, and still less in the developing countries.

More focused antigrowth sentiment, however, clearly influences patterns of economic activity in particular localities, the acceptability of certain kinds of industry, and the related requirements for environmental protection. The potential for more pervasive effects on growth rates lies in two fields: (a) changes in attitudes toward job satisfaction which might radically reduce productivity and require a basic restructuring in the organization and conditions of work; and

(b) a faith in the potentials for energy conservation carried so far as to prevent the development of any kind of workable alternatives to the dwindling supplies of petroleum and natural gas.

THE NATURE AND SCOPE OF GROWTH POLICIES

In spite of the conjectural character of several of the factors discussed above, their overall tendency points toward slower growth than in the postwar era and substantial changes in the composition of output—the directions and patterns of growth. Ultimate consumption should reflect changing age structures, saturation with many consumer durable and semidurable goods, upgrading of quality, and a continuing shift toward recreational activities (active and passive), adult education and health care services, and such public goods as improved environmental quality. Labor supply should reflect an increasing "choosiness" concerning working conditions and rewards.

Will such changes in rates and directions of growth lead to happier and more stable societies? The slow growth optimists will answer affirmatively, anticipating more leisure and less "hassle," more satisfaction derived from noneconomic pursuits, and fewer pressures on environmental carrying capacities. Believers in physical resource limitations on growth will argue that such changes are not only welcome but inevitable.

The slow growth pessimists, on the other hand, will see in these changes the prospect of intensified social struggles, since less growth implies more zero-sum games, in which improvements for the poor or handicapped require a positive sacrifice by the more fortunate. Without a "growth dividend" to be distributed, politics will, in their view, become an increasingly stubborn defense of vested interests, with grave risks to the survival of open societies and democratic institutions. The worsening of inflation in recent years seems to confirm the pessimistic view, since one of its causes is the cumulation of group demands for income gains beyond the actual increases in social product. An observer of twentieth century Latin America can testify to the capacity of inflation to tear apart the social fabric and destroy the political consensus which makes democracy viable. The revival of regional separatism in Europe and Canada; the struggles over energy policy in the United States Congress; the recrudescence of industrial protectionism in many countries; these and other signs all seem to point toward what one cynic has called the "Latin Americanization of the industrialized world." And within Latin America, the social and political effects of years of slow growth in Argentina—once the unquestioned leader of modern development in South America—should serve as a sober warning to the slow growth optimists.

The optimists will have to become correct at some future time, but that time is not yet with us or in sight within the horizons relevant to present policymaking.

On the contrary, for the now foreseeable future, the factors depressing growth rates will place policymakers under heavier pressure to secure maximum results from the resources available. Far from being able to rest on their oars, they will be pressed to give closer attention to efficiency, to innovation, and to dynamism. There will be much more complex trade-offs—more concern for externalities previously neglected, for quality in goods and services, and for accountability. And there will be increasing interest in fitting short-term measures into some sort of framework of long-term objectives—in short, for the development of structural growth policies.

That assertion implies the taking of a position on the spectrum of policy choice between long-term planning and reliance on market forces as the determinants of economic structure—the decisions on what is to be produced, where, and how. In its extreme versions, that choice is illusory. Even in the citadels of communist economic doctrine, there are no longer any advocates of centralized planning of everything. And while the universities and political parties of the Western world harbor many articulate and dogmatic advocates of pure markets, no society in practice accepts their prescriptions literally.

If one simply compares the results in the two major groupings of so-called open-market and centrally-planned industrial countries (the "capitalist West" including Japan, and the "communist East"), the superiority of the former is evident in economic terms and even more evident in political terms for believers in civil liberties and political freedoms. The value systems of the West emphasize freedom, innovation, spontaneous initiative, decentralization, and limits on social controls whether exercised by governments, businesses, religious orders, trade unions, or others. Their experience suggests that market forces—in spite of all the limitations summarized above—are comparatively efficient allocators of resources and means of linking together an enormous web of economic activities. Yet, it is hard to believe that they are a sufficient guide to the longer-term processes of structural change in which all societies are involved.

In fact, even in countries were market forces are most widely respected, the scope of public policies influencing structural change is very wide. It includes social, health care, and educational policies affecting the size and quality of the labor force; provision of physical infrastructure; urban services; regional policies; environmental protection; natural resource development; promotion of scientific and technological research; support for agriculture; a variety of "industrial policies"; social security and welfare programs; and intervention in foreign trade and international investment.[10]

Only in a few cases, however, are these policies guided by a coherent image of the desired future shape of the society, or even subjected to processes of coordination to minimize internal inconsistencies or mutual frustrations. Where such attempts have been made, as in the successive versions of the French national plans, observers find them increasingly irrelevant to actual policies.

A respectable argument can be made that the cardinal aim of structural growth policy should be to provide *resilience*—the capacity to adapt smoothly and rapidly to unexpected change, whether its sources are shifts in preferences, technological innovations, the curtailment of foreign markets or supplies, or new competition from some unexpected source. There is an analogy with planning for national defense by a country which harbors no aggressive designs. Its task is to canvass the array of possible threats to the national security, assign to each some estimate of probability and of severity if it occurs, and then organize the defense forces (and related diplomatic efforts) accordingly. In some aspects of economic security, there are precise counterparts to this strategy of resilience: stockpiles of food and imported raw materials, and reserve capacity in power and water supplies and in systems of transportation.

In this author's view, resilience is a necessary but not sufficient component of planning for structural change. He would complement it with affirmative guidance for certain strategic sectors, using market forces to the maximum for the implementation of those objectives and retaining a very broad area for untrammeled initiative and competition, including real rewards for successful innovation. The strategic sectors of mainly domestic concern include the educational and health care systems, transportation, urban settlement, local environmental protection, social security, and minimum income maintenance. The strategic sectors with both domestic and international implications include population policies (immigration and emigration as well as fertility levels, family planning, and settlement patterns); energy; food; regional and global environmental protection; and the broad shape of industrial adjustment and evolution. The shorter-term aspects of such strategic guidance would be related to macroeconomic demand management, while its longer-term aspects would need to be appropriately institutionalized at both national and international levels.

At the national level, that implies an intimate connection between long-term guidance for the strategic sectors and the critical central agencies of government, the budgetmakers and monetary authorities, without letting longer-term objectives become lost under the well-known formula that "the urgent drives out the important." But institutionalization is not only a matter of bureaucratic organization. Unless its concern also extends to the political level and the political discourse of the society, it is not likely to have much influence. The time horizon of electoral politics is often much shorter than the lead time involved in policies for structural change. Yet mobilization of environmental interest groups has demonstrated that long-term concerns can be brought into active political discourse—even if not always on an intellectually disciplined level.

THE INTERNATIONAL DIMENSION

In the United States, in contrast to countries more dependent on foreign trade, the traditional procedure in making economic policies has been to formulate them first in purely domestic terms and only later, if at all, to see what modifications might be needed to take care of their international aspects. For the second group of strategic growth policy sectors just summarized, that procedure is no longer adequate. The move to interdependence is no mere rhetorical cliché; it is a fact of life.

The main features of this change are well known. Internationally traded goods and services now account for 10 percent of our GNP compared with 4 to 5 percent in the years before and after World War II. Since the early 1960s, exports have risen from 3 to 7 percent of manufacturing production, and much more for some categories. About one-quarter of all farm output is exported; for wheat, the export share is over one-half. Although the United States is more self-sufficient in minerals than most industrial countries, the tendency is toward greater import dependence, conspicuously led by the case of petroleum. Income from investments abroad provides the foreign exchange for about one-quarter of our imports. The linkage with the world economy was dramatized in November 1978, when defense of the external value of the dollar became a crucial criterion of domestic monetary policy.

These trends bring American experience closer to what has long been customary in Europe and Japan, even though public psychology and political discourse have not yet caught up with the new realities. A similar intensification of interdependence extends to most of the developing world, with examples of extreme isolation, such as Burma or Cambodia, all the more remarkable for their exceptionality. Even the two principal communist countries, the USSR and China, whose continental dimensions have made relative self-sufficiency possible and whose doctrines have professed it to be desirable, are moving conspicuously toward greater involvement in international trade, investment, and technology transfer.

It follows that the international dimension now has to be incorporated from the start into the formulation of growth strategies. In many cases, such strategies can be made effective only with some measure of international cooperation. This point can be illustrated in a summary review of the principal sectors involved.

Population Strategy

While policies designed to influence birth rates and urban settlement patterns are prototypically domestic, immigration and emigration are inherently international. With no large unsettled areas of the world inviting large-scale immigration, that kind of contribution to nineteenth century development in the Americas, Australia, and New Zealand cannot be duplicated—a regrettable fact since these earlier migrations probably helped to raise living standards in both the sending and receiving countries. But migration on a smaller scale has, until recently, supplied an important fraction of the work force in northern Europe and the thinly populated oil-producing countries of the Middle East; it is also a major issue in relations between the United States and Mexico. Migrant remittances contribute substantially to the balances of payments of several countries. Migration also involves international issues of "brain drain," of political asylum, and of human rights.

It would lead us too far astray to open up here the social and ethical aspects of migration policies. But it is directly relevant that many European leaders have come to wonder whether a policy of moving capital and labor-intensive industries abroad might not have been wiser than the importation of guest workers on the scale practiced in the 1960s. A similar question might apply to a part of the American immigration from Mexico, although many of the migrants are not seeking industrial employment, and the disparity in wage rates will keep that frontier under pressure for a long time to come. In any event, this aspect of growth policy would benefit from efforts to guide its general magnitude and to regularize the scale and conditions of migration through international agreement.

Food and Agriculture

In the field of food and agriculture, every country already has some kind of growth strategy, either in the form of explicit goals or in the form of implicit targets in systems of price supports, market guarantees, or income supplements to farmers. At the Rome World Food Conference of 1974, there was an apparent worldwide consensus on international cooperation to expand food production in developing countries, create an internationally guided system of security reserves for the major cereal grains, and strengthen the mechanisms for emergency food relief. A few years of good harvests have undermined the Rome consensus and thwarted its translation into effective measures. Meanwhile, consumption has inexorably continued to expand with rising populations and improving living standards, China has entered world grain markets as a regular large-scale buyer, and another crisis period like the mid-1960s or the early 1970s is bound to occur within a few years.

What appears called for in this sector is a follow-through of the principles agreed at Rome, ultimately complemented by long-term projections of production expectations and efforts at harmonizing the various national strategies of the principal food exporters and importers. The gains from such harmonization could be substantial, but the obstacles should not be underestimated.

Energy

The importance of energy policy in relation to growth strategies in all countries—even those fortunate enough to be large exporters of energy materials—has become obvious to the public and policymakers everywhere. The issues become less confusing if sorted out into their major categories, each of which requires national action but could also benefit enormously from international cooperation. In a more extended discussion recently published elsewhere, the author has identified four categories in the following terms:[11] (a) completing adaptation to the economic consequences of the price discontinuity of 1973-1974 (to which might be added the further consequences of price increases in 1979); (b) protecting against the security dangers of supply interruption and of nuclear proliferation; (c) defining and moving to some form of longer-term energy sources to replace cheap oil and gas; and (d) avoiding, minimizing, or managing any oil supply and price "crunch" en route to the long-term alternative.

Although different forms of energy supply are suited to particular uses, there is sufficient overlap so that all kinds of demand and supply now interact, finally coming to focus on residual demand from those residual oil suppliers in the Middle East (notably Saudi Arabia) who combine unused production capacity with earnings beyond their current needs for imports. Actions to conserve energy anywhere, to discover new sources of conventional fossil fuels anywhere, or to develop economical means of exploiting unconventional energy sources—whether fossil fuel, nuclear, or solar—all affect the global energy balance, pricing policies, and the possibilities of supply interruptions. Both coal and nuclear fission—the only two large-scale energy sources which have the assured ability to replace depleting reserves of oil and gas—involve health risks and other major environmental hazards. Uninterrupted energy supplies at tolerable (and reasonably stable) prices are so critical to all forms of economic activity that the effort to secure them can give rise to major international tensions, including armed hostilities. Even among trusted friends and allies, such as the NATO member countries, fears that one or another is getting an unfair "free ride" are major sources of friction. No other set of issues of growth policy is so fraught with potential for worldwide negative or positive-sum games on such a large scale.

In each of the four categories specified above, there are strong grounds for cooperative international action, although the relative weights of national and international considerations vary from area to area. In adjusting to higher energy prices, each oil-importing country must seek its own means of confining the inflationary impact without precipitating excessive deflation, recession, and unemployment. The central international task is to ensure that the new pressures on balances of payments do not lead to beggar-thy-neighbor policies of competitive currency devaluations or trade restrictions.

As pointed out at the beginning of this chapter, the world economy proved more resilient in face of the "oil shock" of 1973-74 than was feared at that time. Even though the scale of price increases in the second (1979) round was smaller, they took place against a background of much greater fragility—higher unemployment and inflation rates in the industrial world and much larger debt servicing burdens in the developing countries. The international task of monitoring and supplementing the recycling of petrodollars and guarding against renewed trade protectionism is now correspondingly more severe. For the poorer developing countries among the oil importers, substantially greater volumes of aid will be essential to avoid serious setbacks in their development prospects, unless the oil exporters can be persuaded to favor them through some kind of two-tier concessional pricing.

Security against sudden interruption of supplies is also mainly a national responsibility, for which diversification of sources and creation of emergency stockpiles are the principal instruments. International arrangements for emergency sharing can provide some reinforcement, as well as a deterrent against politically motivated denials of individual importers. The limitation of nuclear weapons proliferation, on the other hand, is an international undertaking by its very nature. With the diffusion of nuclear science and technology, it has become evident that success in nonproliferation efforts requires the voluntary cooperation of the whole group of countries with a serious interest in nuclear energy, reinforced by a system of sanctions agreed among all but enforced collectively by the suppliers of nuclear equipment and services. The studies called International Nuclear Fuel Cycle Evaluation point toward new kinds of rule-making for supply assurances and new forms of multinational management for spent fuel, waste disposal, and control of plutonium. Developments of this kind would build on the foundation of exceptional authority already granted to the International Atomic Energy Authority.

The longer-term regime to replace low-cost petroleum is likely to differ from country to country, depending on available local resources, the structure of demand, and the capacity to pay for imports. As in the case of food, governments have a natural tendency to seek self–sufficiency in energy supplies, but that goal is physically unattainable for many countries and economically undesirable for many others. The transition will be spread over several decades and will also differ among countries.

While energy supply strategies will be formulated at the national level, they will be strongly affected by international considerations. That is obvious in the cases of oil and nuclear fission, but also extends to such questions as the prospects for a greatly enlarged coal trade, transportation and marketing arrangements for liquefied natural gas, and technological cooperation on novel sources ranging from efficient windmills to photovoltaic cells, ocean thermal generators, biomass conversion, and nuclear fusion. The interaction of energy markets creates a strong convergence of interest in the discovery and exploitation of new fossil fuel sources, especially in the developing countries. Because exploration has been impeded by mutual suspicion between host governments and multinational energy companies, the World Bank has embarked on a fossil fuel development program which may provide a more favorable environment for their cooperation.

In the short run, the strongest compulsions for international energy cooperation arise from conditions in the oil market itself: the world's dependence on internationally traded oil; the prospect that global oil production will pass its peak within the coming decade or two; and the organized control of world oil prices by OPEC—a group of thirteen developing countries. Countless man-hours have been devoted to the construction of elaborate mathematical models of the world oil market, in the hope of descrying the likelihood and timing of a possible supply-price "crunch." But a difference of 2 or 3 million barrels per day can convert a buyers' market into a sellers' market, as demonstrated in late 1978; and there are many unpredictable factors on both the supply and the demand sides which can easily vary by more than those amounts. These circumstances constitute a classic case for maximizing resiliency, for organizing contingency planning, and for erring on the side of anticipating the ultimately inevitable transition to other energy sources. The alternative is to risk a prolonged period of scarcity, in which supplies would be allocated either by price levels far higher than the costs of the ultimate substitutes (squeezing out all but the most affluent consumers), or by political alliances, or by the threat or use of military force.

The industrialized countries have taken some cooperative steps toward enhanced resiliency through the International Energy Agency, with its emergency supply sharing arrangements, pressure for parallel measures of conservation, and promotion of cooperation in the development of new energy technologies. The Tokyo seven-power summit agreement (June 1979) to accept specific nation-by-nation oil import ceilings marked another substantial step forward. But there is no forum for organized cooperation among the oil-importing developing countries, which have been fearful of antagonizing their OPEC allies on broader issues of North-South relations. The same tension between desires for greater bargaining power and fears of confrontation has impeded effective three-way cooperation between the two groups of oil importers and the OPEC exporters.

If the United Nations system were now being designed *ab initio*, it would surely include a specialized agency for energy, parallel with those for food and agriculture, health, education, and labor. That kind of institutional development is not now in the cards because of a widespread (and easily understood) reluctance to create new bureaucracies and because much of the terrain is occupied by others, including the World Bank, the International Monetary Fund, and the International Atomic Energy Agency. There is a strong case for adding a low-key, small-scale International Energy Institute, focused on assisting developing countries with energy assessments, data improvement, technical training, and application of new technologies. But the most provocative idea for institutional development would be a petroleum commodity agreement, in which consumers' acceptance of stable or slowly rising prices (in real terms) would be balanced by producers' commitments to supply continuity and floors under production levels. Short of a formal agreement, one can imagine many possibilities of tacit understandings. All would be facilitated by some organization on the part of the oil-importing developing countries and by the reestablishment of a forum for the three-way dialogue broken off in 1977 with the termination of the Conference on International Economic Cooperation (CIEC).[12]

Other Raw Materials

No other raw materials are even remotely comparable with energy in their economic importance, capacity for worldwide disruption, urgency of technological transition, or environmental impact. Nor is there any serious danger of economic growth rates becoming constrained through the exhaustion of physical mineral supplies on a time scale of decades or even centuries.

Affirmative policies for raw material development, however, are an important component of growth strategies for several reasons. The long-run trend toward stable or declining relative costs can be maintained only by continuous technological improvement in methods of discovery, extraction, and utilization. For many developing countries the production and marketing of raw materials, and the terms and conditions of their development and sale, are crucial to prospects for adequate rates of growth and to hopes for modernization and diversification. At the same time, investment in nonfuel raw materials, like investment in oil exploration, is being hampered by mistrust between multinational mining companies and host country governments.

This is one field in which the often raucous North-South dialogue of recent years is beginning to show some signs of convergence on workable lines of policy. The elements include international monitoring of raw materials markets, with formal commodity agreements to limit price fluctuations in selected cases; broader arrangements to compensate for fluctuations in earnings from raw material exports; improvements in sophistication of mining laws and their administration; and development of mutual confidence between mining companies and host governments through new forms of concession contracts, joint ventures, participation of international financial institutions, or agreed procedures for the settlement of disputes. The opportunities for positive-sum games in this area are so large as to give hope that continued development on these lines will gradually create a kind of case law of effective and internationally respected practices.

Environmental Protection

In terms of their impact on growth strategies, most environmental protection activities tend to be local or national, or confined to geographical regions involving only a few countries. The broader international issues have been well-defined in recent years, aided by the educational role of the United Nations Conference at Stockholm in 1972 and the subsequent work of the UN Environment Programme. The global issues are potentially of the greatest importance; they affect the great ''commons'' of the oceans, the atmosphere, and outer space. The record on cooperation to reduce ocean pollution is a poor one so far, and the effort to secure a comprehensive new legal regime for the seas has been even more frustrating.

Practical scientific cooperation in these fields, on the other hand, has continued to show promising gains, including collaboration among politically hostile countries. That spirit may change when more direct economic interests are perceived to be at stake, as will be evident from reflection on the idea of controlling fossil fuel combustion to avoid carbon dioxide overloading of the atmosphere. For the present, however, environmental protection does not appear to call for entirely new modes of cooperation or major institutional innovations.

Industrial Adjustment

In contrast, the issues of structural change in the industrial sector are of consummate importance in growth strategies. They are at the core of tensions in North-South relations. Their effective management poses some of the most difficult issues in both domestic politics and international cooperation. Unlike the case of energy, they are too diffuse to threaten the international order directly, but their indirect and longer-term consequences are fundamental to prospects for peace and for human dignity.

These strong statements reflect the conviction that industrialization is an indispensable component of successful development for the poorer nations of the world, notwithstanding the widespread discussions of "alternative development strategies" and "basic human needs." The particular modes of industrialization pursued by some developing countries in the 1950s and 1960s were certainly misdirected—especially those focused on replacing imports of limited categories of consumer durable goods. With few exceptions, the dominant strategies of that period also neglected the urgent need for expanding agriculture and increasing agricultural productivity—policies which could have helped to raise incomes of the poorest segments of the population, reduce the pressures for excessive migration to the cities, and avoid the shift from net food exports to net imports experienced by much of the developing world. In addition, methods of industrial and agricultural production were frequently inappropriate to available productive resources, notably in overusing scarce capital and underusing plentiful labor.

Those warranted criticisms, however, apply to the kinds of industries selected and the kinds of production methods promoted; they do not detract from the fundamental importance of economic diversification and a major shift from primary production to industry as a necessary means for the improvement of living standards. If basic human needs are to be met through real opportunities to produce and earn, and not as some kind of dole, industrialization is an essential prerequisite to that goal as well.

But industrialization which uses resources efficiently and which proceeds rapidly enough to keep up with burgeoning populations requires linkage with the already industrialized world. Linkage is needed for technology transfer and the rapid acquisition of skills, for supplies of imported capital equipment, and for markets to earn the means to pay for technology and capital. (It need not be repeated that the technology needs to be adapted to local conditions rather than applied blindly, and the capital equipment should be suitable for higher labor intensity than in the industrial countries.) The more successful developing countries as defined by conventional measures—whether income distribution is highly uneven (as in Mexico and Brazil) or relatively even (as in Korea, Taiwan, and Singapore)—have all followed that course. There is no other in sight or on the drawing boards of econometric modelers which promises effective results.

That course calls for a receptive posture on both sides toward a continued expansion of North-South trade, investment, and technology transfer, and for the poorest developing countries an expansion of Northern aid. It implies a readiness to accept internal adjustments in the North even when economies are beset by unemployment, inflation, a general slowing of growth, and a loss of confidence in all forms of governance—all factors conducive to growing protectionism.

The kind of solution to this dilemma sometimes suggested as part of a "new international economic order," namely, a "planned redeployment" of labor-intensive industry from North to South, is chimerical. The North would not accept that kind of industrial planning for any purpose. Since experimentation and innovation are largely concentrated in the industrial sector, the objections to generalized planning are well founded. At the same time, the progressive reshaping of the division of industrial labor between North and South, including new forms of the intraindustry trade which has been so fruitful in North-North relations, should aid the North in its search for new categories of high technology and "knowledge" industries and new ways of securing a more efficient allocation of resources.[13] The fear of coping with the "absorption of ten or twenty additional Japans" should be mitigated by the realization that each new Japan can become a partner in the absorption of its followers, as Japan itself has been doing to some extent in relation to its developing country neighbors.

Since the Bretton Woods system was conceived in the mid-1940s, the preachments of international economic policymakers in the leading industrial countries have seemed to favor absolute free trade, in industry if not in agriculture. The practice has been much more qualified, but it has also moved surprisingly far in that direction, especially among the principal members of the General Agreement on Tariffs and Trades (GATT). The industrial integration of additional countries into the system, however, will almost certainly require some systemic reforms. The main features of those reforms concern the spreading and the timing of adjustment impacts: sharing by the more industrialized countries in the displacement of obsolescent industries, and slowing the process sufficiently to permit its accommodation into the other on-going processes of domestic structural change. The OECD has wisely called for policies of "positive adjustment"—so far mainly a phrase, but one which points in the right direction.

For heavily capitalized industries subject to severe cyclical fluctuations, international cooperation could help in the projection of global regional marketing prospects, the avoidance of excessively bunched investments, and the accelerated elimination of obsolete capacity. For other industries, the subjection of protective safeguards to orderly international review would improve the prospects for a shift toward positive adjustment policies. The realistic alternatives are backward looking—unilateral protectionism and a spiral of retaliatory measures from which the more advanced developing countries would probably suffer most.

THE POTENTIALS AND LIMITS OF INTERNATIONAL MANAGEMENT

The kinds of policy evolution suggested in this chapter will be disappointing to those who seek a radical reform of the international economic order, to those who favor a quick transition to zero economic growth or to completely altered patterns of growth, and to those who are satisfied simply to carry on with "business as usual." That is inevitable with a diagnosis and prescription which combine some elements of planning for strategic sectors with a strong belief in the value of market forces, in resilience, and in the preservation of large areas of economic activity for the interplay of free competition and innovation.

The prescription does encourage the institutionalization of serious thinking about alternative visions of the future and deliberate choice of general directions, while retaining humility about the capacity to predict with certainty and a strong distaste for efforts to confine the future in a straitjacket. It accepts the condition of continuous change as inherent in dynamic societies enjoying or seeking high living standards, without counterposing such processes of change to the preservation of what is most valuable in the natural environment or in mankind's diverse cultural monuments and traditions.

The concept of affirmative growth policies for strategic sectors implies some kind of institutional evolution at the national level in most countries and in various constellations of international groupings, from the regional to the global. The national level deserves priority attention, since many of the issues can be managed at that level and also because more effective national organization is a prerequisite to international action. For the most part, international action will take the form of exchanges of information among national authorities and transnational organizations, coordination or harmonization of national policies, and development of codes of national behavior with varying degrees of international monitoring. Only rarely will there be need or possibility of direct international or supranational management. And, given the exponential relationship between the numbers of countries involved and the difficulties of decision making and administration, institutional design should search earnestly for means of decentralization to regional or functional groupings or for ways of representing large numbers in smaller bodies.

Behind these apparently modest lines of thought concerning both substance and procedures, there lie two subjective judgments, one optimistic and one pessimistic: (1) the international order as it has been evolving does offer substantial opportunities to all nations for pursuing their developmental objectives, and that it is by no means static or wholly resistant to change; and (2) the enormous inertia of attachment to the nation-state and to the imagery of autonomy and sovereignty—however maladjusted to the needs or realities of the times—makes incremental reform the only possible mode of change short of a global war or a catastrophic economic crisis. But it should be emphasized that incremental reform, which works by building on existing institutions instead of sweeping them away and replacing them, is not necessarily slow or small. In such sectors as energy, incrementalism can be afforded but gradualism can not.

These two judgments are complemented by a third (also optimistic and subjective): the conviction that incremental reform is the best path to durable change. In economic arrangements, as in other spheres, radical reform has often followed the classic pattern of political revolutions: Euphoria; Terror; Thermidor; and Restoration. But even incremental reform does not take place without leadership. On that score the world's condition has been worsened by the passing of American international economic hegemony, even from the viewpoint of many who complained about its exercise. No other single nation is in a position to assume that role; it must be replaced by new forms of collegial leadership. For the whole array of strategic sectors identified in this chapter, a necessary, although not sufficient, condition of successful collegial leadership will be the active participation of the United States.

NOTES

1. Harlan Cleveland and Thomas W. Wilson, Jr., *Humangrowth: An Essay on Growth, Values, and the Quality of Life* (New York: Aspen Institute for Humanistic Studies, 1978), p. 9.

2. The most extensive effort of this type to date is the three-year project at the OECD known as "Interfutures," whose full title is: "Research Project on the Future Development of Advanced Industrial Societies in Harmony with that of Developing Countries." The final report of the Interfutures project, together with many of the background papers, appeared in August 1979 as *Facing the Future: Mastering the Probable and Managing the Unpredictable* (Paris: OECD, 1979).

3. The terminology in *Humangrowth* on this point resembles the argument in the Second Report to the Club of Rome, which contrasted "undifferentiated" with "organic" growth and raised similar difficulties. See Mihajlo Mesarovic and Eduard Pestel, *Mankind at the Turning Point* (New York: E. P. Dutton, 1974), pp. 1-9.

4. Fred Hirsch, *Social Limits to Growth* (Cambridge, Mass.: Harvard University Press, 1976).

5. Mancur Olson, "The Political Economy of Comparative Growth Rates," Unpublished paper presented at a conference at the University of Maryland, February 1979.

6. The dramatic events in Iran in 1978-80 suggest the complexity of this kind of question.

7. The classic discussion of these issues is to be found in Harold Barnett and Chandler Morse, *Scarcity and Growth: The Economics of Natural Resource Availability* (Baltimore, Md.: The Johns Hopkins University Press, for Resources for the Future, 1963). See also V. Kerry Smith (ed.), *Scarcity and Growth Reconsidered* (Baltimore, Md.: The Johns Hopkins University Press, for Resources for the Future, 1979).

8. The Energy Next Twenty Years Study Group, *Energy: The Next Twenty Years* (Cambridge, Mass.: Ballinger), p. 20.

9. Olson, "The Political Economy of Comparative Growth Rates," p. 80.

10. In this connection, Dr. Burkhard Strumpel has commented that affluent societies may develop an increasingly important third category of economic activities, guided neither by market forces nor by governments. He sees signs of such trends in the "informal economy" observed in several countries, including barter, "do-it-yourself," and small group cooperation. This kind of third category is of course paramount in primitive societies. The author is intrigued by the suggestion, but has too little sense of the magnitudes involved to evaluate their importance to economic growth patterns as a whole.

11. See Lincoln Gordon, *Growth Policies and the International Order* (New York: McGraw-Hill for the 1980's Project/Council on Foreign Relations, 1979), p. 110 and Chap. 6 *passim*.

12. Further discussion of these will be found in Harlan Cleveland, ed., *Energy Futures of Developing Countries* (New York: Frederick Praeger, and the Aspen Institute for Humanistic Studies, 1980).

13. The increasing interest of the communist regimes in international trade and technology transfer suggests that East-West relations will also become of growing importance in the international division of labor.

Chapter 10

WORLD SECURITY AND THE GLOBAL AGENDA*

Thomas W. Wilson, Jr.

There was an explosive expansion of the agenda of global issues in the decade of the 1970s. Suddenly governments were seized with an extraordinary range of pressing problems at the international political level that they were accustomed to think of as domestic issues—and, in any case, as matters for "experts" to cope with. It can now be seen that the emergent global agenda is made up of two extremely broad, fundamental, and interacting classes of problems:

- *Issues bearing upon the state of the planet.* During the 1970s, governments met at the plenipotentiary level to consider man-made stresses on global environmental systems, world water resources, pervasive desert encroachment, and—in seven long and painful sessions—management of the vast ocean systems that cover most of the planet.
- *Issues bearing upon the human condition.* During the same period, governments gathered to contemplate world population growth, world hunger, the human habitat, the changing roles of women, global employment, basic health delivery systems, rural reforms, and the use of science and technology to ease the burdens of poverty.

The United Nations Conference on the Human Environment, held in Stockholm in 1972, foreshadowed the emergence of this broad new agenda of global-level issues. Between the Stockholm Conference and the end of the decade, the

*This chapter is an adaptation of a paper originally prepared for the Presidential Commission on World Hunger.

United Nations gathered its still-burgeoning membership in a series of meetings to start talking about a range of issues that affect all nations but are beyond the reach of a single nation acting independently.

The tangible result of these conferences is a mixed bag of successes, failures, and postponements; of rhetoric, posturing, and maneuver; of declarations, agreements, and actions. The least that can be said is that they have helped raise awareness of the existence of problems that previously had not been on the international agenda at all. They also opened up an alternative and more flexible political procedure for doing international business outside the formal structure of international bureaucracies.

But there is little or no evidence that national political leaders have yet begun to perceive the full potential of the new agenda for behavior of nations. No governments are known to be building national policies around perceptions of transnational affairs. Responsibility for management of the global agenda remains largely unassigned in national capitals, while political leaders are preoccupied with what they deem to be more urgent aspects of world politics. And this obscures a crucial convergence between the interests of national security and the capacity to manage transnational problems of a nonmilitary nature.

THE GATHERING OF GLOBAL ISSUES

Problems beyond the reach of national governments are, of course, not new. The International Postal Union was established in the last century for the simple reason that an internationally agreed system was needed to assure that mail sent from one country would be delivered in another—just as common rules were needed for maritime navigation to reduce the likelihood of collisions at sea. Other international agreements and institutions came into being as needs arose to cope with transnational problems: allocation of radio frequencies, communicable disease control, weather reporting, and satellite communication systems are familiar examples.

Indeed, the steady expansion of international organization and agreements and regulations for dealing with a lengthening list of transnational problems has been a distinguishing feature of international life since the end of World War II. The whole system of United Nations agencies stands in evidence—as does a robust growth in nongovernmental organizations and the inauguration of major scientific research projects like the International Geophysical Year.

Some observers of this steady growth of international activities came to believe that a system of world governance eventually must emerge from the ever growing practical need for technical management of functional problems. In this view, such technical arrangements would come to dominate political relations.

Functional cooperation, however, has been treated by governments of the traditional major powers as a minor appendage to the main body of international political relations, located somewhere on the outer periphery of day-to-day diplomacy. For them, the perceived mainstream of interstate relations remained the traditional political-military questions, plus the pursuit of other national interests, mainly economic, as these interests were defined—usually in competitive terms—in national capitals. Technical cooperation, in practice, does not seem to dampen political conflict; indeed, cooperative technical programs and projects can turn out to be among the first casualties of political crises, as we have just seen in the wake of the Soviet invasion of Afghanistan.

But what happened precisely in the 1970s is that the new agenda of global issues was gathered in from the fringes of diplomacy and injected into the mainstream of world politics. Many a matter once thought to be reserved for "domestic" attention was *internationalized*, and many a question once labeled "technical" was politicized within the UN system. In that process, the substance of day-to-day international politics was revised decisively. Issues bearing on the human condition and the state of the planet can no longer be relegated to the realm of functional problems, nor delegated to experts. And diplomacy will never be the same again.

Yet there should be no surprise in any of this. For these global issues are the direct political fall-out of the most conspicuous trends and events of contemporary times; indeed, they collectively describe a large part of the present global predicament.

On the one hand, the integrity of natural systems suddenly is an issue because expanded economic activity by a swollen population has increased exponentially the total human impact on global systems, and raised state-of-the-planet problems that are beyond the realm of national power or influence.

At the same time, the human condition has intruded upon international affairs because the collapse of the European system of empire brought a three-fold increase of accredited members—and a new majority—into the world political arena. The new majority is committed to sustained economic growth to escape the grip of pervasive poverty. And one thing all the new members can agree upon is that the international economic system, in which they are caught up inextricably, was not designed with their needs in mind and does not now function equitably from their point of view.

Small wonder, then, that modernization issues have been moved from the sidelines to center stage at the United Nations, and that a "North–South dialogue" has been opened on demands for the creation of an international economic order better designed for sustainable growth in an interdependent world. The wonder is that this has received such low political priority from governments in the industrialized world.

It is especially surprising because there is no chance whatever that the issues on the new global agenda will go away or diminish from benign neglect. The internal dynamics of the trends that brought some of these issues to the surface in the 1970s guarantee that things will get worse before they get better—hunger, unemployment, and soil erosion are examples—even if vigorous measures are taken now to reverse the trends. And from a political point of view, it is scarcely thinkable that the new majority in the world forum will lose its consuming interest in modernization in the immediate or mid-term future.

Of course, there is no way to force the United States or any other government to pay serious attention to the new agenda. We and other industrialized nations could keep our eyes fixed on the traditional agenda and adopt what is known in the trade as a "damage limiting posture" when forced into debate or negotiation on the new class of issues. This is roughly what has been happening. And this is to say that the world is becoming hungrier, dirtier, more crowded, more deprived, more incapacitated, more quarrelsome, and more prone to conflict from year to year.

WHAT COULD HAPPEN

As things stand now, we are failing by wide margins to cope with the contemporary transnational problems placed on the world agenda during the 1970s. Continued failure could bring the creeping catastrophe of a general crisis in the world's capacity to deal with its most pressing problems. Obviously, it would be foolish to guess at details or dates for such a scenario; the variables are endless in their potential permutations.

But there is no law against looking at what is going on under our noses. And it requires no gift of prophecy to imagine that, in the years just ahead, governments in the industrial world might fail to adapt soon enough to the evident end of an era of cheap and abundant energy; might not come to grips with their problems of inflation, stagnation, and unemployment; might fall into divisive internal conflict between "developers" and "environmentalists"; might yield to protectionist and isolationist pressures; might find themselves simply unable to meet the combined, inflated costs of military establishments, sophisticated technologies, infrastructure maintenance, social welfare programs, decontamination of toxic wastes, and conversion of the economy to sustainable patterns of production and consumption—with the result that policymaking machinery would remain immobilized while once-manageable problems would degenerate into unmanageable crises.

It is not difficult, either, to imagine that governments in the developing world might fall short in their efforts to contain the population explosion; might fail to absorb the floodtides of young people now entering the labor market; might be forced to default on foreign debts, lose access to capital markets, and see their development programs grind to a stop; might fail to resolve internal conflicts between forces of modernization and forces of traditionalism; might not invest enough resources in rural sectors to prevent a continuing spread of the deep poverty symbolized by spreading hunger—with the probable result that political and social institutions would become increasingly vulnerable to repetitive collapse and frequent overthrow in an atmosphere of endemic violence.

Nor is it difficult to envision, in the world arena, that nations, large and small, might continue to dissipate massive resources on military arsenals; developing countries might fail to translate their demands for international reforms into specific, negotiable proposals; the industrial societies, sorely tried by their own problems and weary of being blamed for all of history's ills, might dig in their heels against reform of the world economy; U.N. proceedings might become characterized by extreme rhetoric, political arrogance, double standards, and a diminishing capacity to act or even generate consensus; The North-South dialogue might remain in a state of stalemate—with the result that international relations would degenerate further into competitive struggles over maldistributed resources while natural systems continue to deteriorate at accelerating rates.

The likely price of political paralysis would not be a quantifiable catastrophe or a sudden collapse but, rather, deepening darkness, degradation, and nameless danger—through drift and delay, standoff and stalemate, indecision, inertia, and a bankruptcy of political innovation as structure crumbles, order unravels, violence is more and more taken for granted—and opportunities proliferate for totalitarian scavengers of domestic or foreign origin. The notion of sustainable growth would go glimmering for nations at all stages of development.

Belfast, Beirut, Tehran, Phnom Penh, and Kabul may or may not be models of things to come. But evidence of a failing capacity to cope with modern problems—to make policy, to take decisions, to act in time—is to be found on all sides and at all levels. And perhaps the greatest danger is that the agenda of global issues identified in the 1970s could become the abandoned agenda of the 1980s, neglected through failure to perceive the connection between the state of the planet, the human condition, and world security.

THE DANGER OF POLITICAL PARALYSIS

The security implications of state-of-the-planet issues are not difficult to grasp. Already there is a dawning awareness of a security component in man's rising capacity to bring about disasters analogous to major natural calamities—or, for that matter, conventional wars—without resorting to weapons at all. We can begin now to understand that the burning of fossil fuels might trigger a change in the global climate with an impact on the United States comparable to a major military defeat. We can comprehend that man-made damage to the ozone layer could produce long lists of casualties. We can start to glimpse the potential consequences of headlong destruction of tropical forests, or the steady deterioration of environmentally crucial coastal zones. And we can hear rising voices from the scientific community warning of potential threats to human survival implicit in the rapid disappearance of animal species and the accelerating loss of genetic diversity in the world of plants that provide energy for life on Earth. Military defense is clearly not the only security game in town.

Indeed, one is tempted to wonder what national security can possibly mean for inhabitants of a living planet whose basic biological systems—croplands, pastures, forests, fisheries—are deteriorating steadily under man-made pressures. These clear and present dangers to the security of Planet Earth cannot be met by doubling or redoubling the defense budget or by deploying military forces abroad. Manifestly, our notions of "vital national interest" need a drastic overhaul; and, in that process, protection of the Earth's basic systems must become an integral part of a modern concept of security for this and other nations.

But there is another, more subtle and insidious threat to enduring security in the world today—and it rivets the new global agenda directly to the issue of world peace and security. The name of this menace is political paralysis. It is a much more credible threat to the future viability of this world than is instant disaster from either military or nonmilitary sources. And it is largely the product of a failure to manage the range of problems bearing on the human condition that are now part of the global agenda and at the heart of North-South relations.

As things stand now, both sides have painted themselves into corners. The developing countries agree that they want to participate in economic decisions that affect their interests, but they cannot agree on just how the system should be redesigned or just where to take hold of the problem. Their demands, therefore, are cast in general terms in order to maintain the "solidarity" they perceive to be their only source of bargaining strength.

The market-economy industrialized nations complain that the new order demanded by developing countries is a slogan without recognizable substance; and that, whatever it means, economic reform is not an event to be declared by majority resolution but is a process of adaptation to be negotiated on a technical basis over a period of time. In this defensive posture, most of the developed countries have managed to appear as stand-pat defenders of an international economic system that, it is plain to see, no longer works in the interests of either rich or poor. For their own reasons, the socialist bloc nations remain on the side lines.

So, in effect, nothing happens. This is the anatomy of political paralysis. It poisons an international environment already polluted by ancient feuds, strategic conflicts, and recurrent violence. It accelerates the drift into a general crisis in the world's capacity to handle contemporary problems. And it reinforces disintegrative tendencies that undermine world peace. This is the context in which political paralysis can be seen most clearly as a dangerous threat to national security—and every bit as "real" as military hardware in the hands of unfriendly nation-states.

"SECURITY" IS A WIDENING CONCEPT

No one of sound mind would deny that military capability is an indispensable part of any plausible concept of national security in a disorderly world that is armed to the teeth. Nor does anyone argue that security rests exclusively upon military force.

But the ascendancy of a new political agenda of global problems poses what looks at first glance to be a paradoxical imperative: the need to reexamine concepts of national security in the light of transnational problems. So far, this has been resisted effectively in a near-obsession with the strictly military aspects of the security problem. Governments appear virtually oblivious to the security implications of a failure to cope with the new global agenda. "Defense" issues are still kept in one compartment of policy analysis and though*—along with "strategy" and "vital interests"—while "development," "topsoil," and "hunger" remain in their segregated realms of perception, expertise, and action. This, after all, is how we have been taught to organize and to think about our affairs.

Now this way of thinking has itself become a threat to security. For an extended failure to take positive action on global problems can only contribute to a breakdown of that minimal state of order essential to the peaceful conduct of human affairs—and, hence, to the security of nations and peoples. Indeed, an obsession with the military dimension imperils security on three counts: First, it is limited to perceptions, policies, and modes of behavior that have led, during this century, to two world wars, a nuclear balance of terror, the survival of traditional warfare, unheard-of levels of armament, and a conceptual trap from which there seems to be no escape and no outcome save war or the tyranny of massive armament for the indefinite future. Second, it withholds resources—material and human—from urgent tasks in defense of planetary security, without which the very meaning of national security is called into serious question. Third, near-exclusive concern with the military aspect of security contributes to political paralysis and, hence, to a dangerous neglect of those disintegrative forces that pull the world toward the unmarked boundary between a state of peace and a state of war.

CONCLUSIONS

In a nutshell, then:

- The traditional correlation between military strength and national security has been attenuated over recent decades by many factors, including a pervasive diffusion of military capabilities, the heightened vulnerabilities of an increasingly interdependent world, and the emergence of new sources of power and influence in international affairs.

- Demographic, economic, political, and environmental world trends have combined in recent years to create a qualitatively distinct class of unavoidable world–level problems that were virtually unknown to traditional diplomacy; that are beyond the reach of national governments; that cannot be fitted into accepted theories of competitive interstate behavior; that are coming increasingly to dominate world affairs; that cannot be wished away; and that are indifferent to military force.
- The emergence of these issues raises unfamiliar dangers: on the one hand, physical threats to planetary systems that support all life; and, on the other hand, dangers of an irreversible slide into anarchy, tyranny, violence, and even war, through a political paralysis induced largely by preoccupation with traditional military concepts of national security.
- A draft agenda of some of these global problems was identified during the 1970s, but national governments have not yet recognized the far-reaching implications of world-level problems for peace, security, and the conduct of international relations.

In the final decades of the twentieth century, national security is inconceivable without global security which clearly requires an active defense of planetary life support systems, a positive strategy for breaking out of a political paralysis that abandons the field to disintegrative forces in world affairs, a much strengthened capability for managing an ascendant agenda of transnational problems, and the management of sustainable growth worldwide.

Part Four

THE LIMITS TO GOVERNMENT

Chapter 11

THE GROWTH OF GOVERNMENT

Murray L. Weidenbaum*

THE ARGUMENT IN BRIEF

The basic limits to government are not quantitative. Society's capability to bear the burdens imposed by government essentially is subjective, being primarily a question of willingness or attitude. Yet, some types of government activities may interfere significantly with the ability of a nation's economy to function and to generate the resources to finance these activities.

The economic limits to government may be an important inhibiting factor, especially over a period of time. It is not so much a question of the absolute or even relative size of government, but of the pervasiveness of the public sector's activities. Government intervention in business on an increasing scale has interfered with the ability of the typical business enterprise in the United States to perform its basic economic functions. Yet, there is little evidence of truly basic malaise—as contrasted with specific shortcomings—in the capitalist system.

The political limits to government may be more visible and dramatic—the defeat of a local property tax or the enactment of a constitutional limit on state government spending. In contrast to the "failures" of the market system (the "external" costs generated by business activity which is a standard justification for government involvement), the government apparatus itself can produce

*The author is grateful to Kenneth Shepsle and Ronald Penoyer for numerous helpful suggestions.

"failure." The bureaucracy may develop a set of "private" goals—such as enlarging the size of a bureau in order to justify a promotion or more staff.

Citizens are increasingly looking to new and unconventional ways of controlling government activity. Some of the mechanisms advanced—such as a mechanical approach to balancing the budget—may reflect frustration with conventional approaches to limiting government.

In the final analysis, a new way of thinking about government is required: because society's resources are limited, government cannot attempt to meet every demand of every group. These resources, moreover, are not merely economic or financial. As has been demonstrated so dramatically in recent years, there are severe limits to what government can accomplish. Organizational and managerial ability in the public sector, as elsewhere, is in short supply.

The policymaker needs to be conscious of an unintentional bias—looking instinctively at government for dealing with the problems of society, while ignoring the capacity of the private sector and market competition to deal with many of these questions, at least as effectively as government. Far too frequently, it has been government intervention in private markets that has generated exaggerated problems of rapid inflation and high unemployment which have, in turn, generated pressures for another round of government involvement. Thus, the goal which good public policy needs to strive for is not the maximum limit but the *optimum size* of government in a modern society.

The serious problem we now confront is not how to defeat the growing and at times misinformed public sentiment for limits to government, but to discover ways of responding to it sensibly and effectively.

HOW MUCH IS ENOUGH?

In theory, the state can own all the instruments of production and consumption. In such circumstances, all income and output flows would be recorded in the government's budget which would, in effect, also be the nation's budget. In the limiting case, therefore, the very question of the limits to government could not arise.

In practice, however, every nation has reserved some economic activities for private decision making. A small private sector exists in the USSR, for example, ranging from private purchases of vegetables in farmers' markets to the paintings created by individual artists.[1]

In capitalist nations, in contrast, citizens believe that their welfare is enhanced by a greater decentralization of power in the society. Although production of goods and services occurs primarily in the business sector, many organizations, public and private, profit-seeking and nonprofit, perform some economic function. Even in a capitalist country such as the United States, a large public sector exists. It performs a variety of functions, both sovereign and proprietary.

Adam Smith described three key duties of government—protection against external aggressors, maintenance of order at home, and erecting those public institutions and work "which, though they may be in the highest degree advantageous to a great society, could never repay the expense to any individual."[2] As Paul Samuelson has noted, a liberal enough construction of the last of Smith's three duties could, perhaps, be compatible with a vast expansion of the role of government.[3]

It would thus appear that the limits to government depend on where in the political spectrum a society is located. Socialistic or totalitarian states can and do operate with a dominating public sector, depending on the willingness of the citizens to put up with the centralization of power that results and the likely loss of economic efficiency that appears to accompany such political arrangements. Yet, there would seem to be at least a subjective limit to the expansion of private activity: the accomplishments of a strong and growing private sector—rewarding risk bearing, initiative, and extra work—could well undermine the public support for continuing a basically socialist or communist society. Therefore, the governmental authorities are pressed to limit sharply the private sector in such an economy.

A capitalist society, on the other hand, clearly needs government to enforce contracts and provide the infrastructure and external protection that Adam Smith wrote about. In another sense, government fills a vital role in the market-oriented societies. By correcting, or at least attempting to assuage, "market failures" and other by-products of private production and consumption, the actions by government preserve the essential public support for a predominantly capitalist society. Government taxes and expenditures soften some of the extremes in the distribution of income and wealth. Government curbs some of the external costs of private production and consumption, notably by a variety of environmental and safety regulations. If nothing else, government provides a safety valve for the expression of opposition by those disenchanted with the operations of the private enterprise system.

But how much is enough? In earlier years, some economists wrote about the limits of taxable capacity. Colin Clark, for example, maintained that, when government revenues reach 25 percent of the national income, the resulting reduction in business' profits and ability to expand would constitute a noticeable drag on the economy.[4] More modern economic analysis views the matter in a broader perspective and with less arithmetic precision, giving consideration to both the taxes levied and the resultant expenditures. Richard Musgrave, for instance, has urged that the concept of taxable capacity be discarded in favor of a concept of an optimal budget. Musgrave notes, however, the need to take account of the indirect and more or less hidden effects of budget operations under the efficiency of performance in the private sector. He warns of the possibility of government affecting the supply of work effort, pointing out that an income redistribution policy may reduce the level of output being redistributed.[5]

The fundamental question concerning the future role of government as the American society approaches the decade of the 1980s is whether those departures from private decision making have become so massive and pervasive that they weaken the basic capitalistic structure and, in the process, reduce citizen welfare to the extent that they arouse opposition to the continued growth of government.

THE GROWTH OF GOVERNMENT
IN THE UNITED STATES

Simultaneous with the rapid growth of government activities, public disenchantment with governmental institutions in the United States has become widespread. Political figures in both major parties regularly run for office pledging to cut back on government spending and/or taxes. In conjunction with this political phenomenon, a detailed survey of American public opinion, taken by Opinion Research Corporation in the late fall of 1978, reported that only 20 percent of the public believed that it is getting its money's worth for the federal income taxes it pays; 71 percent said they paid too much. Not too surprisingly, 73 percent favored legislation to put a limit on federal tax increases and 75 percent supported legislation to put an overall limit on the amount that the federal government can spend on all of its programs.[6]

Certainly, government revenues and disbursements (and often deficits as well) have been rising substantially. Federal outlays crossed the half-trillion-dollar point during fiscal year 1979, a doubling from fiscal 1973. In 1973, budget outlays equaled 20 percent of the gross national product. By 1979, the ratio was 22 percent. During the same six-year period, federal receipts rose substantially, but not at the same rapid pace. Consequently the budget deficit reached $40 billion in fiscal 1979, a substantial claim on the nation's supply of savings available for investment. The budgets of state and local governments have been rising at somewhat more rapid rates; the increase in public employment in the last two decades has been almost entirely at state and local levels (see table 11.1).

Perhaps of most important political significance, the tax burden on the typical middle-income family (earning $16,000 in 1977) has been increasing very substantially. The tax burden almost doubled from 1953 to 1977, rising from less than 12 percent of income to over 22 percent. As can be seen in figure 11.1, families with above average incomes also have seen a rising share of that income going to pay federal personal income taxes, social security taxes (OASDHI), and state and local taxes, but the rates of increase in the tax burden have risen at a bit more moderate pace (the total amount of taxes paid does tend to be a rising function of income during the entire period).

From a different perspective, we might consider the fact that the basic limits to government may not be *quantitative* at all. Society's capability to bear the

Table 11.1. Some Yardsticks for Measuring the Growth of Government, 1949-78

	1949	1978	Percent Change from 1949 to 1978
Public Expenditures (in billions)			
Federal	$41.3	$461.0	1016.2
State-Local	18.0	224.0	1144.4
Total	59.3	685.0	1055.1
Public Expenditures as a Percent of GNP			
Federal	16.0	21.9	36.9
State-Local	7.0	10.6	51.4
Total	23.0	32.5	41.3
Public Sector Employees (in thousands)			
Federal	2,047	2,850	39.2
State-Local	4,156	13,000	212.8
Total	6,203	15,850	155.5
Public Sector Employees per 1,000 Population			
Federal	13.9	13.2	−5.1
State-Local	26.1	56.8	117.6
Total	40.0	70.0	75.0

Source: Advisory Commission on Intergovernmental Relations, *Significant Features of Fiscal Federalism*, 1978-79 edition.

burdens imposed by government essentially is a subjective matter, being primarily a question of willingness or attitude. Nevertheless, as we will see, some types of government activities may interfere significantly with the ability of a nation's economy to function and to generate the resources to pay the various costs of government.

Though dichotomies are notoriously and treacherously simpleminded, it may be useful first to discuss the economic, in contrast to the political, limits to the role of government. The political limits, to which we will turn a little later, may be more visible and dramatic since these events make news headlines.

But the economic limits, albeit not so readily apparent, may be an important inhibiting factor, especially over a period of time. Government has powers to affect the economy and the society in many ways. No single statistical measure can hope to encompass all the effects that flow from the exercise of those powers

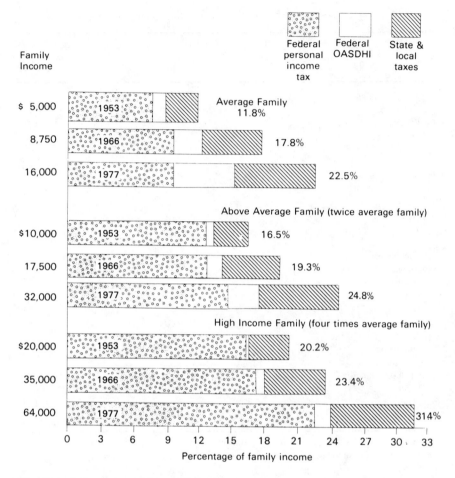

Fig. 11.1. A Comparison of Direct Tax Burdens Borne by Average and Upper Income Families, Calendar Years 1953, 1966, and 1977

Note: These estimates assume a family of four and include only federal personal income taxes, federal OASDHI, state and local personal income and general sales taxes, and local residential property taxes.

Source: Advisory Commission on Intergovernmental Relations, *Significant Features of Fiscal Federalism*, 1978-79 edition.

by a variety of federal, state, and local agencies in the United States. Totals of public expenditures omit an impressive array of off-budget agencies. Reports on revenues collected do not take into account the government's many credit programs. Tabulations of government personnel omit the large and growing numbers of ostensibly private workers performing public tasks under contracts with the government. The impact of deficit financing by governments may be dwarfed by the activities of the monetary authorities. And no budget statement can adequately assess the impacts of the regulatory burdens imposed on the private sector.

Thus, to be blunt, most analyses of the role of government in the United States have missed the central point. It is not so much a question of limiting the absolute size of government, although that is not a trivial concern and is worthy of some of our attention. Rather, it is a matter of understanding the pervasiveness of the thousand ways in which government enters the life of the average person and then reacting to that more complex condition.

ECONOMIC LIMITS TO GOVERNMENT

Far more important than the absolute or relative size of the dollar flows is the fact that government is intervening, on an increasing scale, in the daily lives of its citizens. Decisions by one or more government agencies alter, influence, or even determine what we can buy, how we can use the goods and services that we own, and how we go about earning our daily living. It is no exaggeration to say that governmental decisions also increasingly affect what we wear, what we eat, and how we play. There are very few items of consumer expenditures that escape regulation by one or more federal, state, or local government agencies.[2] The following is a representative sample:

Category	Regulatory Agency
Air travel	Civil Aeronautics Board
Automobiles	National Highway Traffic Safety Administration
Bank deposits	Federal Deposit Insurance Corporation
Boats	Coast Guard
Bus travel	Interstate Commerce Commission
Cigarettes	Public Health Service
Consumer credit	Federal Reserve System
Consumer products generally	Consumer Product Safety Commission
Consumer products containing chemicals	Environmental Protection Agency
Cosmetics	Food and Drug Administration
Credit union deposits	National Credit Union Administration
Drinking water	Environmental Protection Agency
Drugs	Food and Drug Administration
Eggs	Agriculture Marketing Service

Category	Regulatory Agency
Election campaign contributions	Federal Election Commission
Electricity and gas	Federal Energy Regulatory Commission
Firearms	Bureau of Alcohol, Tobacco, and Firearms
Flood insurance	Federal Insurance Administration
Food	Food and Drug Administration and Department of Agriculture
Housing	Federal Housing Administration
Imports	International Trade Commission
Land	Office of Interstate Land Sales Registration
Livestock and processed meat	Packers and Stockyards Administration
Meat and poultry	Animal and Plant Health Inspection Service
Narcotics	Drug Enforcement Administration
Newspaper and magazine advertising	Federal Trade Commission
Pensions	Internal Revenue Service
Petroleum and natural gas	Department of Energy
Physician fees	Professional Standards Review Organizations
Radio and television	Federal Communications Commission
Railroad travel	Interstate Commerce Commission
Savings and loan deposits	Federal Home Loan Bank Board
Stocks and bonds	Securities and Exchange Commission

The nature of regulation varies by product or service and by government agency. In some cases, the focus of control is on the price charged to the consumer, as in the case of the Civil Aeronautics Board and the Interstate Commerce Commission. In other instances, such as the Food and Drug Administration and the Consumer Product Safety Commission, the emphasis is on safety in product characteristics. Still other agencies regulate the method of marketing the product (the Federal Trade Commission and the Office of Interstate Land Sales Registration) or financing the sale (the Federal Reserve System). On occasion, regulation is aimed at restricting sales of goods to certain types of buyers (Bureau of Alcohol, Tobacco, and Firearms) or limiting the amounts of money that we can spend (Federal Election Commission).

Numerous government agencies are involved in regulating the working lives of Americans in their capacities as employees and employers. One or more government agencies are involved in the selecting, training, promotion, and retirement aspects of the job. Whether an individual is hired or not and whether he or she is retained by the employer can be influenced in many ways by actions of government agencies.

The National Labor Relations Board (NLRB) conducts elections governing union representation of workers. Under a "union shop" agreement, workers must—as a condition of continuing employment—join the certified union within a certain time period and pay dues to it regularly. The Department of Labor carries out the compulsory minimum wage laws as well as the statutes on "prevailing" wages on government-financed construction projects. The Equal Employment Opportunity Commission (EEOC) investigates charges of employment discrimination on account of race, sex, or national origin; and files charges on behalf of individual employees and groups of employees. If the company is a government contractor, it must follow an approved affirmative action program for designated categories of "disadvantaged" employees, including blacks, women, disabled people, veterans, and persons with Spanish surnames.

The work environment, likewise, is subject to regulations or influence of the Occupational Safety and Health Review Commission (OSHA), EEOC, and the NLRB. Compensation and fringe benefits are affected by each of these agencies. In addition, the company's pension fund is subject to approval by the Internal Revenue Service. The power to disallow employer contributions to pension plans which do not meet federal standards as tax-deductible business expenses is, of course, a powerful weapon.

From the management's view, basic aspects of operating authority have been taken over, in whole or in part, by various governmental agencies. In some industries, the government must give its approval before a company can go into business. In such specialized areas as nuclear energy, we might expect to see the abrogation of the normal process, whereby individuals and private groups decide who will enter a line of business and market forces determine who will stay. However, on a broad basis, governmental assumption of the traditional role of the market has become commonplace.

An even more representative example is that of interstate trucking. A potential new entrant must convince the Interstate Commerce Commission (ICC) that another company is "needed" before it will grant the necessary authority to operate on a given route. As would be expected, existing competitors usually contend, when they are invariably asked by the ICC as part of the approval process, that they are adequately meeting the public's needs and that no additional competition is necessary.

In the case of other regulatory activities, a company must obtain the approval of a government agency before it can sell a new product. This has become the standard practice in the pharmaceutical area and, as a result of the passage of the Toxic Substances Control Act, will soon become the order of the day in the chemical field as a whole.

When a company can proceed with producing a new product, it soon finds that a variety of agencies is involved in the production process. OSHA has established a multiplicity of standards governing an almost infinite variety of production activities and working conditions. Affirmative action programs require redesigning the work place to make job opportunities available for disabled workers. The Environmental Protection Agency (EPA) is concerned that production processes do not pollute the environment. The Department of Energy is involved in minimizing the use of natural gas and petroleum. And, on occasion, that agency requires companies to shift to coal where previously the same companies were forced by EPA to switch from coal to petroleum.

The multiplicity of regulators and regulations has, of course, increased the likelihood that an individual firm will be caught in the crossfire of conflicting directives. Many types of businesses increasingly find themselves between the proverbial "rock and the hard place." One company attempted to install an elevator for its handicapped workers as a part of its affirmative action program, but the company was delayed for six months by the EPA's insistence that it file an environmental impact statement. Federal food standards require meat packing plants to be kept clean and sanitary. Surfaces which are most easily kept clean are tile or stainless steel, but these are highly reflective of noise and may not meet OSHA noise standards. Coal must be combined with lime in order to remove the sulfur and thereby reduce air pollution, but this process generates large quantites of waste calcium sulfate, the disposal of which contributes to water or surface pollution.

Many other government requirements generate impacts that are much more visible to all of us, ranging from the seat belts that must be installed in the passenger automobile to the mandatory scrubbers on electric generating plants. But it is the indirect effects that are far more costly—the vast amount of corporate time and energy poured into paperwork and the subsequent slowdown of productive effort as the firm tries to meet the myriad regulations imposed on its operations. And on top of all that are the induced or third-order effects—the cumulative negative impacts of regulation on industry's pace of innovation and development, on its ability to finance growth, and on its capability to perform the basic function of producing goods and services.

Even more important than these inefficiency effects are the circular, "feedback" impacts of government intervention in economic activity. When government policies raise the inflation rate by increasing the cost of private production, they create pressure for greater government involvement via wage and price controls. When government regulation of business reduces the ability of and incentive for private enterprise to engage in technological innovation, the economy suffers a further reduction in its capability to achieve such important national objectives as greater job opportunities, rising standards of living, and an improved quality of life.

And when government policies sharply curtail the ability of the private sector to generate adequate savings to finance economic growth, not only is the government looked upon as the banker of last resort, but the basic vitality of the business system is called into question. Public dissatisfaction with business performance is increased in the process. In turn, that often sets the scene for another round of government involvement.

It is difficult to turn to a matter of concern to the average citizen without implicating some government action as a cause, at least in part, of the problem. For example, we have seen news stories about striking independent truck operators, long lines at gasoline stations, and high inflation. The relationship between government—especially deficit spending—and inflation is becoming clear to much of the public. With reference to the "shortages" of gasoline, many economists have been writing for years about the tendency of price controls to result in inadequate supplies in relation to consumer demand at the controlled price.[8]

But even the truckers' strike is related to government policy. A key issue is the inability of the truck owners to pass on the rising costs of diesel fuel—a situation which arises because of the rate restrictions (price controls) established by the ICC. The case of the truckers may be typical in two ways: (1) the public does not appreciate the extent to which government policies have created the problem in the private sector, and (2) the suggested solutions are virtually all proposed in terms of further government action. In this case, the public pressures seem to be directed at altering and expanding the government's efforts to allocate energy supplies within a system of partial price control and partial rationing.

Some conservative analysts believe that the many inroads by government which were described above may have fatally weakened the private corporation as a functioning economic unit.[9] As any detailed examination of the impacts of the rapid expansion of government regulation reveals, government intervention on an increasing scale has interfered with and inhibited the ability of the typical business enterprise in the United States to perform its basic economic functions.[10] In spite of this, however, there is little evidence of truly basic malaise or malfunctioning—as contrasted with specific shortcomings—in the capitalist system.

Therefore, deep pessimism about our economic system is not warranted. By its very nature, private enterprise will not merely endure, it will prevail. Albeit with various modifications, the private business firm is likely to continue to be the dominant economic institution in the United States for the foreseeable future, generating the incomes from which government secures its revenues. In the short run, therefore, it is more likely that effective pressures to limit the size of government will flow from political rather than economic sources.

POLITICAL LIMITS TO GOVERNMENT

From time to time, we may need to be reminded that, despite its many imperfections, the American body politic is directed, at least on fundamental matters, by the will of the people. The direction that will takes can be an important warning. The public opinion polls that report inflation to be the number one problem on the mind of the average citizen also show that government is closely implicated. According to the Opinion Research Corporation cited earlier, 57 percent of Americans stated that government was most responsible for inflation—compared to only 29 percent blaming labor and 15 percent holding business responsible.

But in spite of this citizen concern with the size and power of government, it is easy to predict the public reaction any time the Congress or the White House takes a specific action to close a Navy yard or an air field because it is no longer needed. Howls of anguish arise from the locality in which the military installation is located. A solid phalanx of business, labor, and civic groups in the community bitterly oppose this "blow" to their local economy. Sure, they are for economy in government, they respond, but why pick on them? And the unneeded bases usually remain open.

Similar examples of public opposition to cut-backs abound in civilian programs. Virtually every president since Harry Truman, Democrat and Republican alike, has tried to cut back the "impacted" school aid program—which, it turns out, is a subsidy from the general taxpayer to some of the wealthiest school districts in the nation. The result of the attempts is predictable: more howls of anguish and opposition to cut-backs. It is easy, of course, to berate the inconsistency in public attitudes. No one wants to volunteer to reduce his or her share of the national "pie." The basic cause of this state of affairs, however, lies in the differential nature of the distribution of expenditure benefits and tax burdens.

Beneficiaries of any specific government program tend to be very highly concentrated in specific segments of the population. Thus, a defense contract awarded to a firm in Los Angeles will generate employment in Southern California. A subsidy to sugar beet growers will generate income to farmers in Idaho. ICC barriers to new trucking firms will protect the profits of existing truckers, and so on and on.

In striking contrast, should any of those government activities be reduced, the saving to any individual taxpayer or consumer will be extremely modest, and usually not worth the major effort needed to obtain the result. Thus, a strange type of benefit/cost analysis tends to dominate government decision making: Government programs that generate more costs to the nation than benefits are enacted and maintained because these programs generate more benefits than costs for the concentrated pressure groups which are the beneficiaries. Moreover, the persistent notion that it is the duty of members of the legislature to obtain those special benefits for their constituents is as blatant as it is widespread. In June 1979, for example, a member of the House of Representatives—angry that other legislators were attempting to eliminate a pet project in his district—suggested the establishment of a Pinocchio award. The idea would be to ridicule a legislator who "sticks his nose" into another's business. In the Congressional debate, the very right of a member to oppose a project in another district was challenged and the specific efforts to eliminate low priority pet projects were defeated.[11]

This situation, of course, is precisely the opposite of the typical justification for government intervention in the private economy. The "market failure" justification for many governmental undertakings argues that uncontrolled private activity may generate private benefits greater than private costs, to the extent that government involvement is necessary because the social costs may be greater than any social benefits (the classic case being environmental pollution). But, far too frequently, "government failure" itself generates an excess of social costs over social benefits.

Although citizens have been aware of the limits of government as well as its immense size, scholars are now examining those shortcomings of governmental activity in a comprehensive way. As Peter Schuck has pointed out, however inadequate consumer information may often be, information in the political marketplace (where there is no Federal Trade Commission to police claims) is probably worse.[12] Charles Wolf, Jr., has noted the various shortcomings of government activity, including the "internalities" it generates. In contrast to the "external" costs generated by business activity (which is a standard justification for government involvement), the government apparatus itself develops a set of "private" goals—such as enlarging the size of a government bureau in order to justify a promotion, a fancier title, and more staff.[13] The point, however, should not be pushed too far. For the most part, government employees may believe deeply in the objectives of their programs and contend that they contribute more to the general welfare than if the resources remained in the private sector.

But belief and sincerity may only make the problem of change more difficult. These sentiments and good intentions only complicate any effort to achieve the public desire for more effective and less burdensome government. Unlike a private business faced with unwilling customers, modern government does not seem to have much capacity to correct itself. Government officials have tended to avoid tough choices and decisions. New government priorities have been superimposed on old programs; unlike old soldiers, old programs neither die nor fade away. For example, neither the proponents of national health insurance nor the advocates of subsidies for energy development have coupled their proposals with any planned reductions in existing government programs.

The result is the condition described earlier in this chapter—a great increase in both the size of government budgets and in those "off-budget" costs that are imposed indirectly on the private sector. Despite the procedural improvements that were made by the Congressional Budget and Impoundment Control Act of 1974, it is difficult to demonstrate that the institutional changes to date have resulted in any significant reduction in the growth of the federal government or even in any major rearrangement in the allocation of resources (e.g., in any shifts in priorities). At best, the flow of authorizations for new programs has only been slowed down.

Under the circumstances, then, it is not surprising that citizens across the nation are increasingly looking to new and unconventional ways of controlling government activity. For instance, the overwhelmingly favorable vote to cut property taxes in California was more than a "straw in the wind"; it was a true harbinger of things to come. On the same Tuesday that Proposition 13 was enacted, taxpayers in Ohio turned down 86 out of 139 school tax levies. In November 1978, thirteen states had on their ballots measures designed to restrict government taxation and/or expenditures. Ten out of the fifteen measures passed (see table 11.2). The successful ones ranged from a property tax limit in Alabama, to a California-type property tax rollback in Idaho, to a lid on state spending in Texas.

The rising attention to bills in the Congress to alter the Constitution to mandate a balanced budget or to otherwise constrain government is another indicator of citizen arousal. Table 11.3 shows the array of alternatives currently being considered. These political measures take three general forms. The first strategy, seen in the tax and expenditure control bills and in the balanced budget control bill, aims at limiting government spending through the imposition of procedural changes in the Congress that would increase the political accountability of elected officials. Each of these proposed bills—the one for indexing the personal income tax to prevent increased taxation generated by inflation; the "sunset" legislation to evaluate the effectiveness of existing federal programs; and the proposed requirement that both the President and the Budget Committees submit balanced budgets—requires legislators to take publicly recorded stands on issues that involve the continued expansion of government.

The second strategy is statutory and would impose formal legal restraints on federal taxation and spending. The two Kemp-Roth proposals would require by law a 10 percent reduction in income tax rates for three consecutive years, and annual reductions in federal expenditures based on percentages of the Gross National Product. The Balanced Budget Proposal would require that the congressional resolutions on the federal budget be in balance, unless they are overridden by a two-thirds majority in both houses of Congress.

The third strategy involves direct constitutional control of the budget, subject to an extraordinary two-thirds majority rule in the Congress. One proposed amendment would limit federal taxation to 15 percent of the GNP. Another would prohibit federal spending from rising faster than the GNP and would reduce its growth rate in times of inflation. Another proposal would constitutionally require that congressional budget resolutions be in balance. The philosophy behind the constitutional strategy, not unlike the statutory proposals, is that Congress is too responsive to special interests, and that limits to its spending must be set forth in law.[14]

Table 11.2. 1978 November Ballot Measures

State	Constitutional Amendment	Statute	Passed	Failed	Type
Alabama	X		X		Property tax limit.
Arizona	X		X		State spending lid, can be exceeded by ⅔ vote of legislature.
Colorado	X			X	State and local spending lid, can be exceeded by majority vote of people.
Hawaii	X		X		State spending lid, can be exceeded by ⅔ vote of legislature.
Idaho		X	X		Jarvis-type property tax rollback, new local taxes need ⅔ vote of people; and new state taxes, ⅔ vote of legislature.
Michigan Headlee Amendment	X		X		State and local spending lid, can be exceeded by majority of voters.
Tisch Amendment	X			X	Property tax and income tax reductions.
Missouri	X		X		Gives legislature authority to rollback property taxes.
Nebraska	X			X	Local spending lid, can be exceeded by majority vote of the people, ⅘ vote of the legislature.
Nevada	X		X		Jarvis-type property tax rollback—new local taxes need ⅔ vote of the people, new state taxes ⅔ vote of the legislature—must pass again in 1980.
North Dakota		X	X		Income tax reduction.

Table 11.2 (cont.) 1978 November Ballot Measures

State	Constitutional Amendment	Statute	Passed	Failed	Type
Oregon Measure 6	X			X	Jarvis-type property tax rollback, new local taxes need ⅔ vote of the people, new state taxes ⅔ vote of the legislature.
Measure 11	X			X	Major property tax reduction with state assuming 50% of property tax bills.
South Dakota	X		X		Any increase in state taxes must have ⅔ vote of legislature or majority of people.
Texas	X		X		State spending lid, can be exceeded if legislature declares an emergency.

Source: Advisory Commission on Intergovernmental Relations

Table 11.3. An Outline of Proposals for Fiscal Restraint

Proposal	Basic Approach	Implementation Strategy	Description
Indexation Bill HR 365 Rep. Willis Gradison	Tax control	Strengthen political accountability	Indexes the federal individual income tax to offset inflation.
Sunset Bill S 2 Sen. Edmund Muskie	Expenditure control	Strengthen political accountability	Establishes a ten-year schedule for reauthorizing all federal programs.
Budget Reform Amendment to HR 2534 Sen. Russell Long	Balanced budget control	Strengthen political accountability	Beginning with FY 1981, Congress must consider a balanced budget resolution.
Kemp-Roth Income Tax Proposal S 33	Tax control	Statutory	Individual income tax rates would be reduced annually by 10 percent for three consecutive years, with indexation in following years.
Kemp-Roth Spending Limitation Proposal S 34	Expenditure control	Statutory	Would require that federal outlays be held to 21 percent of GNP in FY 1980, 20 percent in FY 1981, 19 percent in FY 1982, and 18 percent in FY 1983.
Balanced Budget Proposal S 13 Sen. Robert Dole	Balanced budget control	Statutory	Would amend the Congressional Budget Act to require that Congressional resolutions on the budget be in balance; provision for override by ⅔ majority of both houses.
Tax Limitation Proposal HJ Res 43 Rep. Tennyson Guyer	Tax control	Constitutional	Would amend the Constitution to provide that total taxation by the federal government not exceed 15 percent of GNP; provision for override by ¾ majority of both houses; also has balanced budget provision.
National Tax Limitation Committee Proposal SJ Res 56 Sen. H. John Heinz, III	Expenditure control	Constitutional	Would amend the Constitution to prohibit federal spending from rising faster than GNP; also reduces growth margin when inflation is greater than 3 percent; provision for override by ⅔ majority of both houses.
Balanced Budget Proposal SJ Res 4 Sen. Richard Lugar	Balanced budget control	Constitutional	Would amend the Constitution to require that Congressional resolutions on the budget be in balance; provision for override by ⅔ majority of both houses.

Source: Advisory Council on Intergovernmental Relations

325

Although the notion of mandatory budget balancing—whether by constitutional amendment or otherwise—is a perennial favorite of advocates of smaller government, that approach might not necessarily result either in smaller government or lower taxes. Eliminating the massive deficits of the federal government would help alleviate inflationary pressures. However, should a requirement for budgetary balance be enacted, one result might be new pressure for tax increases on the part of the traditional proponents of big government.

Also, the "escape valves" inserted in many of the proposals for compulsory budget balancing can cause problems. If the Congress can readily violate the intent of the amendment by declaring an emergency, then the measure would be seen as little more than pious rhetoric. Alternatively, however, if no "loopholes" are provided, then the possibility would arise of substantial economic damage in the event of a real emergency, such as a large-scale depression or a major war. Thus, less dramatic but more realistic approaches need to be considered to achieve what Paul Samuelson has termed the "disciplined and informed choice by a democratic people of their basic economic policies."[15]

A more direct way of attempting to control the public sector may be based on the concept that taxes are the lifeblood of government bureaucracies. To be sure, deficit financing and off-budget transactions provide some leeway. But, by and large, the flow of revenues into public treasuries is the key determinant of the ability of government agencies to expand their activities. Perhaps even more to the point, it is via the payment of taxes that the individual citizen feels the burden of big government most directly.

Therefore, an effort to trim the size of government might well focus on tax policy as the prime mechanism. At the state level, a lid on the overall tax burden (defined as tax payments as a percent of personal income) may be the most fruitful approach. Moreover, because revenues and expenditures (at least from 'own" sources) are closely connected, a lid on spending might, in effect, achieve the same purpose. Of the successful state measures shown in table 11.2, four were of that general variety; Arizona and Hawaii passed limits on spending which can only be circumvented by a two-thirds vote of the state legislature; Michigan enacted a spending ceiling which can be exceeded by a majority of the voters; the Texas lid on spending can be breached if the legislature declares an emergency.

At the federal level, the most direct method for setting limits to taxation and spending may be a sustained, across-the-board reduction in income tax rates. The key point in such a measure is to put substantial tax reduction at the top of the governmental agenda. Therefore, with an anticipated lesser flow of revenues into the national government, planning of government spending would have to be more modest than it has been. But to date, taxes have been cut only after budget appropriations have been passed, and thus the practical result tends to be larger budget deficits. In short, limits to government tend to expand rather than contract.

A fundamental change could occur in government thinking if the revenue cuts preceded the determination of budget appropriations for future spending. Rather than concentrating on the still further expansions in government that could take place (which is the traditional approach), the White House and the Congress—and their state and local counterparts—would be forced into a new way of proceeding. They would have to ferret out old and obsolete programs that are no longer worth doing under the new fiscal restraint imposed by the cuts in revenue. Moreover, the tax reductions might also force the legislative and executive branches of government to pay serious attention to the mechanisms available to reform government.

For specific government programs, those involving making expenditures as well as using rule-making authority, economics offers a straightforward reform mechanism for assisting decision makers: benefit/cost analysis. Although looked upon by many as a "green eyeshade" device, the benefit/cost approach can take on more substantive significance when properly used. The basic notion is not a fascination with accounting details but, rather, the idea that government decision makers should carefully examine the disadvantages as well as the advantages of their proposed actions and then weigh one against the other before they act.

That examination should be made from the perspective of the society as a whole, rather than from the limited viewpoint of any specific government agency or private interest. Moreover, in the course of the analysis, the most economical and effective means of achieving the public purpose should be identified. Thus, properly performed, benefit/cost analysis is a useful tool in raising the sights of public policymakers and in assuring that—at least to the extent that the effects can be measured—government does more good than harm in the policies and programs it undertakes.

We need to recognize, of course, the counterarguments of some advocates of the status quo, who contend that specific government undertakings are so worthy—e.g., national defense or social regulation—that they should not be subject to the scrutiny of "green eyeshade" accountants and economists. But, hopefully, a more informed review of each government program will, in many cases, yield ways of achieving higher levels of effectiveness from a given outlay of resources for each of these purposes. Such improvements will not be possible so long as major shares of government activities are treated as "sacred cows," immune to analysis of the various impacts—good and bad, intended and unintentional—that flow from them.

Another sensible change is for all government activities to be subject to a "sunset" mechanism; Senator Muskie's proposal, in table 11.3, is an example. Each agency would be reviewed by the Congress on a timetable to determine whether it is worthwhile to continue it in light of changing circumstances, or whether the "sun" should be allowed to "set" on its existence. Many government programs tend to be prolonged far beyond their initial need and justification.

In a world of limited resources, the only sensible way to make room for new priorities is periodically to cut back or eliminate older, superseded priorities.

In the final analysis, however, a new way of thinking about government is required: in essence, the notion that, because society's resources are limited, government cannot attempt to meet every demand of every group within the nation. Those resources, moreover, are not merely economic or financial. As has been demonstrated in both military and civilian areas in recent years, there are severe limits to what government can accomplish. Organizational and managerial ability in the public sector, as elsewhere, is in short supply. And since society has given government many important responsibilities—ranging from maintaining the national security to providing a system of justice—it is important that government do well those tasks that it is charged with performing. More important, however, we need to take note of the fact that society has given government *too many* important responsibilities. So if we find that government is too big and too powerful, then we need to remember that it has become that way, to a large degree, as a response to public pressures. If we find that government has too much control over our lives, then we must realize that it is our choice and our duty to regain that control and to reassume the responsibilities we have delegated. That is likely to occur only if a proper distinction is made between the truly important need for governmental intervention and the relatively trivial matters that are of passing public concern.

A final thought along these lines is that the public policymaker needs to be conscious of what often is an unintentional bias—looking instinctively at government for dealing with the problems of society, while ignoring the capacity of the private sector and market competition to deal with many of these questions, at least as effectively as government. Far too frequently, it has been governmental intervention in private markets that has generated or exacerbated problems of rapid inflation, high unemployment, low productivity, reduced private initiative, and slowdowns in innovation which, in turn, have generated the pressures for another round of government involvement. Thus, the goal which good public policy needs to strive for is not the maximum limit but the *optimum size* of government in a modern society.

The relationship between the government of a society and the private sector is a delicate one, involving difficult balancing of many important considerations. Clearly, polar alternatives are unrealistic in modern Western nations. Laissez-faire is hardly a sensible response to the needs of a modern, high-technology society. People do need to use the powers of government to help members of society achieve important objectives that they cannot attain on their own.

Yet, as we look around the world, it appears all too easy to move far beyond that position toward that other polar alternative—the totalitarian state—where the government winds up directing the details of the day-to-day lives of the individual members of society and substituting public for private values. Some caution, then, does seem to be in order. That the sentiment for Proposition 13 has not yet led to constitutional restraints on government should not be misinterpreted. More carefully fashioned bills have been gaining passage in several states and, at least for the time being, the national pressures are resulting in a slow-down in the growth rate of government at all levels. The serious problem we now confront, therefore, is not how to defeat the growing public sentiment for limits to government, but to discover ways of responding to it sensibly and effectively. The "invisible hand" that Adam Smith spoke of may or may not help find those ways, but it is our own quite visible hands that must carry them into action.

NOTES

1. Frederic L. Pryor, *Public Expenditures in Communist and Capitalist Nations* (Homewood, Ill.: Richard D. Irwin, 1968), p. 27.

2. Adam Smith, *An Inquiry Into the Nature and Causes of the Wealth of Nations*, Modern Library edition (New York: Random House, 1937), p. 681.

3. Paul A. Samuelson, "Aspects of Public Expenditure Theories," *Review of Economics and Statistics*, November 1958, pp. 332-38.

4. Colin Clark, "Public Finance and the Value of Money," *Economic Journal*, December 1945, pp. 371-89.

5. Richard A. Musgrave, *The Theory of Public Finance* (New York: McGraw–Hill, 1959), pp. 51-52.

6. Harry W. O'Neill, *The Mood of America: A Post-Election Survey*, a paper presented before the Executive Briefing on the New Outlook for Government Regulation of Business, Washington, D.C., November 30, 1978, pp. 13-19.

7. See Murray L. Weidenbaum, "Government and the Citizen," in *Dilemmas Facing the Nation*, edited by Herbert Prochnow (New York: Harper and Row, 1979).

8. See Kenneth J. Arrow and Joseph P. Kalt, *Regulating Energy Prices: The Case of Oil*, Discussion Paper Number 715, Cambridge, Mass., Harvard Institute of Economic Research, 1979.

9. Michael C. Jensen and William H. Mckling, "Can the Corporation Be Saved," *MBA*, March 1977.

10. See Murray L. Weidenbaum, *The Future of Business Regulation* (New York: Amacom, 1979).

11. Ward Sinclair, "Meddling Members of Congress Warned Off Others' Pet Projects," *Washington Post*, June 16, 1979, p. A9.

12. Peter H. Schuck, "Regulation: Asking the Right Questions," *National Journal*, April 28, 1979, p. 711.

13. Charles Wolf, Jr., "A Theory of 'Non-Market Failure': Framework for Implementation Analysis," *Journal of Law and Economics*, April 1979.

14. For details see, John Shannon and Bruce Wallin, "Restraining the Federal Budget: Alternative Policies and Strategies," *Intergovernmental Perspective*, April 1979, pp. 8-14.

15. Paul A. Samuelson, "Two Views of the Budget-Balancing Amendment," *AEI Economist*, April 1979, pp. 4-6.

Chapter 12

THE REVIVAL OF ENTERPRISE

Theodore J. Gordon

Enterprise is defined as the willingness to venture on bold, hard, and important undertakings with energy and initiative. In our culture, we save the term for many of our most heroic images and myths: ships that sailed against the Beys of Tripoli, aircraft carriers, Captain Kirk and his starship, the space shuttle, and the go-getters (individual and corporate) who shaped our nation. *Enterprise* captures the zeitgeist of an active society moving to improve its lot against odds, sloth, and random choice. It describes not only an organization but the way it functions. As used here, it includes purposeful productivity.

Yet *enterprise*, in our time, has a strangely negative pall. The heroic ships meant war; the go-getters went too far; the starships are for another age. We are modern, somehow postenterprise; our myths have led us astray. Horatio Alger found Watergate, drugs, and the tragedy of the commons. Our American system of enterprise lies in modest disarray, encumbered by regulation, and by diverse and conflicting views about what constitutes the public interest. To the developing countries, the Western model of enterprise is at one time the cause of unfilled aspirations and the cure for poverty.

Adam Smith wrote of enterprise in *The Wealth of Nations*. In Smith's view, the production and exchange of goods can be stimulated, and the standard of living thereby improved, only through the efficient operation of private, industrial, and commercial enterprise, with minimum regulation and control by government. While this view has changed since the mid-eighteenth century, it provides important perspective for economic thought today. Each person working toward his own ends, improving his state of affairs through his actions—his enterprise—produces a society that is best not only for the individual, but for society as a whole. A nation, acting according to such principles, improves not only its standards, but the standards of other nations as well. This is Smith's great perception of the *invisible hand*: a society in which each individual works toward selfish ends resulting in the highest order of total well-being.

Competition is the balance wheel. Where competition is unimpeded, manufacturers and suppliers of services cannot exploit a strong demand and mark up their prices because, in concept, new competitors would step in to satisfy the market at lower prices. Similarly, the company that pays less than prevailing wages loses its labor force to those who pay fully. Thus, socially desirable behavior is encouraged by competition. Galbraith adds to this idealized self-regulating system the idea that market power can be exercised by strong buyers against weak sellers as well as by strong sellers against weak buyers. Organized labor is one source of this countervailing power; consumerism is another.

We have long since attempted to guide the invisible hand by defining acceptable limits of behavior for individuals, corporations, organizations, and nations. Yet these attempts to shape economic order bring with them as many risks as advancing technology and are viewed by many as unwarranted interventions. In terms of the operations analyst, the invisible hand involved suboptimization; that is, only subelements of a larger system are optimized. Operations analysts can demonstrate mathematically, and judgment provides intuitive evidence, that suboptimization does not lead to optimum systems. In other words, a society composed of individuals acting according to Smith's invisible hand, working to improve their own individual situation, will not necessarily be optimum in the aggregate.

So on one hand exists the notion of enterprise: the challenge to be met and the life to be fashioned. Conventional free market economic theories since the time of Adam Smith support this view. On the other hand, enterprise, unbridled, is apt to be socially chaotic, unjust, and destructive. Out of this tension comes all mainstream political and economic thought, from Marx to Friedman. Out of it also come all explanations of our present state and its implications for the future, from Shumacher and Meadows to Kahn.

There is a feeling in this country that enterprise is dead, or at least seriously ill. Reference is made to diminishing productivity growth, the decreasing rate of new product introductions, and loss of competitiveness on the international market. Causes offered include overburdensome regulation, excessive demands of labor, "topping out" of previously fecund technological domains, and loss of national will and spirit. But,

- Was enterprise ever as real and intense as we remember, or was it only a myth, a chimera, more vital in recollection that reality?
- If it was real, have things changed; has enterprise diminished as we suspect?
- If enterprise has diminished, why?

- Given the reasons for shift, can it be revitalized?
- Even if it could be revitalized, should it be; do we really need enterprise in the years ahead or has its time, like Victorian moralism, gone by?

These five questions constitute the framework for the inquiry contained in this chapter.

WAS ENTERPRISE EVER REAL IN AMERICA?

Over the last 200 years in our country, there was a spectacular confluence of forces that gave rise to unparalleled economic and social growth. Markets were created by the growing numbers of people in our country, their inexorable move toward the West, the entry of the United States into world trade, and the spectacular rise in affluence by any measure—quantitative or qualitative—of the average American. These markets could be served because of a vital labor force—sometimes imported—the ready availability of inexpensive raw materials, and the creation of a national capital base. These forces came together in an era of invention that ran continuously from the first beginnings of electromagnetic theory to large-scale integrated circuitry; from phlogiston (a "heat fluid" used in the erroneous explanation of thermodynamics) to the internal combustion engine; from the birth of atomic theory to nuclear reactors; from the telegraph key to the communications satellite; from the frontiers of western Massachusetts to the frontiers of space. And the invention was not confined to mechanical devices; it pervaded the social arena as well.

We saw in this same interval the transition from an agricultural society to an urban culture; unprecedented experiments in freedom of life-style from generation to generation; the emergence of the political system in America; the growth and demise of imperialism; the advent of bipolar military stability; the abolishment of slavery; development of the corporate form and the regulatory structure that shaped it; the shift from exploitation of the environment to its preservation and restoration; changes in authority and structure of almost every institution from the family to the church and school, even changes in what constitutes the "good life," what a citizen owes his country, and what a country owes its citizens.

The weave of this fabric of enterprise and growth is extremely complex and it is very difficult to separate cause from effect. Without innovative technology, the labor, capital, and raw materials would have been devoted to mundane improvements, not the fireworks of technological revolution. Without social innovation, all would have been a rigid exercise in the preservation of archaic and static values. But the whole was dynamic, vital, discontinuous, explosive; and the consequences are undeniable. In the space of a few generations in our country, we saw the transition from an agrarian, rural nation to an industrial, urban nation. The number of people employed in agriculture today is about equal to the number employed in agriculture in 1840, but, in that same time, the size of the domestic population served by the agricultural labor force rose from 17 million to over 200 million. Since 1939, the number of business enterprises has grown from 1.8 million to 14 million; of this number, corporations have grown from 400,000 to almost 2 million. Exports have increased since 1939 from about $4 billion to about $180 billion, and imports in the same interval from about $3 billion to about $190 billion. Even in constant dollar terms, affluence has grown tremendously. In 1951, only 2.3 percent of American families had incomes over $15,000 (1967 dollars); by 1970, 18 percent were in this bracket. Since 1960, median family income has increased more than 30 percent in constant dollars. In 1974, 84 percent of households owned at least one automobile and 33 percent, two or more. Ninety-seven percent of households owned at least one TV set; 61 percent, a color TV. Some 32 percent had room air conditioning; 20 percent had central air conditioning. Ninety-nine percent had a refrigerator; 72 percent a washing machine. The list of statistics could be much longer; but it all points to this time being the age of affluence.

Of course, such statistics do not reflect quality, only measures of state. Quality of life, some argue, has been diminishing along with this growth and these changes are also undeniable: in the domains of product reliability, the environment, and violent crime, to name a few that come easily to mind, conditions have deteriorated.

The complex mesh of forces that led to these huge changes was driven not by fiat and design, but largely by individual innovators, entrepreneurs, and social revisionists. The names that come immediately to mind are Edison, Ford, Armour, Birdseye, Pullman, McCormick, Whitney, Hartford, Singer, Howe, Rockefeller, and Morgan: organizers, entrepreneurs, go-getters carrying ideas and concepts to the market, creating markets. Theirs is the history of enterprise. With the advantage of current perspective, we see some of that enterprise as destructive, and no one would wish those excesses to be imposed again. But the changes they brought made our world as we know it, for good—and there's a lot of that—and bad.

As Daniel Boorstin, Librarian of Congress in Washington, said:

> The century after the Civil War was to be an Age of Revolution—of countless, little-noted revolutions, which occurred not in the hall of legislatures or on battlefields or on the barricades, but in the homes and farms and factories and schools and stores, across the landscape and in the air—so little noted because they came so swiftly, because they touched Americans everywhere and everyday. Not merely the continent, but the human experience itself, the very meaning of community, of time and space, of present and future, was being revised again and again; a new democratic world was being invented and was being discovered by Americans wherever they lived. [1]

The innovations were mechanical and institutional, and often one or the other in a counterpoint so intimate it was difficult to tell where one began and the other ended.

A little more than 100 years ago, Edwin L. Drake pumped the first oil well at Titusville, Pennsylvania. According to all accounts, it was a flash of inspiration that led him to drill; until that time, petroleum was collected from seepage pools on the surface or from deposits floating on streams. Its principal use was medicinal, although through privately funded research it was known that petroleum (or ''rock,'' as it was known) could be used in lamps with much less odor than kerosene (known as ''coal oil'') and was an excellent lubricant. The dogma of the time was very strong and well drillers that Drake contacted thought he was insane. Yet the first well, at a depth of about 70 feet, yielded oil. That was on August 27, 1859. By 1865, western Pennsylvania was in the midst of an oil boom. The situation was chaotic; oil was being produced but storage area was in short supply. Furthermore, no systematic means existed for refining the product and bringing it to market as an illuminating fuel. John D. Rockefeller, sensing the opportunity, became a partner in a refinery in 1863 (he was then 24) and bought out his partners two years later. By the end of 1865, his venture had grossed more than $1.2 million. His refineries survived the depression of 1867-1868 and, by 1872, his company was refining 10,000 barrels per day. The position of Standard Oil, his company, was solidified in a ''great monopoly.'' This organization was ruthless with respect to competition and set prices in monopolistic fashion; it was the basis for the Sherman Anti-Trust Act of 1890, which made such trusts illegal.

The market for the products fashioned from petroleum would have declined early in this century as electric illumination became popular were it not for the coincident popularity of the automobile. The story of petroleum involved the inspiration that led to drilling for oil, the organizational genius that led to its massive refining and use, and the coincidence of emerging markets.

The inspiration of Richard Sears in 1886 was to set up a mail-order watch business, promoted with clever advertising. By 1893, the business had grown into Sears & Roebuck and the form was one we recognize as quite modern. One million copies of the catalogue were being printed by 1904 and three million by 1907. Woolworth took another tack: ''price-lining,'' a system in which goods are sold, not on the basis of their brand, but on the basis of their price. More recently, we have seen the growth of franchising of business; easy credit enhanced by credit cards; electronic funds transfer; the agglomeration of markets by TV; easy travel by automobile, bus, air; huge proliferation of printed materials; dissection of markets into smaller and smaller specialized segments; and the rise of rapidly changing fashion.

We forget that even government institutions can be innovative. Take the Post Office. Before 1847, the recipients of letters had to pay for their postage. This meant that letter carriers had to find the recipients in person to collect the fees. This slowed communications. In 1847, the postage stamp came into use—with a lot of unhappiness because it required prepayment by the sender and obviously, it was argued, the recipient was the person who received the service—the information—and thus should pay. The mail carriers at the time were not federal employees; most people still went to the Post Office to collect their mail. In large cities the confusion was unimaginable. In 1863, Congress provided government salaries for mail carriers, with no extra postage for city dwellers. Free delivery to the farm—RFD—came later, and it was a political battleground. Congressmen argued about the possible impacts of RFD without much data—an early example of technology assessment—but some saw in the new system the prospect of lessening the isolation of the farm. Congress passed a resolution providing for this service in 1893 and, by 1898, RFD was available to groups of rural dwellers who petitioned their Congressmen for the service. The mailbox design was also a focus of controversy. When the design was published by the Postmaster General there were charges of monopoly, but the RFD system was operating well by 1906. According to Boorstin:

> This was the least heralded and in some ways the most important communications revolution in American history. Now for the first time it was normal for every person in the United States to be accessible by cheap public communication. For the rural American (more than half the nation's population by the census of 1910), the change was crucial. Now he was lifted out of the narrow community of those he saw and knew, and put in continual touch with a larger world of persons and events and things read about but unheard and unseen. RFD made these everywhere communities possible. From every farmer's doorstep there now ran a highway to the world. But at the price of dissolving some of the old face-to-face communities.[2]

Parcel post did not exist before 1913. Before that time, the maximum weight for domestic mail was four pounds. Packages were carried by private express companies: Adams Express, American Express, United States Express, and Wells Fargo. Although it seems strange to us now, opponents of parcel post warned that it would "change fundamentally our conception of government."[3] Within a month after the service began in 1913, the Postmaster General called parcel post, "the greatest and most immediate (success) ever scored by any new venture in the country."[4] Within a year, 300 million packages per year were being mailed. And with the introduction of parcel post, the big mail-order houses of Sears and Wards thrived and the old general store began its decline. When we think of the Post Office now, losing money every year, faced with the challenge of private carriers and the technological revolution of electronic mail, it seems hard to believe that the institution was a carrier, not only of mail, but of social innovation and bureaucratic revolution.

Americans, at least until recently, seemed to have had a frontier mentality. The frontier is where the action is; that is where careers are made, fortunes are to be found. The first American frontiers were geographic, pushing West until the nation was filled. The next frontiers were markets based on technical and institutional innovation, as well as inexpensive resources. Scientific frontiers pushed back the boundaries of the unknown, and the technological frontiers pushed forward the boundaries of the doable. Space, more recently, rekindled the old instinct, and novel social programs quite appropriately were called—at one stage—the new frontier.

Certainly, in all of this there were excesses and abuses. The line between enterprise and exploitation was often blurred and vague and changed as modern values emerged. The common environment was used without question as a source and a sink for production. Child labor was not only acceptable at one time but considered a virtue. Trusts, clearly seen as exploitive in retrospect, were the economic engine for accumulating capital which, reinvested, promoted growth and social development. In short, the virtue of enterprise and the evil of exploitation could not, cannot, be defined in any absolute sense established by measurable specifications which are unchanging in time. Rather, they are concepts evaluated in terms of the morality of the moment.

The morality of enterprise has changed over our history. Four epochs are apparent: the emergence of laws affecting operation of business from an antitrust standpoint, the creation of regulated monopolies, the rise of laws defining the rights of labor, and, finally, the initiation of laws protecting the consumer and the environment.

The first major law affecting the general conduct of business was the Sherman Anti-Trust Act passed in 1890. This act prohibited contracts or business combinations that were formed for the purpose of restraining trade or commerce between states. Any person found to be monopolizing interstate trade or commerce would be guilty of a misdemeanor. The Clayton Act of 1914 prohibited local price discrimination, exclusive selling agreements, tying agreements, holding companies, and interlocking directorates.

The Sherman and Clayton Acts were beneficial in helping to restrain the monopolist powers of the trusts. Other federal legislation complemented these acts; for example, in 1914 the Federal Trade Commission Act created a board charged with the responsibility of preventing the formation and continuation of illegal trade combinations. In 1936, the Robinson-Patman Act forbade price discrimination and eliminated many of the practices used by chain stores to secure price concessions from their vendors. Although the definition of what constitutes restraint of trade has changed with time, the spirit of these early laws still is very much in evidence today.

The second force for change in business practices was the rise of regulated industries. The policy of promoting competition could not apply when competition was not deemed to be in the public interest. In these instances, the government chose to permit limited monopolies, controlled through federal regulation. The first major act dealing with direct regulation was the Interstate Commerce Act of 1887. It was aimed specifically at the railroads; and major portions of the law were designed to provide that all rates be just and reasonable, prohibit personal discrimination in the form of special rates and rebates, and create an Interstate Commerce Commission to hear complaints of violations. In 1906 pipelines came under government regulation; in 1920, water transportation;

motor carriers in 1935; and air transport in 1938. The Public Utility Holding Company Act of 1935 and the Natural Gas Act of 1938 did much to regulate the utilities industry. The Federal Communications Commission was formed in 1934 to regulate the communications industry. These acts illustrate two major trends of government intervention: first, the acceptance of additional monopolies and the efforts to regulate them; and second, the desire to prevent monopolies and encourage competition in the rest of the economy.

Labor unions provided a third source of pressure for change in American business practices. American labor unions had been attempting to organize almost since the signing of the Constitution. The first craft unions were confined to the Northeast and North Central sections of the country and involved only a few industries: ship workers, printers, and shoemakers. Although unions in the 1830s showed a substantial growth rate, the panic of 1837 hurt their cause. Employers typically resisted union demands and, in many instances, successfully sought court protection. Early court cases held unions to be illegal conspiracies under common law.

After the panic in the early 1850s, a resurgence in national prosperity brought workers back to the unions in great numbers. More than 400 strikes were staged during 1853 and 1854. In 1869, the Knights of Labor came into existence. Formed by a small band of Philadelphia tailors, it was originally a secret organization. By 1886, its membership had risen from 9,000 to 700,000. Poor leadership eventually led to its demise, but another craft group, the American Federation of Labor (AFL), formed in 1886 under the leadership of Samuel Gompers, survived and prospered. By 1900, AFL membership was 870,000.

Because the AFL was craft-oriented, there was a void for industrial workers. In 1935, that void was filled by the Congress of Industrial Organizations (CIO) which organized the powerful steel and auto workers. There were other important union movements during this period; the railroad brotherhoods, started just after the Civil War, and the United Mine Workers, headed by John L. Lewis, made especially significant impacts on their respective industries.

By the late 1800s, unions and management seemed to be in a continuous battle. When the initial laws to regulate commerce from 1886 to 1914 were passed, most businessmen failed to recognize the trend. The strength of the union movement accompanied by supporting legislation was proof that important changes were taking place.

The Railway Labor Act was passed in 1926. Although it applied only to a limited number of employees and was confined to the railroads, it provided a precedent for the establishment of collective bargaining machinery. The stock market crash of 1929 and the depression that followed provided a great boost for union power in America, giving birth to the Norris-La Guardia Act of 1932. That act stated that a union could enjoin a threatened violation of a contract by an employer. Norris-La Guardia protected the most powerful tool of the union, the right to strike.

In 1935, a new height of union power was reached with the passage of the National Labor Relations Act, commonly called the Wagner Act. It was the first broad declaration of labor policy enforced by the federal government. The act stated that employees would have the right to engage in concerted activities for the purpose of collective bargaining or other mutual aid or protection. It also listed unfair labor practices of employers; these included discriminating in conditions of employment against the employees for the purpose of encouraging or discouraging members in any labor organization. The Wagner Act also provided for establishment of a National Labor Relations Board to prevent unfair labor practices and to conduct employee elections for the purpose of selecting a bargaining agency. The elected agency, usually a union, would then possess the collective bargaining power of all workers. Between 1930 and 1945, union membership rose from 3.5 million to approximately 15 million. This growth is largely attributed to the Wagner Act.

The labor laws presented management with a second front. Businessmen were restricted not only as to trust and other areas defined as unfair competition, but now they were also forced to recognize and bargain collectively with unions. Today it is no longer the right to organize that preoccupies the unions; it is job security, guaranteed income, and a safe place to work.

In short, *enterprise* was real in America and its fruits are abundantly evident. But what was meant by *enterprise* has changed over time as values have changed.

HAS THE LEVEL OF ENTERPRISE DIMINISHED?

Have things changed? Productivity and enterprise are not synonymous. Productivity in the economist's sense is unit output per man-hour. Enterprise, in the terms of this chapter, involves bold, hard, and important undertakings. Thus, not all productivity is enterprising, nor enterprise productive. Consider the development of an effective plan for cutting health care costs. This would certainly be enterprising, but might not result in an increase in economic measures of productivity. Similarly, more efficient production of unimportant trivia might increase measures of productivity but is not enterprise. Yet, the two are surely linked. What is characteristic of enterprise promotes productivity: employment of workers at their highest skills, invention of new products and processes, and use of innovative management techniques. In a society suffering from lack of enterprise, a drop in productivity is probably symptomatic of the basic malaise. Since productivity can be measured and enterprise cannot, the next two sections focus on productivity and then extend the conclusions to enterprise.

Productivity growth in the United States is down, in comparison with both its historic rate and other developed countries. The Council of Economic Advisors (CEA) estimates that growth of output per labor hour averaged 3.3 percent per year in the period from 1948 to 1966. Between 1966 and 1973, output grew

2.1 percent per year. For the most recent interval, between 1973 and 1978, growth rate declined to 0.9 percent per year. While productivity typically declines in a recession, the recent deterioration has taken place in the midst of an economic recovery; for example, productivity increased only 0.5 percent in !978. Furthermore, taking 1967 as a reference year, labor productivity in the United States has grown only 27 percent compared to 106 percent in Japan, 70 percent in West Germany, 62 percent in Italy, 43 percent in Canada, and 27 percent in the United Kingdom.

The decline in innovation and productivity can be measured in other dimensions as well. Government spending for research and development rose from about $10 billion (1972 dollars) in 1955 to about $30 billion in the late 1960s. Since that time, total research and development expenditures have remained virtually constant. In 1964, R&D outlays totaled 3 percent of the GNP; in 1977, R&D spending dropped to about 2 percent of the GNP.

Contrary to the trends in federal government expenditures, private industry R&D has about doubled in constant dollar terms over the last 20 years. Real growth rate of industrial R&D expenditures has been about 2 to 3 percent per year. While this growth has been important, analysts have found that the R&D projects have generally moved away from a search for new products and processes and toward the upgrading of existing products and processes, in particular, as necessitated by compliance with government regulations. Today, private industry and the federal government both fund R&D at about the same level. Typically, federal R&D, particularly in basic research, has focused on longer-term prospects than industrial R&D; therefore, these trends suggest that R&D activities focused on the long term—that is, activities that will "pay off" in new products and processes within the next 20 years—have diminished in relative, and probably absolute, terms.

While expenditures on research and development have diminished in the United States, they have increased in other countries. For example, in West Germany and Japan R&D expenditures averaged about 1.3 percent of GNP in 1960; currently West Germany spends about 2.2 percent of GNP, and Japan, 1.8 percent of GNP, on research and development. Government R&D spending in the USSR as a percentage of the GNP has risen to about 3 percent.

These adverse trends also are reflected in the number of patents issued by the United States. In 1960, approximately 40,000 patents were issued to U.S. residents by the Patent Office. By the early 1970s, about 55,000 patents were issued per year to U.S. residents. The number declined in 1978 to less than 50,000 patents. At the same time, the number of U.S. patents issued to foreign citizens rose from about 7,000 in 1960 to about 22,000 per year currently, Patents are not, of course, a precise indication of innovation because many firms prefer to use other means for the protection of proprietary concepts. Furthermore, a simple number count of patents does not necessarily indicate their quality or potential contribution. Nevertheless, these numbers indicate that something is changing. This view is confirmed by studies done by the Stanford Research Institute (SRI) and elsewhere; in the 1950s, about 80 percent of all new products were developed in the United States but, by the late 1960s, only about 55 percent of the new products were of American origin.

There are other indicators of the decline in innovation as well. Capital investment is growing more slowly in the United States than in other countries. Recently, growth has averaged about 14 percent in the United States, 30 percent in Japan, and 20 percent in Germany.

Small business, historically, has been responsible for a great deal of innovation. Yet, over the last ten years, small business has had difficulty in attracting venture capital. In 1969, there were about 550 public offerings of stock from companies with less than $5 million net worth; these offerings raised a total of about $1.5 billion. In 1975, there were only 4 offerings which raised a total of $16 million. This represents not a decline, but essentially an elimination of public stock offerings as a means for raising capital by small businesses.

There are no indicators to the contrary: things have changed, the level of enterprise has diminished. In terms of innovation, productivity, competitive position, or even a qualitative feeling of vitality, the trend is apparent everywhere.

While measures of change are relatively simple for manufacturing industries, measures in the service sector are much more difficult. How does one measure, for example, productivity in government? Can the productivity of a police force be measured in terms of its traffic citations per officer? Economists have been able to develop implicit measures of productivity of the service sector. It is important to attempt to construct such measures since the growth in nonmanufacturing activities as measured by relative employment has been rising at a rate more rapid than employment as a whole. The implicit productivity measure is constructed by removing the statistical contribution of manufacturing from a productivity index that ostensibly measures all private, nonfarm activity. This index involves the difference of two relatively uncertain numbers and, therefore, has great uncertainty itself; nevertheless, the index shows productivity gains running below those of manufacturing and, as in the case of manufacturing productivity, a recent diminishing of growth rate. Despite these difficulties, there are some indications that innovation and productivity have also diminished in the service sector of the economy. The number of employees in the federal government has grown from 2.4 million in 1960 to 2.8 million in 1977. While this growth is appreciable, nonetheless it represents a relatively constant percent of the work force as a whole. However, in state and local government, the numbers are quite different. In 1960 there were 6.4 million employees in state and local government. This number had increased to 12.6 million by 1977, representing a growth in terms of percentage of the labor force from 9 to 13 percent. Expenditures at the federal level rose from $166.9 billion in 1970 to $289.7 billion in 1976, and at the state and local level from $131.3 billion in

1970 to $256.8 billion in 1976. Public reaction to this growth in government has been expressed in rather direct terms through "proposition 13" type referenda.

In the time that we recognize in retrospect as creative, would these same statistics have shown the vigor of enterprise? Certainly, to some degree. But missing from the numbers would be the spirit of the time, telling those who dare that innovation is not only acceptable but desirable; that creativity is not a disease; that aspiration is not an aberration but the norm; that achievement, even individual achievement, is possible and to be revered; that institutions can work, and work in the interest of those who are served. These sentiments, to varying degrees, are seen as obsolete today, and that is the strongest indication that enterprise has diminished.

WHY HAS PRODUCTIVITY DIMINISHED?

Given that enterprise has diminished in its vigor, what are the reasons? The answers are many and varied and range from and include economic, political, and social factors. Here are some of the most important explanations that have appeared in the literature.

Productivity growth is down because of rapid energy price escalation. Recent energy price increases have lowered productivity by decreasing the rate of growth of capital stock (through diverting of capital that might have been used to increase productivity), by making energy-intensive processes unprofitable and some capital stock obsolete, and by making labor substitutes for energy more feasible than was previously the case. (But note that price escalation also acts as an enormous stimulant to innovation.)

Productivity growth is down because of inflation. For over a decade, inflation has increasingly characterized the U.S. economy, coexisting both with economic growth and with recession and unemployment. Rising price levels have been the result of structural shifts in the United States and world economy, as well as of the shock effect of increasing energy prices. The underlying causes of endemic inflation include the ever-increasing role of government in the economy (now accounting for 36 percent of GNP), increased deficit spending and the consequent increase in the money supply, and the growth of government regulation. The attempt to absorb the cost of the Vietnam war without raising taxes, the Russian wheat sale, the decision to fight unemployment rather than inflation in 1976 and 1977, and uncertain oil supplies have also taken their toll in inflation. Recent government policies that have raised the cost of living include farm price supports, steel reference pricing, increased Social Security taxes, and the rise in the minimum wage.

The longer inflation persists, the more institutionalized it becomes. All sections of the economy adjust to it in various ways, and it becomes increasingly difficult to cut into the spiral. Federal Reserve Bank Chairman Miller estimates that gradual deceleration of the growth of the nation's money supply—the goal of the Federal Reserve Bank—may take five to seven years, and the voluntary guidelines of the Carter administration appear to be foundering on the shoals of cost-push pressure.

A wide spectrum of the economy is now indexed to inflation. Social Security benefits, civil service pensions, and wages of many federal employees are now adjusted to the cost of living, and over 60 percent of workers with major union contracts are covered by cost-of-living adjustment clauses. More contracts of all types now stipulate inflation factors, and there has been some indirect indexing of taxes by cuts in federal tax rates. Through the various ties to the Consumer Price Index (CPI), even temporary price rises become built into the upward spiral. Financial instruments have also adjusted to inflation, making it more difficult to apply monetary restraints. Housing has been insulated from disintermediation by the six-month certificates, and by the pooling of mortgages with "pass-through securities." Various other interest-rate ceilings have also been raised, eliminating previous mechanisms that cut lending in a booming economy.

The effect of inflation on productivity is complex. Clearly, diminishing productivity growth adds to inflation; less obviously, inflation contributes to diminishing productivity growth. In the presence of inflation, uncertainty is higher and long-term commitments to industrial research are diminished. Furthermore, as the recent federal R&D budget has illustrated, the tendency is to keep funding at a constant level and, due to inflation, today's R&D dollar buys less than it used to. Finally, inflation acts to deter venture capital, equity investment in entrepreneurial activities, and even long-range corporate projects, since the return-on-investment equation always requires some future estimate of income. Venturesome projects reach "break-even" some years hence; the changing value of the dollar in the interim can make otherwise attractive projects quite uncertain. Many investors argue, rationally, that investment should be made in real property that appreciates with the dollar as a hedge against uncertainty. Inflation alters the rate of return for new investment that businesses can accept. In the presence of inflation, businesses seek faster recovery of capital costs. Inflation adds to the cost of capital—contrast this with the lowering cost of capital in both Japan and Germany.

Productivity growth has diminished because after-tax profits have declined. The facts are clear enough: after-tax profits of corporations in the United States averaged about 8 percent of sales during the mid-1960s and currently average around 4 percent. The lowering of after-tax profit impedes the cash flow of American industry, and restricts capital availability for new ventures by investors. Double taxation of corporate dividends and long periods of capital recovery for investment in plant and equipment tend to amplify this factor.

Productivity growth has diminished because of the changing work force composition. Over the last three decades, the labor force has grown more than 50 percent. A major portion of this growth has occurred as the result of the entry of young people into the labor force (the post–World War II baby boom) and the entry of women into the labor force. The influx of less skilled, less experienced persons, combined with earlier retirement of skilled adult men, is presumed by some to have reduced overall labor productivity. Furthermore, while the educational attainment of the labor force has been increasing, it has been slowing in its rate of increase. Today, about 20 percent of the labor force has completed four or more years of college; another 40 percent are high school graduates. As rate of educational attainment slows, productivity increases from this source slow as well.

Productivity growth has diminished because of the dynamics of the shift of labor from agriculture to industry. The percentage of the labor force involved in agriculture has been steadily dropping and, today, stands at a low of about 3.5 percent. This seems to represent a minimum level, and there is reason to doubt that it will drop lower. Thus, the transition, measured by this agricultural participation rate, has been completed. Since productivity in agriculture is generally below productivity of other economic sectors, the shift from agriculture to other sectors generally represented a trend in the direction of increasing productivity. Now, with that shift completed, this source of productivity growth has ended.

Productivity growth has diminished because technological progress in the mid-1960s was exceptional. In many ways, the mid-1960s were unusual. The spin-offs from the space program were coming to market; integrated circuitry was finding product application; American technology was a stimulant to international trade. As pointed out earlier, R&D spending peaked at about 3 percent of GNP in the mid-1960s. Since then, the number of "leading edge" technologies has diminished and R&D spending has dropped to some 2.2 percent of the GNP. One good measure of the contribution of technology to productivity is the age of capital goods, since new capital equipment generally embodies the latest technological advances. In the period from 1948 to 1966, the average age of capital goods declined about three years; between 1967 and 1972, the decline was about one year. From 1973 to 1977, there was essentially no change at all in the age of capital equipment in the United States.

Productivity growth has diminished because of increasing emphasis on antitrust actions. The essential antitrust question of distinguishing restraint of trade from successful business practice remains unresolved after a century of litigation, and antitrust scrutiny is being expanded in areas of corporate mergers, professional sports, the learned professions, and regulated industries. Conflicts often exist between antitrust objectives and other social goals. For instance, the patent laws reward creativity and technological innovation in contrast to the competitive demands of antitrust. One requires disclosure, the other secrecy. Large corporate mergers present classic questions of efficiency versus competition. In general, antitrust actions—or the threat of them—tend to increase uncertainty and promote a short-term focus. Momentous developments in antitrust are currently being decided. The future of IBM, AT&T, and possibly the major oil companies is currently being determined; the resolution of these cases will set the philosophic tone for the antitrust field for many decades to come.

Productivity growth has diminished because of excessive expenditures on weapons. This view, developed in detail by Seymour Mellman, asserts that decline in productivity growth results from excessive expenditures in the military sector.[5] U.S. expenditures are high, averaging 5.4 percent (constant 1975 dollars) of the GNP in 1976. By contrast, Japan's military expenditures averaged only 0.9 percent of the GNP in 1976. Thus, according to this argument, Japan's high productivity growth is, at least in part, due to miniscule expenditures for defense. Furthermore, since both defense contracts are cost-plus-fixed-fee, there are no rewards, and perhaps penalties, associated with improvements in efficiency.[6]

Productivity growth has diminished because of new product liability concepts. As recent lawsuits and awards have shown, injured persons have the right to reimbursement for damages incurred as a result of defective products. In addition, large awards have resulted even though the product was misused in a manner that could not have been easily foreseen. The liability associated with these products extends over long periods of time and "ripples" back from the manufacturer of the product to lower echelon suppliers. Recalls are triggered by the new accidents or injuries which indicate potential danger to large numbers of people. These recalls can be very expensive. Firestone, for example, recalled 10 million tires over a period of several years at a cost of $135 million. The Ford Pinto gas tank suit resulted in an award of $128 million for a single event; this award was subsequently reduced on appeal to $6,600,000. Product liability insurance, which reduces the risk of product recalls and product liability, is available, but the cost has increased tremendously in recent years. The threat of continuing liability of uncertain size, evoked by conditions of use not under the control of the innovator, probably leads to less innovation.

Productivity growth has diminished because of worker dissatisfaction. According to this interpretation, diminishing productivity growth stems from worker dissatisfaction and alienation; worker frustration stems from bureaucratization and segmentation of the work process, of work processors, and over-qualification for jobs. McConnell quotes Arthur Burns, former chairman of the Board of Governors of the Federal Reserve System:

> It is not at all clear that people actually perceive that lessened work effort inevitably must be reflected in the material benefits we the people can enjoy. That linkage was inescapably evident earlier in our history—when, to a much greater degree than is now the case, men and women could literally see what their individual effort yielded in consumable products; but the linkage has been blurred as our productive and distributive mechanisms have grown in complexity.[7]

The other side of this coin involves the activities by organized labor that are designed to preserve jobs irrespective of the possibility of technological improvement. This is known as "job security" and is directed against the possibility that technological improvements, such as automation, could eliminate jobs, or that jobs would be eliminated as a result of moving production facilities to another geographic location, particularly offshore. Over the past several years, job security has emerged as a major bargaining issue. Previously, it tended to be cyclical, with increase in job security clauses after economic recession. The current emphasis is based on more than cyclical trends and, therefore, cannot be expected to abate quickly.

Some forms of job security are already included in a significant portion of labor contracts in the United States. For example, the Bureau of Labor Statistics, in a 1975 survey, found guarantees of work or pay in 12 percent of all major contracts, guarantees of severance pay in 32 percent, and supplementary unemployment benefit plans included in 15 percent. In 1978, the Bureau of National Affairs (BNA) reported that technological changes were restricted in 17 percent of the contracts it sampled, with 27 percent of those requiring retraining. Thirteen percent of the contracts sampled had plant shut-down or relocation limitations, and half of these provided for transfer rights.

Productivity growth has diminished because of egalitarian drives that make entrepreneurship less rewarding. The egalitarian bent of our society, the argument goes, requires that gains be distributed—but entrepreneurship by its nature is pursued to achieve unequal gains. When such gains are perceived as intrinsically evil or disproportionate, or their pursuit is perceived as "greed," inhibition is inevitable.

Productivity growth has diminished because limits to growth are being approached. In this view, growth inevitably results in world catastrophe and, in the course of that collision, productivity must diminish as the quality and amount of input material drop. Near the limits to growth, one must work harder and harder simply to maintain a position. Technological progress itself is subject to diminishing returns; it no longer offsets declining labor productivity. Furthermore, in the past, society used easily available resources. These were cheaply obtained and processed. Now, it costs more in terms of capital and human effort to obtain a unit of output.

Productivity growth has diminished because of inappropriate government intervention in the free market. In this view, diminishing growth in productivity stems from tampering with the free market system. The domain of tampering is extensive and includes excessive individual taxation that lowers work incentives, regulations that require corporate expenditures for pollution control and occupational health and safety, and a degree of unionization. The growth of the public sector both causes and reflects inefficiency in the economy. Wage-price guidelines result in misallocation, shortages, and uncertainties that reduce, directly and indirectly, production efficiency. Finally, as pointed out earlier, inflation itself adds greatly to uncertainty and, therefore, limits investment, particularly long-term investment.

Productivity growth has diminished because of the adverse regulatory climate. Appreciable investment is required to comply with pollution abatement, safety, and health requirements. GNP computations include estimates of these investments when they occur; thus, investments in capital equipment designed to meet regulatory requirements do not adversely affect productivity estimates, at least initially. However, when the equipment is used later on to produce improvements in the environment and worker health and safety, the "output" cannot be captured numerically and, therefore, does not appear as productive investment. In effect, there is no way to account for cleaner air and water and improved worker health and safety in the current economics of productivity. Thus, the loss in productivity growth may be a myth, an artifact of the way we account. If we could include the improvements in the environment in the productivity equation, perhaps growth would still be increasing.

With respect to the conventional estimates, however, regulation has lengthened the period before new developments can become profitable. Most importantly, inconsistency of government regulations, supervised by different agencies, changing at unexpected times, places additional uncertainties and complexities on planning by businesses that must cope with such redundancies. Finally, regulatory requirements divert capital expenditures from other operations. This is an appreciable amount; expenditures for antipollution devices involved 10 percent of capital spending in 1977.

As an example of the difficulties produced by regulation, the 1976 Toxic Substances Control Act (TOSCA) expands federal regulation of the chemical industry by granting EPA the authority to require the testing of chemicals and to regulate chemical production, use, and disposal. TOSCA approaches the huge task of regulating chemicals by categorizing them according to risk, a priori, and requiring premarket notification and testing of potentially dangerous chemical products. The chemical industry is classed as high technology with extensive product innovation. This industry has found it critical to maintain the confidentiality of its trade secrets. The fund of trade–secret information generated serves as an essential base for further technological innovation. The ability to preserve confidentiality under TOSCA regulations may be severely compromised. Thus, regulation not only adds costs and complexity, but may affect the proprietary nature of information, as well.

Productivity growth has diminished because of satiation. We have got what we want, most of us. Entitlements guarantee it whether we work hard or not. Why push?

Everybody sees in the situation a reflection of their own prior biases, from Marxist to Keynesian. Perhaps they are all correct, to some degree.

CAN ENTERPRISE BE REVITALIZED?

Productivity is *in* this year. President Carter has asked the Department of Commerce to suggest legislative initiatives that would have the effect of stimulating innovation and productivity in the country: these suggestions will go forward from the White House to Congress later this year. This activity is known as the Domestic Policy Review (DPR). The Council of Economic Advisors (CEA) is also probing productivity and Congressional committees are holding hearings on the topic. The reasoning behind these activities is that productivity growth in the United States is waning, lagging behind Japan and West Germany. This puts us at a disadvantage in world markets, hurts balance of trade, and is inflationary. Public policies can induce increased innovation, and innovation will stimulate productivity. Improved production will lower prices and make our products more competitive on the world market. Thus, it is anti-inflationary and job creating.

The work being conducted under the Assistant Secretary of Commerce for R&D involves a number of subcommittees to consider environmental, health, and safety regulations; patent and information policy; research and development; industry structure and competition; labor; economic and trade policy. The recommendations of these groups are far reaching and include

- changes in corporate tax policy to encourage investment and R&D. The tax bill of 1978 reduced corporate tax rates and liberalized the investment tax credit. Clearly, government policy is already shifting toward investment incentives. These might include more rapid write-offs for R&D expenditures, and tax credits for individuals and corporations for contributions to research-oriented activities.

- promoting commercialization of inventions made under government contracts or in government laboratories by, for example, permitting the inventor to have proprietary rights to the invention, and reserving for the government a royalty-free license.

- improving the support of research and development associated with generic technology, that is, technology that can spawn a large number of downstream products and processes. Similarly, improved support of basic research represents a long–term strategy for improving innovation and productivity.

- improving incentives for small business, improving incentives for investment in small innovative businesses. Historically, small businesses have contributed innovations at a greater rate than business as a whole; therefore, incentives favoring small business should be quite efficient. These incentives might include stock option time limitations for founders and key personnel, postponing tax on income derived from sale of shares by founders, allowing start-up losses to flow free to founding investors, allowing tax-free rollover of equity investment in small businesses, etc.

- promoting international trade, particularly export activities of American firms. This would involve intensifying current efforts to supply American business with information about markets in foreign countries, providing low interest loans for the development of foreign business, amending tax laws to encourage these activities, etc.

- organizing information relevant to the innovation process, and making it available in a timely manner to potential entrepreneurs. This includes information about markets, technology developed under government contract which is available to small businesses, regulatory impediments and means for coping with them, etc.

- reducing inconsistency of regulation or uncertainty of regulation standards. Inconsistent regulations may exist within a single agency, or conflicting regulations may exist between agencies. Furthermore, means for measuring compliance often are not clearly understood at the outset; these factors cause uncertainty and delay innovation.

- focusing regulations on standards of performance, not the processes used to achieve the standards. By concentrating on the ends rather than the means, the regulations themselves might stimulate innovation.

- studying the consequences of innovation before implementation. Many new regulations have many unintended consequences; for example, clients with new regulation have caused many smaller firms to close and larger firms, as a result, have increased their concentration in various markets. This was often accomplished through acquisition, a consequence certainly not intended by the original regulation.

- amending antitrust policies to reduce their effect on innovation. In today's environment, the introduction of a new technology might give the innovating firm a substantial market share which would then attract the attention of the Justice Department. Antitrust considerations also are involved if two firms were to engage in cooperative research pertaining to a new product or process. Antitrust policies might be amended to permit such activities where they seem to be in the national interest.
- where product safety is involved, employing product performance standards rather than detailed design standards. This would allow maximum opportunity for innovation.
- limiting product liability to a specific number of years after a product is introduced in the marketplace and introducing a statute of limitations to limit the number of years from discovery of damage to suit. Other means also should be sought to evaluate the products in terms of the technologies and standards that existed at the time they were designed and manufactured rather than present-day technology and standards. Some authorities have suggested that our system of contingent legal fees is a considerable incentive to sue and seek very large awards. Almost every country in Europe as well as Australia forbids such contingencies and considers them unethical.
- amending patent timing. Where regulatory approval is required of a new product before its introduction to the market, patent issuance should coincide with the approval date. For example, new drugs must receive FDA's approval. This often takes a number of years and reduces the effective time of patent coverage if a patent were issued earlier. To provide the innovator with a. least a "standard" 17 years of protection, the beginning of the elapsed time should coincide with FDA approval.

The social currents that produced consumer, worker health and safety, energy, and environmental legislation are not temporary. Despite the costs of these activities to the consumer and taxpayer and the difficulties they create for business, the pressures will continue and, in some instances, intensify. The reasons for this view are many:

- The media display the values behind these laws as modern and desirable, and opposition to them as reactionary.
- Government bureaucracies dedicated to these goals are in place and bureaucracies tend to be self-perpetuating.
- Some sectors of industry and academia produce goods and services for the new bureaucracies; hence an industry–government complex has been formed that gives these activities considerable momentum.
- The values behind these pressures are intensely held by some groups who are willing champions, and these values are generally popular.

- Some of the enterprises that are necessary to accomplish national objectives are too costly or risky for private business to undertake (e.g., large-scale energy demonstration plans).
- Most importantly, the regulatory activity has brought changes which, for many people, are worth the cost.
- While inflation is high, unemployment is low and, therefore, public pressures to change the situation are, at best, marginal.

The government will continue in its role as promoter of national economic development; regulator of activities that have seemed to enhance the public interest; contractor for goods and services required in the achievement of national goals; provider of social services designed to reduce personal risk and improve security; and a guardian of the shared resources including air and water. Further action by government in these areas will result in additional regulation of business and in moving some activities currently performed by the private sector into the public sector (e.g., health care).

Business leaders ask for less regulation of their activities for two reasons. First, the social goals behind the regulations often seem inappropriate; and second, the means for achieving the goals may seem too costly or overly burdensome. While there is apt to be little retrenchment in goals, there are likely to be important attempts to improve the efficiency of attaining them. One such attempt may be in the area of regulatory reform. President Ford created a national committee on regulatory reform. The purpose of this committee was to reduce the unnecessary burdens placed on business by regulation and to help restore confidence in government. The program involved selective reductions by federal agencies in their paperwork requirements and revamping of rules in regulatory agencies. Deregulation of airlines and communications industries are early examples of the efficiency improvement sought.

The public is mistrustful of both big government bureaucracy and big business. Many people are reluctant to accept the word of government as absolute truth or all of its actions as necessary; as for business, many people feel that corporations generally act in their own interest (even when such action endangers the health and safety of their workers and customers), and that they distort the truth as well. In short, there is a loss of confidence in both government and business and their ability to act for the public. This expressed lack of confidence is a mark of our time.

Perhaps paradoxically, the public looks to government to provide a still–growing list of "entitlements": jobs, housing, health care, education, financial security, and clean air and water. The growth in the demand for job security is another manifestation of this "entitlement" attitude. In many instances, business is the instrument by which government accomplishes these ends; if business can't, or is profiting unduly, the prevailing attitude is that the government should perform the function. Because of mistrust, changing values, and pressing societal issues of unprecedented complexity, urgency, and importance, a new contract is being forged, piece by piece, between government, business, and society.

To revitalize production is not as simple as putting things back as they were. We can't return to that seemingly simpler time before double-digit inflation, the growth of regulation, and the era of entitlement. Many of the policies suggested by the Domestic Policy Review and the Council of Economic Advisors, undoubtedly, will have the effect of promoting investment and perhaps even stimulating innovation in the sense of encouraging the introduction of new products and processes.

These are necessary but not sufficient results; they do not assure a return to enterprise—a willingness to venture on bold, hard, and important undertakings with energy and initiative. Perhaps we as individuals and the institutions we have created are beyond that point. Why, after all, should we do hard things when "entitlements" let us survive? In the end, to revitalize enterprise requires attitudinal shifts reminiscent of President Kennedy's exhortation: We must, and our institutions must, ask what needs to be done, not what's coming to us. This requires a means for *searching our goals* and making our measures of achievement coincident with goal attainment, not raw production. This goal reward should work at all levels.

For the nation, this means creating incentives for institutions and individuals which reward contribution to unambiguous national goals. For business, this means profit-sharing. For labor, this means tying compensation to productivity. For society, this means relating "entitlements" to needs.

SHOULD WE SEEK TO REVITALIZE ENTERPRISE?

Granted that innovation and productivity, or even enterprise, could be stimulated, should we make the attempt? To what end? Some people would argue that any encouragement of enterprise represents only a last-ditch effort to preserve the old order in which decay is becoming increasingly evident. But I believe the problems which we have now or which appear on the immediate horizon *require* revival of enterprise; without it, these problems will be more severe and lasting. The problems requiring hard, bold, important actions include those which flow from the increasing number of people who, without accelerated innovation and productivity, would be doomed to crushing poverty; and the "problematique," that confluence of global issues, largely stemming from population and economic growth, including adequacy of food, energy, resources, and the environment.

Other difficult issues requiring enterprise could certainly be cited: the control of nuclear proliferation, repairing the frailties of the world monetary system, dealing with terrorism, urban renewal, immigration, conservation without stagnation, improving agricultural production and the balancing of inflation and employment within the United States and other countries. World population and the "problematique" are used as examples in the discussion that follows. These issues cry for sensitive, intelligent, inspired enterprise; with it, these issues may be successfully faced; without it, the risks are magnified.

There are about 4.3 billion people in the world today. Population growth rate currently is on the order of 1.8 percent per year and falling slowly. This decline in population growth rate represents a very optimistic trend. However, since the women who will become mothers within the next generation or so have already been born, this trend will have little effect on total world population within the next two decades. Even if world population growth rate continues to drop, population by the turn of the century will be approximately 6 billion, give or take a few hundred million. Thus, within the next 20 years or so, world population will grow by 50 percent.

Suppose for a moment that we could measure every facet of existence—e.g., mobility, education, food consumption, housing, and shelter. To prevent such things from getting worse on a per capita basis, during the next two decades everything would have to increase by the same amount as population; that is, 50 percent. This is a staggering image: to avoid decreasing the value of such measures on a per capita basis over the next two decades, everything will have to increase by 50 percent.

Technology might help, somewhat. For example, during this period of population growth, to keep the per capita level of education at least constant, communication satellites might substitute for brick and mortar schools. To keep food per capita at least constant, new technologies might be employed to improve agricultural productivity. To keep shelter per capita at least constant, new technologies might be employed to use structural plastic foams or to discourage private and encourage multi-unit dwellings. To keep mobility per capita constant, new, less energy-intensive technologies might be employed, or mass transit or communications might be substituted. In other words, without innovation in social and physical technologies, things are likely to get worse for most people in the world.

Most population growth occurs in countries that are currently poor; by the end of the century, approximately 80 percent of all people of the world will live in poor countries. Poor people won't settle for keeping per capita measures only constant. The difference in consumption levels between the rich and poor countries is a factor of 6 or 7. Any argument designed to only preserve the status quo would be taken as a denial of the right of most people to pursue the affluent life-styles enjoyed by only a relatively few today. Taking the two factors together—population growth and rising expectations—the implicit growth required during the next two decades is essentially an order of magnitude. These aspirations are probably impossible to satisfy by any means, particularly in the presence of high energy costs.

"Problematique" is a term that has come to be applied to the confluence of global issues that confront us now and promise in their complex interactions to change the social, political, and economic order in the years to come. These issues abound; their domains include: food, the environment, energy, inflation, resources, finance, productivity, equity, poverty, and the potential for war. None of these issue-elements are independent; none can be addressed singly; all promise long-lasting, perhaps permanent, change—much of it undesirable. Because past national, corporate, and personal policies seem to have brought us to this point, these policies now seem flawed, and the old goal structures on which these policies were based seem inadequate and short-sighted. Hindsight is usually marvelous but fails us here; if we ask what would we have done differently to diminish the potency of the "problematique," the answers are even now less than precise. When we ask what we might do differently tomorrow morning to slow the cresting and diminish the consequences of these issues, the answers are even more obscure. And yet, the search for workable, practical concepts that address such problems is, above all else, an immediate and intense need, for failure implies suffering on a tremendous scale and stagnation of human potential and spirit.

Whether or not there will be enough food for all the people in the world over the next two decades remains uncertain. The technological potential for increasing the amount of land or water used in raising food, and for increase in the productivity of that land or water, is indeed great and is described later in this section. However, most of these technologies are expensive and the nations that need food are poor. Therefore, it is not at all clear that the free market will be robust enough to convert these technological opportunities into life-sustaining food.

In developing countries, agricultural production and population growth have barely kept pace over the last decade. In the rich countries, food production has grown faster than population. This suggests that an important aspect of the world food situation will be one of distribution, from high–income to low-income countries.

The growth in world population—as well as the drive for increasing economic output—requires continually increasing the rates of raw material used. In the 20 years prior to the oil embargo, world production of important minerals had increased by a factor of about 5. Growing population and affluence certainly indicate a continuation of this trend. A shortage has developed in materials for which indigenous but untapped reserves exist. These deposits will undoubtedly be tapped. However, developing these resources will have significant environmental and political ramifications. The environmental implications are obvious; as for political, we already hear dissatisfaction expressed by a state rich in one or another mineral, which feels exploited when it must "export" that mineral at the expense of its environment to satisfy the needs of another state. The emerging raw material situation will stimulate not only the development of marginal resources but also the search for substitutes and acceptable recycling techniques. The use of recycling or substitute materials as a strategy to respond to increasing material cost and uncertain availability is, of course, not new. What has changed are the equations by which the economics of such activities are evaluated. Over the next 10 years, there are likely to be increasing attempts to intervene in these economics by stimulating conservation, providing a floor for material prices (to reduce uncertainty), encouraging scientific and techno- logical pursuits which could result in effective substitutes, and by the more efficient extraction of marginal ones.

With respect to energy, it is unlikely in the short term that prices will drop or that abundant supplies will once again be available. Solar energy, geothermal power, and fusion will require research and pilot plant operation before they can reliably produce appreciable amounts of power. At best, solar and geothermal energy may begin to make substantial contributions early in the next century. Fusion power will probably not be a significant source until well after the turn of the century.

In our country, of course, coal supplies are abundant. However, coal is not the energy form most in demand; except through liquefaction processes, coal cannot be used to power automobiles or to heat most homes. Furthermore, there are limits on coal production and use including availability of manpower, en- vironmental considerations—particularly those associated with deep mining and the sulfur content of coal, and safety and health considerations of the workers involved in mining operations.

To pay for imported energy supplies, developing countries have to borrow extensively. These obligations will have significant effects on the domestic economies of many countries, necessitating reordering of priorities, encouraging exports to develop sources of foreign currency, and implementing programs of import substitution to improve balance of trade. The linkage between high energy costs and food production is a dramatic demonstration of the connectedness of world problems. Nitrogenous fertilizer is energy-intensive. Application of ni- trogenous fertilizer increases agricultural productivity. However, in the presence of high energy costs, this strategy for improving food production may not be available to energy-poor countries most in need of the food.

Another demonstration of the connectedness of world problems is in the growth of atmospheric carbon dioxide. Carbon dioxide results from burning; as world energy consumption increases, the amount of carbon dioxide introduced into the atmosphere must also necessarily increase. The amounts added are not significant: from 1960 to 1975, the carbon dioxide levels increased by 4 percent. Energy policies in most countries favor the continued burning of fossil fuels; therefore, the outlook is for increasing levels of carbon dioxide in the decades ahead. The effect of increasing levels of carbon dioxide in the atmosphere is to trap longer wave lengths of incident energy within the atmosphere and, thus, increase atmospheric temperature. This "greenhouse effect" has been balanced by a concomitant increase in the amount of particulate matter in the atmosphere that has the effect of increasing the earth's reflectivity and, thus, tends to lower the temperature. Particulate matter is now being trapped in most developed countries through the use of filters and precipitators. Therefore, the balance between increasing "greenhouse effect" and increasing albedo may be upset. Finally, since growing plants remove carbon from the atmosphere, plant life production—particularly of forests—acts as a sink for the carbon dioxide. Deforestation, which is accompanying increased population levels, has adverse consequences for this issue.

Within the United States, improved enterprise will provide an antidote to inflation. As pointed out earlier, inflation is increasingly the result of structural changes in our economy and, therefore, will resist conventional policies. Almost nothing is anti-inflationary except improved productivity. The situation is, to a degree, self-correcting. As improved productivity diminishes inflation, uncertainty caused by inflation will be diminished as well. This encourages investment which has the effect of improving productivity.

These examples illustrate the need for a special spirit of enterprise. Coincidentally, there is quite a range of nascent technologies which, through planning, can be utilized to advantage. These include:

- *The microelectronic revolution* will manifest itself in robots, computers, electronic books, electronic photography, flat wall television screens, listening and speaking machines, artificial intelligence, appliances that reason, games that are fun, human-like creative machines, and new processes and their control. This technology is relatively inexpensive and will help provide education, communications, and organizational efficiency.
- *Genetic technology* offers the possibility of designing plants and animals with desired properties. Before too long, this technology will find application in agriculture, animal husbandry, medicine, mining, environmental control, water purification, aging research, botany, and control of chemical processes by providing new means of catalysis. This technology promises profound impact on many industries. Two examples: it is possible to imagine the development of new microorganisms for digestion of waste cellulose to produce either sugar or alcohol, thus influencing availability of both food and energy. Secondly, the commercial production of pharmaceuticals through biological processes is not far off; the gene for producing insulin has already been moved to microorganism.
- *Nutrition* is a kind of black art today, but improved epidemiological analysis may help identify the health consequences of various diets and environments. We are on the frontier of making nutrition a science by finding out how various things we eat and do influence health, disease, and vigor. This science will provide the basis for more rational environmental and work place regulation.
- *Biomedicine* is likely to undergo significant improvements in the years immediately ahead. Since the turn of the century, life expectancy at birth has

increased dramatically. In the same interval, however, life expectancy at middle age and early old age has hardly improved at all. This is because diseases of childhood have been essentially conquered and diseases of middle and old age—cancer, heart disease, stroke—have remained the major killers. There will be many innovations that have the effect of diminishing death rate from these causes at least until extreme old age and, as a result, life expectancy at middle age will begin a dramatic climb. Vigor in middle age and early old age will improve. This will have important consequences for retirement and the nature of work.

- *The psychology revolution* will add to our understanding of how the mind functions, how thought is stored and recalled. New fundamental knowledge in this field seems likely, and with it efficiency of education will be improved; the ability to communicate rationalized; and behavior more easily understood.

- *Agricultural technologies* will add to the world's arable land, increase productivity, provide new sources of protein, and improve the quantity of nutritious food which is finally delivered. These technologies, include new uses of the oceans, such as domestication of fish, raising of biomass in the form of aqueous plants, cultivation of shellfish; inland fish farming; the development of self-fertilizing plant strains through genetic engineering; development of new plant strains that can be irrigated with brackish or salt water; controlled-environment agriculture utilizing large hothouses; drip irrigation in which small quantities of water and plant nutrients are delivered directly to the plant root structure rather than over wider areas; new means of preservation to reduce post-harvest spoilage; development of single-cell protein which digests petroleum or other organic materials; and new means of forecasting harvest size in order to obtain early warnings about crop failure, through such means as the Landsat. As mentioned earlier, it is not at all clear that the promise of these technologies will be realized since the market incentives that might have brought them into large-scale use may not be present. This points out the basic need for infrastructural development. Furthermore, many of these technologies are capital–intensive, and the countries that must improve their indigenous agricultural systems may need to develop highly efficient, labor–intensive systems.

- *Communications and information technologies* make practical the storage and retrieval of large amounts of data almost instantly at trivial cost, present essentially any program to small audiences with great flexibility, and increase the potential for direct education.

This list is obviously incomplete and, in retrospect, we will find other technologies which have had even greater impact on the world scene two decades hence. Yet, even with these omissions, the technological opportunities on which enterprise can build seem profound indeed.

Enterprise grows, not only from technology "push" but from "market pull" as well. There are huge market needs on which enterprise can focus. There is an express need for:

- very inexpensive water desalination;
- a plethora of energy-related devices: solar heating, photovoltaics, waste utilization, tertiary recovery, waste heat storage, low-grade heat amplification, cost–effective gasohol, coal gasification and liquefaction, safe nuclear waste disposal;
- nuclear reprocessing cycles that eliminate the possibility that the materials being processed can serve in weapons;
- a means for making tropical rain forests agriculturally productive by, for example, controlling hardening of lateritic soils;
- substitution in important manufacturing processes of plentiful materials for materials likely to be in short supply;
- development of processes for in situ leaching of low–grade ores;
- technique for improving agricultural production and food preservation and distribution; and
- the construction of efficient and livable urban environments.

This list is also incomplete but illustrates the enormous scale of opportunities which exist.

So the answers to the questions that form the framework for this chapter are: Yes, enterprise did once flourish in our country and has now diminished in intensity. The reasons for its current lack of vigor are hard to pin down exactly, but include satiation, diversion of resources from "productive" to "unproductive" pursuits, and above all, increased uncertainty occasioned by inflation and regulation. Uncertainty results in a short-term perspective. There are many proposals for stimulating innovation and productivity but their effectiveness is not certain, by any means. Even if these proposals work as intended and they stimulate innovation and productivity, they may not rekindle enterprise. After all, innovation may be channeled to trivial pursuits and our definition of enterprise requires hard, bold, and important action. Finally, the opportunities and needs for enterprise abound and, in some instances at least, there seem to be few alternatives.

How then might sensitive and intelligent enterprise be revived? Certainly, the proposals directed toward stimulating innovation will help, but beyond these, hard, bold, and important action requires another fundamental change. We have lost the belief, or the courage to enforce the belief, that good performance should bring rewards and bad performance should not be rewarded. A waitress offering poor service should not expect a tip; wages should be linked to productivity; incentives should be linked to the achievement of goals. With wages linked to productivity, enterprise is carried to the level of the individual, inflationary pressures subside, and competitive stance improves. With incentives linked to the achievement of goals, the invisible hand writes not to suboptimize individual gain, but to accomplish higher ends.

We have fooled ourselves with inadequate measures of productivity. First, these measures do not include the output—e.g., improvements in the environment—which we have bought too dearly through the diversion of fiscal resources. Second, these measures are tuned to industrial output rathe⁻ than services, and the further we move into postindustrialism, the less adequate these measures become. Clearly, new measurement systems are in order.

We should track the consequences of our policies. It is not enough to debate policies before the fact; it is essential to track the results after implementation, because most policies are necessarily based on inadequate information. As information improves and results are measured, policies can be "tuned" to meet their objectives and to diminish unintended side effects.

But what goals are appropriate? How can diverse interests be balanced? How can public views be generated and more effectively injected into the decision process? The efforts by presidential commissions to set "national goals" have been notoriously unsuccessful. Where is it that we want to go? Knowing this, most else can follow. Harlan Cleveland has stated that planning cannot be detailed regulation by federal agencies, but will be improvisation by each of us on an agreed sense of direction. The revival of enterprise needs that sense of direction. We need planning that identifies issues and ties rewards to their solution.

Some national programs have galvanized action; the .nost inspired of these have organized national purpose, captured imaginations, and harnessed enterprise (for example, the Apollo program). Other national programs have been no more than weak slogans, seen as ineffective by both cynics and supporters (for example, The Great Society). The differences are matters of clarity of purpose, perceptions about need and self-interest in the program, and the charisma of the program's champion. The United States needs new, effective programs in energy, agriculture, ocean utilization, and urban redevelopment. Properly framed, these can stimulate enterprise in the interests of society.

Enterprise is like a giant engine, idling now, but capable of great horsepower. We can, and therefore should, aim it at our problems and turn it on again.

NOTES

1. Daniel J. Boorstin, *The Americans: The Democratic Experience* (New York: Random House, 1973), p. 9.

2. Ibid., p. 132-3.

3. Ibid., p. 134.

4. Ibid., p. 134.

5. Seymour Mellman, "Decision Making and Productivity as Economic Variables: Depression as a Failure of Productivity," *Journal of Economic Issues*, June 1976, p. 230.

6. James M. Faurev, "Profits on Performance of Aerospace Defense Contractors," *Journal of Economic Issues*, June 1976, pp. 399-400.

7. Arthur F. Burns, "The Significance of Our Productivity Lag," Speech, May 14, 1977, in Campbell R. McConnell, "Why is the Art of Productivity Slowing Down?" *Harvard Business Review*, March 4, 1979.

Chapter 13

GIVE THE GOVERNMENT A CHANCE

William S. Sneath

A MIX OF SIGNALS

Newspapers confirm that this is truly the season of discontent in the nation's capital. They reported recently that, of all American cities, Washington has the highest per capita concentration of psychiatrists. Why is this? The newspaper stories offered several reasons. One is the pace and pressure of work in government. Another is the feeling of insignificance that comes with being one among nearly 3 million federal government employees. Another possibility, for some at least, is the fact that job security may have more to do with the outcome of the next election than with their own competence.

But I prefer another theory. I believe that among the people who work in Washington, anxiety levels are high because the level of dedication is high. I believe that most government employees truly wish to serve the nation well, and I have met and worked with enough of them over recent years to have an informed opinion. The question is why such dedication to service should have such traumatic consequences.

Psychologists have long known that a good way to induce symptoms of anxiety in laboratory animals is to set up an arbitrary system of rewards and punishment. They mix the signals, rewarding the same behavior with a piece of cheese on one occasion and with an electric shock on another. The result is that animals, who would ordinarily complete the maze, start running around in circles and climbing walls. It just might be that mixed signals are also the root of the problem in Washington. For some years now, Americans have been calling for more and more from government. We demand more equity, more benefits, more laws and regulations, and generally more attention to our special needs. You will not be shocked to learn that business people have also gone to Washington seeking favors. But now that Americans have a lot more government, we want less. We resent government's expanding presence in our lives. Perhaps Oscar Wilde was right when he observed that there are only two tragedies in this world—one is not getting what you want, and the other is getting it.

No wonder there is anguish in Washington. And no wonder we appear to be on the road to government without limits. Our government has been making a mighty effort to be everything to everybody, a task that obviously offers unlimited growth potential. It is certainly a major cause of the growth of federal government outlays from 2 percent of GNP 50 years ago to 22 percent today.

Americans have become alarmed at the enormous cost of government and dismayed by a pervasive feeling that we do not get our money's worth. We are reacting in clear and unequivocal fashion, sending a message across the land in the form of measures like "Proposition 13" (and other restrictive measures) that would limit taxes and spending in order to call a halt.

Based on my corporate experience, I predict that budget and spending limitations will have a salutary effect. At Union Carbide, for example, we tend to trim the budgets of businesses that are not meeting our performance standards. The result is often rapid cutback of marginal operations, new attention to efficiency and productivity, and more aggressive attention to meeting our planned objectives.

But a business chastened by budget limitations has one formidable advantage over government when it comes to reorganizing and reviewing activities to improve performance. Management knows precisely the business it's in and what is required of that business. That critical element seems to be missing in the Proposition 13 approach to limiting government. There are no guidelines. The only message conveyed is to do less and to do it better.

If we expect to get government off the couch and on its feet, I think we owe it something more. I think we owe it fewer denunciations and clearer guidelines about what we do and do not expect. We owe it a reasonable chance to succeed. And we owe it the wherewithal and support it needs to perform the tasks we require. Ah, but there's the rub. What are the tasks we require, and why is the question of government's role proving to be so divisive?

Of course, there is no last word on the role of government in a democracy. It is different in war and in peace, in seasons of social harmony and in conflict, in good times and in bad. We all benefit from the ongoing debate that reflects society's changing needs and goals. Policies that address the concerns of one season in our history are wrong for another. But there are times when, instead of helping the nation adjust to new circumstances, the debate about government's role does more to disrupt and divide. That is happening now on the question of government intervention in our free enterprise system.

We have become antagonists, out to have government tame corporations or to get government off our backs. Neither slogan adds much to our comprehension of the issues, or to clear thinking about our priorities. Yet it ought to be possible to agree on the broad outlines of a role for government in our mixed system of capitalism and democracy, to create a job description that is short on hyperbole and ideology and long on real world needs and experience.

A JOB DESCRIPTION FOR GOVERNMENT

Obviously, relying on a job description is the approach of a manager, which I am, and not the approach of a social scientist, which I am not. But I would like to suggest the brief outline of such a job description as a basis for discussion of the limits to government.

First, my position description would acknowledge that government is inescapably involved in the American economy. Government makes economic activity possible in the first place. The corporation itself, the main engine of the economy, is a creature of government, given perpetual life so it can better fulfill the public purpose of organizing and using resources to create wealth, income, and jobs. Further, without rules established by government and without government's coercive power to enforce contracts and protect property rights, business activity on a large scale would be chaotic at best, and might not be possible at all. To relate this to our question of limits, I would be careful to spell out the mutual dependencies in this arrangement. The system needs the government; but a vigorous private economy sustains us *and* our government.

Second, our job description should note that government also has a legitimate restraining role on marketplace activities. The marketplace is that peculiar arena of the invisible hand, where self-interest and the public interest come together in a productive amalgam that allocates resources according to the wishes and values of consumers. Its signals call for such products as housing, medicine, tools, and textbooks that are essential to society's well being; and they also call for striped toothpaste and hula hoops. The role of government is not to make value judgments about these goods, but to see that the business of supplying them is conducted openly and fairly; and to see that no one who is able to compete for a share of the market is unfairly kept from doing so.

On the question of limits, I would caution that the job is to give consumers confidence in the market, not to change their verdict. That is the quickest way to defeat its highly useful purpose. In open competition, some companies will succeed and some will not. Some will grow large and some will drop out. Most often, the companies that are large and successful have done something right, not something wrong. It is not part of the job to punish them for their success.

We have been dealing up to this point with the workings of our mixed economy. But a democracy must also consider the results. Our economic system is meant to reward the winners, but it can also mete out harsh penalties to the losers. When the economy is depressed, there are not enough jobs to go around. When older industries are displaced by newer ones, Americans may be left without work. As industrial processes become more technical and sophisticated, there are fewer jobs for the unskilled and the undereducated. These are problems that a democratic society cannot ignore. Since the marketplace is not always self-correcting, government must have a hand in dealing with them.

The task of correcting the deprivations suffered by the disadvantaged and unemployed is among the toughest in our job description for government. It requires judgment about what our society must and can do. It needs deft management to make the most of resources transferred to the poor. Experience has taught us that, when government transfers wealth from one segment of society to another, a lot of it simply evaporates in the process. So the process will always be subject to debate and criticism. But it should be the kind of debate and criticism that is aimed at improving government's performance, and not the kind that denies the need to do the job at all.

Finally, our job description for government needs to take into account the fact that purely private transactions may also affect the lives and values of people who are not involved in those transactions. I can sign a contract to build a new plant, but I still need to find the community willing to have it for a neighbor.

In our complex high-technology society, this opens up another difficult and demanding role for government—a political balancing act—one that requires mediating a host of private concerns and interests to achieve a result that is in the public interest. The issues are familiar ones, many growing out of our need to employ new technology on behalf of the increased growth and productivity on which our standard of living depends.

If we considered *only* economic costs and benefits, we would be further along in the use of nuclear power to meet the nation's future energy needs. We would have had Alaskan oil some four or five years before it actually came on the market. We would not be diverting capital investment from new plant and equipment to the technology we need to protect the environment. We would be making greater use of the nation's coal resources.

Those are facts. But even stating them matter of factly—without suggesting whether our choices have been foolish or prudent—sets the heart racing a little faster. Our values are involved. Third party interests are affected. When these interests are sufficiently broad based to be advanced in the political process, they will sometimes require the government to act and to regulate.

Thus, it makes no more sense to reject a role for government than to dismiss the benefits of technology to producers and direct consumers. Accepting government's role permits us to focus on its performance, on the need for developing greater government expertise in dealing with technical matters, and on the need for assessing both risks and benefits in deciding them.

But my objective here is not to grade government performance. There are few

364   The Management of Sustainable Growth

Americans in or out of government who do not feel it needs to be improved in all of the areas just mentioned. I am simply suggesting that the debate over limits to government could be more productive. It should proceed from an awareness that government's role in mediating the issues that shape our future is necessarily large and important. And it should aim at promoting government's ability to focus on new priorities as the seasons change.

A GOVERNMENT FOR ALL SEASONS

We will continue to need a government for *all* seasons. It can benefit from our constructive criticism, but it also needs our support. The fact is, we are entering a season of slower growth that will put the government and the nation to exceptional tests. We will need new policies and new initiatives to ease the transition and sustain our economic strength. It is a poor time for Americans to divide even further over the question of social values versus economic values. Let us turn instead to finding the means of greater accommodation of both sets of values. A system determined to have efficiency prevail over other human values will one day find itself without support. At the same time, a society with an underpowered economy will fail to achieve either social *or* economic goals.

Sustaining both sets of values does not seem to be an impossible task. By way of illustration, let us look at some of the current concerns about our economic performance.

Our decline in productivity, if it continues, will have serious consequences for all Americans. Declining productivity reduces the output of goods and services and accelerates the rate of inflation. And as the nation becomes poorer, confidence in our ability to cope with social problems, and with the novel and difficult problems of access to energy and other raw materials, will be seriously undermined. Internationally, declining productivity means we will be less able to compete in world trade. According to the American Productivity Center in Houston, if current trends continue, both Japan and West Germany will pass us in overall productivity in the 1980s.

In that context, tax policies that encourage investment in new plant and equipment and in research and development—thereby helping to increase productivity—do not benefit business alone, as some argue. Increasing productivity permits wages to rise faster than prices. It creates new wealth, new jobs, and new markets for American goods and services around the world.

The question of government regulation of the economy is another that is too often distorted by pitting one set of values against another. The enormous cost of compliance with government regulation—which Murray Weidenbaum has estimated will reach $103 billion this year—clearly affects all Americans. It drives up the price of everything produced in our economy. And excessive regulation slows the economy in other ways. The uncertainty of regulatory policy, and the disregard in many cases of cost efficiency in achieving regulatory goals, puts a brake on business investment. Regulation can also restrain competition by making compliance prohibitively expensive for all but the strongest companies.

Again, the result is fewer jobs and higher prices. There is no need to deny the government's essential role as regulator to suggest that this is the wrong season for what Arthur Burns has referred to as "the regulatory frenzy." It is, rather, a season in which we should be applying a stringent cost-benefit analysis across the entire spectrum of government regulation. I might add that the public has some definite ideas about this. An opinion poll Union Carbide commissioned recently shows that a majority of Americans wants both environmental *and* economic impact statements.

It may be difficult to break the habit of turning to new regulation to solve all our problems, but the habit must be curbed to keep regulatory agencies from undermining our elected government's responsibility for addressing these problems. The agencies, which are not subjected to the checks and balances of the political arena, may be too well insulated from the changes of season in our economy. Let us all, therefore, encourage the new interest in making the regulatory process more efficient.

There is plenty of room for improvement in this area and plenty of room for innovation. Agencies are only now beginning to realize that they are better at setting regulatory goals than at telling business how to reach them. There is talk of a place for market incentives in reaching environmental goals, and new willingness to question the efficiency of mandating the use of costly pollution control devices. There is increasing interest in the potential for self-regulation by industry and in providing consumers with enough product information so they can start making more of the product choices that government increasingly has been making for them.

The zero-based budgeting and sunset ideas may be gaining ground as means of reviewing agency missions to see if they are still relevant and still needed. And there is growing interest in increasing the power of the courts to overturn agency regulations.

I am not suggesting that all of these approaches have equal merit. Some may have no merit at all. But I am suggesting that we are obliged to find out. The aim should not be to stop regulation in its tracks, but only to make it more efficient, less costly, and better attuned to the nation's needs. Both social and economic values will be better served.

I do not know how big the government needs to be to meet the test of efficiency in this difficult season, and to meet the responsibilities outlined in my job description. A bloated government will surely move too slowly in changing priorities to meet our new needs. Many Americans believe that this is the source of many of our problems. Our poll shows that six out of ten Americans feel the government is larger than it should be. On the other hand, a government that is undersized and undernourished will also fail to cope. But given the problems we will face in sustaining and increasing economic activity in the next decade, it seems clear that whatever its size, we will need a government with clearer purpose and direction, one that operates with less fumbling and more efficiency, and one that understands that social progress and economic efficiency are goals that we will achieve in tandem or not at all.

There is no doubt that we will need limits on government growth. But I am inclined to hope for a government of self-imposed limits, based on an understanding of when to intervene in the economy and when to get out. When we have such a government, I think those who serve in Washington will all feel less anxious and more secure, and so will we.

Chapter 14

ACTION FOR NEW GROWTH

Maurice F. Strong

The Third was the first of the Woodlands Conferences on Growth Policy in which I have been able to participate. But I have followed with special interest and great admiration the innovative ways in which the McHales and Harlan Cleveland and their colleagues have worked to give the dialogue on growth and sound conceptual and intellectual underpinnings which have enabled us to move increasingly from generalities to specifics. Thanks largely to the enlightened leadership of George Mitchell, the Woodlands series of conferences have become, in a very real sense, the prime forum for the dialogue on growth which *The Limits to Growth* and, I like to think, the Stockholm Conference on the Human Environment did so much to initiate.

The quality of the material presented to the Third Woodlands Conference, and of its participants, moved the growth dialogue another major step forward. But the time has come now to move from dialogue to action—and this dialogue can itself be a launching pad for action.

A CONSENSUS FOR CHANGE

The growth dialogue really began in the United States, the home of the growth culture. This society more than any other has been the prime mover in the creation of the technological civilization which now dominates our planet. It is also the largest and most powerful product of that civilization. And, if real changes are to take place in our technological society, they must first and foremost take root here.

Thus, it is encouraging to see the broadening concern of the American public for growth-related issues. What began as a discussion among a relatively small core of scientists, futurists, and public interest groups now engages the attention of almost the entire American public. This has fed on the personal experiences of most Americans with environmental problems and energy shortages. The public concern has inevitably been accompanied by the fears and resentments of those whose lives have been directly affected by shifting patterns of employment, unemployment, hazards to their own health or safety; and the impediments to the development of enterprises to which their wealth or careers are committed because of government regulation or public reaction. I doubt if there is a community or family in the United States which has not been touched by some aspect of the growing controversy over growth.

So, today, we have succeeded in creating—albeit at a high level of generality—a high degree of awareness. But, we are just beginning to see the kind of difficulties we will confront. For I am persuaded that we are in one of those seminal transition periods in human history. The way in which we deal with the growth issues will determine the direction in which this transition will take us. At this point, I believe its outcome is not predetermined. Most of the basic forces which are shaping our future are the result of human actions and human failures. In a very real sense, then, technological man is today in command of his own evolution.

Whatever may be the result, one thing is certain—the period we have now entered will inevitably be one of greater turbulence, uncertainty, and conflict. We have just begun to see the real and deep-seated conflicts that can be engendered over the growth issue—conflicts over scarce resources, over land and living space, over the control of ocean resources, and even over the use of the air waves. These conflicts will range from the local level to the international level. But whether local or national, the issues must inevitably be perceived within a global framework while being dealt with at the most effectively managed scale and level of decision making.

There seemed to be a broad consensus among the participants in the Third Woodlands Conference that the processes of economic growth which have characterized our industrial societies in this century, and notably in the past 30 years, either cannot or are not likely to be sustained in the future, and that significant changes must occur in the content and dynamics of growth. Most would also accept the premise that our attitudes toward growth and our expectations as to the benefits it can produce must also change. Indeed, such changes are already under way. They will require a strong commitment to conservation, less wasteful use of resources, and measures to minimize environmental degradation.

However, I do not yet see a real consensus emerging as to the fundamental purposes of growth and the basic values which are to underlie our future approach to growth. At one end of the wide spectrum of opinion are those who are convinced that nothing short of the collapse of our industrialized societies will ensure our continued survival. At the other end are those who believe that any changes required can take place in an evolutionary manner without serious disruption of our present system or life-style. There is also no generally agreed set of answers as to what we *can* do, even if we could agree on what we *should* do.

I certainly don't have any pat answers. But, I do believe that we must change some of the basic assumptions underlying our economic activities if man's productive capacities are to be managed so that decent human life will be possible for the majority of people in an increasingly crowded and interdependent world. And, most of these people live in the developing world. Much of my life has been spent living and working with them and their fate continues to be my prime concern. However, I have been spending most of my time recently in the United States, and the reason for this is that I have come to believe that what happens here will have a major impact on the fate of the world as a whole, and particularly the peoples of the developing world. A collapse of our present system, however deficient and battered it may be, would be disastrous for the people of the industrialized countries. It would also, in its immediate impact at least, add to the already heavy burden of problems which afflict the poor of the world. They could survive such a collapse much better than we. Most of the survivors would be in the developing world. And in the process of rebuilding after collapse, their positions relative to that of industrialized society might well be improved.

Why, then, should one who is preoccupied by concern for the disadvantaged peoples of developing countries be preoccupied today with what is happening in the United States? In part this is because this continent is my home, I am a product of North American society, and I am deeply concerned about its future. But, I also prefer to be an optimist, and I believe that it is possible to make, in a peaceful and evolutionary way, the radical transition that I feel we must make. And I believe that the leadership and example which will make this possible must come primarily from the United States. What other industrialized societies do will be important; what the United States does will be decisive.

A COMMITMENT TO "NEW GROWTH"

No one will deny that the technological civilization which now dominates our planet is the direct product of economic growth. The belief that growth in the purely material sense would bring about a corresponding increase in human satisfaction and well-being has been the underlying premise of virtually every modern nation, whatever the political ideology of its government.

The premise has been reality—up to a point. The explosion of our capacity to produce a wide variety of material goods and services which accompanied the Industrial Revolution brought unprecedented benefits to the peoples of the industrialized world. It has also made it technically possible to effect vast improvements in the conditions of life available to the entire human population. Nevertheless, a failure of moral and political will continues to prevent some two-thirds of the world's people who live in the developing countries from realizing a fair share of this potential.

It is through the growth process that we create the economic means for meeting our social needs, that we provide employment and make possible leisure. It would be irresponsible to knock growth per se. It is as necessary and as good for society as it is for individual human beings. But, for societies as for individuals, the wrong kind of growth can be unhealthy and destructive. Patterns of growth which are healthy at one stage of development can become distorted, even self-destructive, if carried too far or continued beyond their time. Thus, the very same growth processes which have produced so much wealth and power for the industrialized societies are also producing widespread destruction of the Earth's natural resource base and new risks to human health and well-being. These are accompanied today by persistent inflation, unemployment, escalating social conflicts, economic disparities, and potential shortages of energy, food, and other key resources. These must be seen as strong signals—symptoms—that something is wrong with the life system of our technological society.

The industrialized societies today are like the person who eats compulsively, who cannot restrain his appetite, until his body begins to degenerate with heart disease, high blood pressure, diabetes, gout, and other debilitating conditions that feed on indulgence and carelessness. Our present growth patterns are based on consumption beyond our needs, on wastage of resources, on rapid obsolescence, on fostering overindulgence on the part of those with the capacity to buy and neglecting needs of those who cannot afford to buy.

This kind of growth is not sustainable. A society which lives by running down and wasting the basic resources which constitute its capital will be no more viable in the long run than a business which does not provide for depreciation, maintenance, and preservation of its capital. But, if the alternative is a society in which people's lives are tightly controlled by an omnipresent bureaucracy, where employment opportunities are scarce and personal options severely restricted, not many of us would opt for it. We would prefer to continue to live dangerously, whatever the ultimate risks.

"No growth" is not the answer. What we need is "new growth"—that is to say, a new approach to growth, to its purposes, its content, and its direction. It must be the kind of growth which makes possible an exciting society, a challenging society, a dynamic society—one which produces significant opportunities for employment and enjoyment. The "new growth" approach must be based on the realization that the most important kind of growth is that which produces the maximum opportunities for self-expression and fulfillment on the part of individual people. This is the kind of real growth to which we should aspire.

There *are* limits to growth in the purely physical sense; limits to the quantities of nonrenewable resources we can use; limits to the amount of land that can be used for producing food, and the amount of food which can be produced from a given unit of land; limits to the quantities of chemicals we can put into the soil, the waters, and the air without unacceptable risk to our own health and well-being; and limits to the total number of people the planet can support at a level compatible with human dignity and well-being.

Even the most optimistic assumptions about population growth in the developing countries lead to the conclusion that their claims on world resources will escalate sharply within the next three decades. The implications are staggering. In order for all of the present population of the world to reach a standard of living equivalent to that of the United States in 1970, it would require extraction or recycling of some 75 times as much copper, 200 times as much lead, 75 times as much zinc, and 250 times as much tin, and increases of similar orders of magnitude in the production of many other basic resources.[1] As for energy, such a standard of living would require the equivalent of seven times as much oil, eight times as much gas, and nine times as much coal as are now produced annually.[2]

Physical limits are clearly important. They establish boundary conditions which we must ultimately respect. This means that they must begin now to influence our planning and our behavior. But the limits we are beginning to confront today are not primarily physical. They are limits of political and social will and of adequate institutional means to assure careful use of the Earth's resources and equitable distribution of benefits and costs resulting from their use.

Our industrialized societies are very much like the physically mature human being. For us to continue to pursue purely physical kinds of growth would be as unhealthy and self–destructive to our society as it would be for an adult person to pursue ways which simply added to his physical dimensions. And it would be just as wrong to say that societies must stop growing when they reach the stage of physical maturity as it would be to say that people stop growing when they stop growing physically. The real growth of our societies in human terms can still be ahead of us. But it demands that we change our ways and adapt to a more mature kind of growth that is less physically oriented and less demanding of resources and the environment.

This change is possible. Each new day sheds light on more opportunities and examples of "new growth" in practice. The case of Georgia-Pacific (Oregon) can become typical. This U.S. forest products company gains 40 million BTU of energy per hour, meets air emission standards, and eliminates disposal costs by burning dust from its sanding machines to provide heat for the drying of veneers and coatings.[3] Many more activities can be carried out with greatly reduced demand on physical resources, energy, and materials, while providing new opportunities for employment. A recent U.S. Federal Energy Agency paper estimates that installation of ceiling insulation and automatic thermostats and retrofitting of furnaces in 32,372 homes would produce 487,000 jobs over a seven-year period.[4] The same agency estimates that a switch to refillable containers in the United States would produce a net increase of 117,000 jobs.[5] The U.S. Environmental Protection Administration monitored the employment impact of its regulations between 1971 and 1976 and found that an estimated 300,000 jobs were created each year as a direct result of pollution control.[6]

A commitment to new growth will clearly change employment patterns, and it will create many new employment opportunities. But even this is not likely to be enough. It must be accompanied by changes in our attitudes toward employment.

In the technological society, capital will continue to replace labor in many of the traditional functions related to production of goods and services. Already a majority of Americans are employed in the field of services—a growing number of them in areas related to the production and use of information. Professor Ward Morehouse of Columbia University suggests that the revolution in microelectronic technology will serve to reduce dramatically the employment opportunities in the service industries. He quotes John Bessant of the Technology Policy Unit at the University of Aston in Birmingham (U.K.) in pointing out that computing costs had already decreased from the equivalent of $200 per logic gate to about a half cent by the mid 1970s and were expected to reach about 1/200th of a cent by 1980. At the same time, increases in computer power, measured in components per chip have, in the same period, risen from one in 1960 to 64,000 by 1975 and 1 million by 1980. He projects that these increases will reach as much as 16 million by the mid-1980s.[7] While these dramatic changes will certainly provide some new opportunities for employment, they will also vastly reduce the number of people required to perform many of the services in which people are now employed.

All of this can liberate people from performing so many of the dull and routine tasks by which industrial society has tended to turn men into machines—but only if new and more satisfying avenues of employment are available. Many of these opportunities will not be in areas directly related to production—or, at least, to present methods of production. People will have the time to rediscover the satisfactions that come from applying art and craftsmanship to production of high quality goods on a human scale. This, too, could be economically attractive as people become more and more prepared to value such quality and pay for it in the marketplace, thus stimulating growth of additional smaller-scale ventures. The ensuing diversification of tax revenues would free many towns and regions from dependence on a small number of significant tax sources to pay for municipal services. Thus, increased opportunities and ideas for businesses, governments, and individuals would encourage more "new growth."

At the same time, we will have to learn to regard as "employed," people who use their time to develop and express their talents and aspirations in the fields of art, culture, sport, and the spirit in ways which do not contribute to production. We will have to recognize more than we now do that such activities do contribute to the good of society by producing better people, even when they do not produce income or add to the GNP in the conventional sense. As capital produces more and more of society's income, more and more people will have to receive their livelihood out of the income produced by capital rather than that produced by their employment. This means that the ownership of capital and the benefits it produces must be much more broadly distributed than they are today.

MARKETS AND VALUES

Most of the managerial, technological, and institutional capacity of our society resides in the private sector. The creative capacity of business and industry can be a powerful ally in effecting the transition to the "new growth" society. Corporations are, in many ways, the most innovative and effective agents of change we have, and we need to harness their energies. The suspicion and antagonism toward business on the part of so many people threatens to undermine and weaken the very institutions the "new growth" society needs most if it is to provide the new economic dynamism and the expanded employment opportunities which must be its hallmark. We simply cannot afford to perpetuate the debilitating and self-defeating conflict which has marked relations between business and the rest of society, particularly over the past 50 years. I know of no priority more urgent or more compelling than the need to forge a new, and positive, alliance between the business community and the other principal actors in our society.

Under our economic system, the free market plays a major role in determining how our research and development efforts will be deployed and our industrial capacity engaged. Thus, it is not enough simply to recognize the need for these new activities and the fact that they are technically feasible. There must be an effective market demand for them. Here, government can help by the way in which it manages the system of incentives and penalties which influence supply and demand in the marketplace. In short, it can use its powers to provide incentives to business, industry, and consumers to do those things that will help bring about a "new growth" society and impose penalties and disincentives to those activities that are wasteful, destructive, or unduly risk-creating for society.

Of course, no one wants to add unnecessarily to the controls government exercises over our economy. The last thing we need is more bureaucratic regulation of our lives. Most governments today are already attempting to do more than they can do well. In order to take on new functions, they must shed functions that can better be performed by other levels of government or the private sector. What is needed is a reorientation of the manner in which governments use the public policy levers which they now deploy to influence the economy.

There is ample precedent for this. Perhaps the best example is the capacity industrialized societies have demonstrated in fighting or preparing to fight wars.

It is not the free operation of the market economy which produces the massive market for war materiels. That market is created by an act of public policy carried out by governments in response to the belief of their people that their security is at stake. If expenditures on war materiels, which are inherently wasteful whether they are used or not used, can be a major stimulus to the economy, surely expenditures on building more livable cities, improved cultural and educational facilities, recreational areas, and opportunities for meaningful leisure can be just as stimulating to the economy, can create new and challenging employment opportunities, while at the same time adding positively to the real capital stock of our societies.

But the capacity of governments to act depends, in the final analysis, on the will of the people. The transition to "new growth" will be made feasible only by basic changes in the attitudes, values, and expectations of people. This will mean, for industrialized societies, a virtual cultural revolution. The dominant culture must be one that places highest values on quality rather than quantity, on conservation rather than waste, on cooperation above competition.

If we consider those periods in history that have produced the greatest and most durable advances in human understanding, learning, and culture—the Golden Years of Greece, of Rome, and of the Renaissance, for example—we find that the dominant ethos was not a commercial one. Of course, the accomplishments of these societies were undergirded by successful economies and by a high degree of respect for commercial pursuits and for the practitioners of commerce. But, the value system accorded greatest respect and highest honor to those who contributed to the advancement of the human experience in the fields of ideas, philosophy, and the arts. It is in these areas, after all, that the highest levels of human development are reached. Our technological civilization has been given an unparalleled physical base which makes it possible for all members of our society to enjoy enlarged opportunities for this kind of human growth. But it will take a conscious change in our attitudes and priorities to translate our unprecedented material success into opportunities for human growth rather than barriers to it.

This requires that we reorder our own priorities so as to place highest value on those activities and life-styles which truly enlarge the opportunities of individuals for self–development. This requires that we learn to applaud and look up to those who adopt life-styles that are modest in the amounts of space they monopolize or the amounts of material and energy they consume. Ostentatious, wasteful, and indulgent living will become socially reprehensible. There must be an acute sensitivity to all activities which create risks of damage to our natural heritage or impair the quality of life for others. The people of industrialized societies, in particular, must again nourish their cooperative instincts and discipline their competitive drives.

There is little evidence at present that we can realistically expect this kind of cultural transition to take place wilfully. The status quo continues to have a powerful grip, even on those who concede that such a transition is necessary.

We owe a great deal to those people in our societies who are already demonstrating that more qualitative, less materialistic life-styles can be not only feasible but highly desirable and rewarding. The Stanford Research Institute has estimated that "between 4 and 5 million adult Americans have transformed their personal life-styles from their former slots in the dominant industrial, consumer economy to a 'voluntary simplicity' approach of frugal consumption, ecological awareness and a dominant concern with personal, inner growth."[8] But it is not easy to pursue such life-styles today, even for those who wish to do so. We must do more to encourage and support experimentation by people in new life-styles, both within existing communities and through the creation of new communities. If more of the leaders of thought and opinion in our society would themselves adopt such life-styles, it would have a powerful exemplary effect on the changes in our value system that are needed to make the "new growth" society possible.

UNMET NEEDS AND REDEFINED NEEDS

No country can make the transition to the "new growth" society in isolation; the United States, more than more countries, has a deep-vested interest in the healthy functioning of the global system. According to former Secretary of State Cyrus Vance, "The U.S. now earns more from its exports to developing countries than it does from those of Western Europe, China and the Soviet Union combined."[9]

It is surely axiomatic that the transition to "new growth" requires the full participation and active cooperation of the two-thirds of the world's people who live in the developing world. The concentration of production and its benefits in the industrialized nations of the world enabling a few people to have great material abundance while the majority are living at the edge of poverty and despair is not a rational or a viable basis for a secure world. As Barbara Ward has written, the poor majority "consume fifteen times less energy, on the average, than do the citizens of developed societies. . . . A third of the municipalities of Latin America have neither sewage systems nor piped water. The proportion in the Indian subcontinent and parts of Africa is higher still."[10] Moreover, "three percent of the land holders in 83 countries control 80% of the farmland."[11]

If we continue to ignore the economic and social conditions which create such a world, if we continue to seek security simply by spending more on arms while refusing to make the changes in the operation of the world's economic and trading systems to give the developing countries a fairer access to the benefits of the technological civilization, we shall be doubly wasteful of the Earth's resources. While the developing countries must rely ultimately on their own efforts, these need not and should not take place within an international system in which the dice are heavily loaded against them. It is widely assumed that assistance to developing countries imposes a burden on the industrial economies which we can less and less afford. But a larger vision of current economic realities would suggest that the vast unmet needs of the developing world represent the greatest opportunity for getting the world economy moving again.

One of the principal reasons for the slowdown in growth in the industrialized world is the lack of real demand for the increasing quantities of products we have the capacity to produce. If the needs of the developing world could be translated into real market demand, it would provide the kind of dramatic stimulus which would start the world's economic wheels turning vigorously again. This can only be done if purchasing power is available to those who have the needs. There is nothing revolutionary in this concept. It is a matter of applying the same basic principles which have undergirded the growth process in our own economies—effecting a broader distribution of purchasing power. Henry Ford did this when he started paying $5 a day to his workers. Governments have done it through the redistributive mechanisms of the tax system. The Marshall Plan provided a vivid demonstration of how the same principles may be applied internationally when it enabled the productive capacity of North America after World War II to be used to rebuild war-torn Europe.

In similar fashion, a massive commitment today to the development of the developing countries will provide the best means of renewing the economic dynamism of the industrialized countries and creating the New International Economic Order which is so essential to the peace and well-being of both rich and poor. This will, in turn, provide both the time and the means for the present industrialized societies to make the transition to the "new growth" which I believe is the key to their future viability.

It may seem paradoxical that I am suggesting, at the same time, a transition to "new growth" in the industrialized societies and use of some of the traditional means for stimulating growth in the developing world. Consider once more the analogy to growth in the human being. The developing countries are by definition at the early stage of their growth where they must concentrate on the physical aspects of growth so as to give their societies the physical and material infrastructure which is an essential precondition to their capacity to meet the basic needs of their people for food, shelter, jobs, and education, as well as their aspirations for greater social, intellectual, and cultural fulfillment. But in pursuing their more physically-oriented growth patterns, they can avoid many of our mistakes; indeed, they can ill afford to repeat them. Already, many of their precious natural resources are being destroyed through soil erosion, deforestation, urban blight, and other destructive consequences of imbalanced, ecologically unsound growth practices. In building their societies they need not slavishly follow our examples. To do so would be, for most developing countries, a dead-end pathway to disaster. They have as much to learn from our mistakes as from the managerial and technological know-how we can make available to them. They have the opportunity to build "new growth" principles right into their own patterns of growth.

It will not be easy for the developing countries to do this. Indeed, they will need our help and understanding if they are to do it successfully. But, in principle, they may be better able to make this adjustment than we are. Our societies are more mature, more wedded to the status quo, and it will not be easy for us to change our ways.

There is the danger in talking about problems on a global basis that we will lapse into thinking that all solutions to them can only be mounted at the global level. There *are* many global problems in the sense that they can only be fully understood and addressed in their global context; however, there are few global solutions. Most of the actions required to deal with these problems must be taken at the local and national levels. One of the problems with the United Nations is that it is not able to concentrate its attention on the relatively few issues requiring global cooperation. There is a great tendency to try to elevate every issue to the global level for its solution. On the contrary, every issue should be dealt with on the specific level in the decision making chain on which it can be managed most effectively.

We must start by examining our own style of life, our own values. We must be willing to manage our own lives as well as our national economy in a less wasteful, more rational manner. We should not underestimate the power of example. My experience in the developing world has brought home to me the fact that they are far more likely to be influenced by our example in opting for technologies and life-styles that are modest in their demands on material resources by what they see us doing in our own countries than by what we are exhorting them to do in theirs. American-style materialism is the dominant culture of our times, and the majority of the people in the developing world still take it as the model to which they aspire. Even the Chinese seem to be jumping on this bandwagon. Yet how sad, and how disconcerting it is to see how other countries are committing themselves to patterns of growth that have little chance of meeting the basic needs of their people, let alone their aspirations, at the very time when the peoples of our more developed societies are beginning to realize the weaknesses in these patterns and the need to change them.

America must start sending out a new set of signals to the world, start setting an example of how more modest life-styles which are less demanding of energy and materials cannot only be viable but superior. If the people of the richest country on Earth can learn to live on less and place greater value on the non-material dimensions of human life, then others would begin to take more seriously our new attitudes toward growth. This requires that we use our technology and capital with greater wisdom and efficiency to reduce the overall demands which the United States makes on the world's resources. We need also to diversify the goods and services we supply, to value conservation as much as we value consumption, and to stop confusing "the good life" with "the indulgent life."

And yet, if we are to make this transition peacefully and willfully, we must start where we are. We cannot suddenly abandon the traditional tools and instruments with which we have fashioned our present society and, as suddenly, take up the new tools and instruments with which we would fashion the "new growth" society. Nor, can we, in one miraculous moment of enlightenment, shed the deeply rooted attitudes and values which now condition our behaviors and replace them with a new set of revealed truths.

I do not minimize the difficulty of changing attitudes, but neither do I minimize the potency of the means of information and persuasion which are now at our disposal. If we have not hesitated to use techniques of mass persuasion to sell consumer products of dubious value, I do not now see why we should not use them in pursuit of a more healthy and humane life-style. People are already choosing to read alternative publications. As Hazel Henderson has noted, *Prevention* now has a U.S. circulation of 2 million; *Organic Gardening*, 1 million; *Rolling Stone*, 1.5 million; *Mother Earth News*, 300,000; in addition to *The Whole Earth Catalog*, *Foxfire*, *Journal of the New Alchemists*, *Co-Evolution Quarterly*, *Rain*, *Science for the People*, *Self-Reliance*, and hundreds of others.[12]

Moreover, in spite of the difficulty of changing attitudes, I think we should take heart from evidence of some recent, modest changes. Within the last few years, and with much less than an extensive and concerted campaign, jogging and marathon running have become pastimes for very large numbers of Americans. A few companies have instituted fitness programs and businessmen jog at noon in downtown urban areas. These are the seeds of an important change, with more people taking more responsibility for their own health.

Also, concern with energy has now gone past the talking stage, though not nearly far enough. Prodded by the rising costs of energy, people are insulating their houses, buying smaller cars, using public transit, and building solar installations in their homes. These are small beginnings, but with more information, with education programs, with higher prices for scarce materials, and with incentives—such as lower mortgage rates on better insulated houses—we can go much further. People will be more and more willing to reexamine their life-styles and to try alternatives.

A NEW GROWTH VENTURE CAPITAL CORPORATION

The scale on which these changes are taking place is still far too small. And it is difficult for individuals who wish to change their personal life-styles to do so in isolation. Certainly, there is much they can do in terms of changing their personal consumption habits. But what happens when an individual in Houston decides he wants to use public transportation rather than his own car? What does an architect or an engineer or a construction worker do when he wants to commit his career to designing and building housing and buildings that are materials and energy saving? What does a business executive do when he wants to use his talent to demonstrate that private enterprise can lead the way in showing that business can combine economic viability with social utility? What does a farmer do when he wants to apply more organic ecological systems to raising food, or a public relations man when he wants to use his talents to sell the new growth ethic rather than cigarettes or Coca Cola?

There are opportunities—but they are relatively few and far between. And I am meeting more and more people who are seeking such opportunities. But they are not easy to find. Equally, entrepreneurs with talent to offer but insufficient capital are ready to stake their careers on "new growth" business enterprises. And there is a growing market for products and services of such people and enterprises. I know two innovative architects who design houses with energy–saving design and insulation which reduce heating costs from $250 in the coldest months of winter to $10. They couldn't get builders to build them and people wouldn't buy them if they couldn't see them. Yet, when they scraped up enough of their own personal money and built them to look conventional within a conventional urban development and to sell competitively, buyers were enthusiastic.

Inertia is a powerful force and it is just as powerful in human affairs as in the physical world. However much we talk about change, our entrenched habits and patterns of life tend to keep us moving in the same old direction. To change direction, we need to break that inertia in which we are all caught up. This calls for some new initiatives and who better to launch such an initiative, in what better place, than the people assembled in a Woodlands Conference on Growth Policy?

What might we do in practical terms? Let me make some suggestions. Collectively, we could set up a mechanism for monitoring what is actually going on—what new technologies are available, where and by whom they are being tried and what will result, what new ideas are being put into practice. Enthusiasm feeds on information, and people doing things are encouraged by the knowledge that they are not working alone as well as by the practical help they get from the experience of others. We could also create a means of recognizing the achievements of individuals and corporations in producing "new growth" ideas and techniques as well as putting them into practice. Here, the way in which the Mitchell Prizes have been used to stimulate the growth dialogue provides a fine example.

But let me concentrate on another, very specific initiative that I believe we can and should launch. It calls for the creation of a new corporation to provide a source of equity and equity-related capital for enterprises which have practical "new growth" ideas that can be profit-making, as well as access to the management and technical capability to carry them out. These could range from energy-saving housing, ecologically sound methods of food production, new production processes that are materials and energy saving, and the development of small, community-type enterprises utilizing the skills that so many people are now developing in the arts and crafts.

I am impressed with the number and the quality of proposals that are coming to my attention these days—far beyond my personal capacity to respond. Many of them come from people who have the technical skill and the will, but often need (in addition to capital) a certain amount of financial and management input. They find the traditional sources of venture capital are not usually attuned to meeting their needs. And this isn't always because of lack of interest; it is more generally because they lack the experience base or access to the specialized knowledge and expertise that would be required to evaluate proposals and to monitor the investments once they are made. Often, too, it is because the capital required is too small to justify the attention of institutions which characteristically have to restrict their attention to investments of a larger size.

My solution to this problem is a "New Growth Venture Capital Corporation" which could enlist as shareholders other financial institutions that are prepared to set aside a certain amount of their own capital for this purpose but aren't yet in the position to provide, themselves, for the kind of specialized management that such investment requires. The participation of such institutions would expose them to experience in "new growth" investments which might lead them to setting up their own programs. The more competition the new venture generates, the more successful it will be.

I believe that the ingredients needed to launch such a new venture are all present in the community of persons professionally and personally interested in "growth policy." There are enough potential investors as well as potential sources of management and technological capability. Surely, those who have helped to engender the growing concern for the need to change our attitudes to growth should be prepared to accept a high share of responsibility for helping to translate these concepts into reality. I am personally prepared to participate in terms of both capital and some part of my personal time and energy. Better to light even one candle than to curse the darkness. As you know, a single candle can start a big fire.

Also, surely, such an initiative is in the best American tradition. The United States is still the leading society of our period of history. It can lead the way into the new era of growth—or "humangrowth" as Harlan Cleveland so appropriately terms it—by its initiative and its example, better than by its failure and collapse. The people involved in the Woodlands Conference process have the capacity to take this initiative. The thinkers have made us all think in new ways about growth. The actors can now, in consequence, take the lead in action. I look forward to discussing further with the Woodlands participants the specific ways in which we might join in addressing this new challenge. It is clearly up to us. Through leadership and example, the United States can best show the way to a new and more enlightened period of human growth for the entire community of man.

NOTES

1. P. R. Ehrlich, and A. H. Ehrlich, *Population, Resources, Environment*(San Francisco: W. H. Freeman, 1972), p. 62.

2. Report of the Workshop on Alternative Energy Strategies (WAES), *Energy: Global Prospects 1985-2000* (New York: McGraw Hill, 1977).

3. Pollution Probe Foundation, "Probe Post," Toronto, Canada, January–February 1979.

4. U.S. Federal Energy Agency, *Conservation Investment as a Gas Utility Supply Option*, Washington, D.C., 1977, in Environmentalists for Full Employment, "Jobs and Energy," Washington, D.C., December 1977, p. 10.

5. U.S. Federal Energy Agency, *Energy and Economic Impacts of Mandatory Deposits*, Research Triangle Park, September 1976, in ibid.

6. U.S. Environmental Protection Administration, in Alternatives Inc., "Conserver Society Notes," Downsview, Ontario, Spring 1978, p. 7.

7. Presentation by J. R. Bessant, Technology Policy Unit, University of Aston at Birmingham, to the Lund Workshop on Technological Change in Industrialized Countries and Its Consequences for Developing and Industrialized Countries, Lund, Sweden, 28-30 May 1979, in Ward Morehouse, "Letting the Genie Out of the Bottle? The Microelectronic Revolution and North-South Relations in the 1980's," Council on International and Public Affairs and Columbia University, New York, June 3, 1979.

8. 1976 Report of the Stanford Research Institute's Business Intelligence Program, Guidelines No. 100, in *Creating Alternative Futures*, edited by Hazel Henderson (New York: Berkley Books, 1978).

9. Vance speech cited in R. Thomas Hoffman and Brian Johnson, *Beyond Oil and the Atom? Energy Cooperation with the Third World in the 1980s*, International Institute for Environment and Development, August 1979.

10. Barbara Ward, *The Home of Man*, (Toronto, Canada: McClelland and Stewart Limited, 1976), pp. 7-8.

11. F. M. Lappe, and J. Collins, "Food First," Inter–Church Committee for World Development Education, Item 4, distributed at Canada Food Week, Feb 11-21, 1977.

12. Hazel Henderson, *Creating Alternative Futures*, (New York: Berkley Books, 1978).

INDEX

Government, 32, 33, 220-21, 264, 340-41; agencies, 315-18; constraint of, 322-28; and education, 182; intervention, 41, 318-19, 346; limits to, 12, 309-15, 319, 321-28; regulations, 316-18, 346-47, 350, 364-65; role of, 142-50, 350, 360-66, 373-74; mistrust of, *see* Public mistrust
Grain, 241, 252, 289
Gravel, Mike, 149
"Greenhouse effect," 354-55
Gross National Product (GNP), 4, 20, 83, 86-88, 126, 130, 138-39, 145, 184, 213, 220, 273, 278, 281-82, 322
Gross World Product, 128
Guggenheimer, Elinor, 38

Halberstam, David, 103
Hamilton, Alexander, 16
Harried Leisure Class, The, 107
Harvard University, 12; Business School, 94
Heilbroner, Robert, 245
Henderson, Hazel, 378
Hirsch, Fred, 85, 107, 111, 275
Hodges, Carl, 257
Houston, University of, vi
Hughes, Dorothy, 40
Humangrowth, 272-73, 274
Humanism, 112-13
Human rights, 11
Hydroponics, 257

Index of Consumer Sentiment, 24
India, 251-52
Industrial nations, 128, 273-74, 275, 278-88, 292, 294-95, 304, 369, 370, 371, 374, 376
Industrialization, 137, 138, 294-96
Industry economy, 118
Inflation, 19, 25, 35-37, 84, 117, 131, 136, 214, 267, 284, 285, 341-42, 335, 364

Information economy, 118-21, 128, 136, 138
Information resources, v, 7
Input rates, 115
Institute for Social Research (University of Michigan), 24
International Atomic Energy Authority, 291
International dimension, 287-97
International Energy Agency, 292
International Nuclear Fuel Cycle Evaluation, 291
Interstate Commerce Act, 336
Iran, 10
Irrigation, 252-53

Japan, 102, 121, 283, 344
Jay Douglas, 94
Joint Economic Committee of Congress (1976), 149

Kahn, Herman, 20, 31, 124, 195, 196
Kaiser Steel, 230-31
Katona, George, 24-25
Kemp-Roth proposals, 322
Kendrick, John, 138
Kerr, Robert, 108
Keynes, John Maynard, 8, 90, 92, 130, 131, 133, 134, 135, 136, 144, 149, 150, 198
Kristol, Irving, 31
Kuhn, Thomas, 133

Labor, costs, 19, 228-29; division of, 97, 98; force, 117, 343, 345; unions, 214, 215, 337-38
Ladies' Home Journal, 191
Land, 132, 246, 251-52
Land Grant College Act, 171
Large Area Crop Inventory Experiment (LACIE), 250
Lasch, Christopher, 207
Las Vegas, 85
Law firms, 225-29

ABOUT THE EDITOR AND CONTRIBUTORS

HARLAN CLEVELAND, political scientist and public executive, was Director of the Program in International Affairs of the Aspen Institute for Humanistic Studies from 1974 to 1980, and served as program consultant to the Third Woodlands Conference on Growth Policy. In August 1980, he became Director of the Hubert H. Humphrey Institute of Public Affairs, and Professor of Public Affairs, at the University of Minnesota. He was a Rhodes Scholar in the 1930s; economic warfare specialist and United Nations relief operator (in Italy and China) during the 1940s; foreign aid administrator (The Marshall Plan), magazine editor and publisher (*The Reporter*), and graduate school dean (the Maxwell School at Syracuse University) during the 1950s; Assistant Secretary of State and U.S. Ambassador to NATO in the 1960s; and President of the University of Hawaii from 1969 to 1974. He served as Chairman of the U.S. Weather Modification Advisory Board during 1977 and 1978. His many writings include *The Overseas Americans* (1960), *The Obligations of Power* (1966), *NATO: The Transatlantic Bargain* (1970), *The Future Executive* (1972), *China Diary* (1976), *The Third Try at World Order* (1977), and *Humangrowth: An Essay on Growth, Values and the Quality of Life* (1978), co-authored with Thomas W. Wilson, Jr.

LINCOLN GORDON, political economist, is Senior Fellow at Resources for the Future, a nonprofit research organization concerned with resource and environmental matters, in Washington, D.C. He earned a D.Phil. degree from Oxford as a Rhodes Scholar in 1936, and has divided his subsequent career almost equally between public service and the academic world. He was on the research staff of the National Resources Planning Board before World War II; an official of the War Production Board throughout the war, ending as its Program Vice-Chairman; on Bernard Baruch's delegation to the United Nations Atomic Energy Commission in 1946; a consultant to the State Department in the development of the Marshall Plan, and subsequently on Governor Averell Harriman's staff in Paris and Washington; Minister for Economic Affairs in the American Embassy in London (1952-55); U.S. Ambassador to Brazil (1961-66); and Assistant Secretary of State for Inter-American Affairs (1966-67). He was on the Harvard University faculty for twenty-five years, serving from 1955 to 1961 as William Ziegler Professor of International Economic Relations. From 1967 to 1971, he was President of the Johns Hopkins University. In 1972-75, he was a Fellow of the Woodrow Wilson International Center for Scholars, after which he assumed his present position. His principal publications include *The Public Corporation in Great Britain* (1938); *Government and the American Economy* (with Merle Fainsod 1941, 1948, and 1958); *A New Deal for Latin America* (1963); and *Growth Policies and the International Order* (1979). He also edited the OECD's Marshall Plan Commemoration Conference proceedings under the title *From Marshall Plan to Global Interdependence* (1978).

THEODORE J. GORDON is President of The Futures Group, a management consulting organization which he founded in 1971. He has been associated with futures research and policy analysis for many years and has participated in the development of several methods of forecasting. Mr. Gordon has directed many studies performed by The Futures Group, covering such topics as life extending technologies, energy, international developments, social change, new ventures, and innovation. He is a frequent lecturer at corporate and university seminars and is the author of several books and more than 100 reports for government and corporate clients. Mr. Gordon was one of the authors of the recently published *Life-Extending Technologies: A Technology Assessment* (1980).

ROBERT HAMRIN, economist, is a professional staff member of the President's Commission for a National Agenda for the Eighties. He has been a college teacher; a staff economist with the Joint Economic Committee, U.S. Congress, where he served as director of the Committee study series, *U.S. Economic Growth from 1976-86: Prospects, Problems and Patterns*; Fellow of The Rockefeller Foundation; and senior policy economist of the Environmental Protection Agency. He is a frequent lecturer at colleges and conferences. His recent writings include: *U.S. Long Term Economic Growth Prospects: Entering a New Era* (1978) and *Managing Growth in the 1980s: Toward a New Economics* (1980).

BERNARD LEFKOWITZ is a freelance writer and a consultant to the Ford Foundation, reporting on social issues. His most recent book is *Breaktime: Living Without Work in a Nine to Five World* (Hawthorn/Dutton). It will be published as a paperback by Penguin in 1980. He also teaches at the Columbia Graduate School of Journalism.

WILLIAM LEE MILLER is director of the Poynter Center at Indiana University, a teaching and writing project dealing with American self-government; he is also Professor of Religious Studies and Political Science. He received his Ph.D. in Social Ethics from Yale in 1958, and taught there from 1958 to 1969. He also taught at Smith College 1953 to 1955. Mr. Miller was a staff writer for *The Reporter* magazine and has written articles for *The New York Times Sunday*

Magazine, The New Republic, The Washington Post, and many other publications. He is the author of four books: *Piety Along the Potomac* (1964); *The Fifteenth Ward and the Great Society* (1966); *Of Thee, Nevertheless, I Sing: An Essay on American Political Values* (1975); and *Yankee From Georgia: The Emergence of Jimmy Carter* (1978). He has been a fellow of the Center for the Study of Democratic Institutions in Santa Barbara, California, and of the Washington Center for Foreign Policy Research of the Johns Hopkins University. He has received a travel and study award from the Ford Foundation, and a fellowship from The Rockefeller Foundation. Professor Miller has also directed two NEH Seminars for Public Administrators, and an Aspen Institute Executive Seminar.

JAMES O'TOOLE is Director of the Twenty Year Forecast Project at the University of Southern California, where he is an Associate Professor of Management in the Graduate School of Business. As a Rhodes Scholar, he received a Doctorate in Social Anthropology from Oxford University. He has served as a Special Assistant to the Secretary of Health, Education and Welfare; as chairman of the Secretary's Task Force on Work in America; as a management consultant with McKinsey and Company; and as Director of Field Investigations for the President's Commission on Campus Unrest. He directed an Aspen Institute project on Education, Work and the Quality of Life. In a recent American Council on Education survey he was selected among the "one hundred most respected emerging leaders in higher education." His primary professional interests are public policy analysis, futures research, government/corporate relations, and human resources development. He has published numerous magazine and journal articles. His six books include: *Work in America, Energy and Social Change* (1979), and *Work, Learning and the American Future* (1977).

WALTER ORR ROBERTS, astronomer and climatologist, is Director of the Program on Food, Climate and the World's Future at the Aspen Institute for Humanistic Studies in Boulder, Colorado. Educated at Amherst College and Harvard University, he was first director of the High Altitude Observatory at Climax and Boulder, Colorado from 1940 to 1960; first president and director of the University Corporation for Atmospheric Research which merged with the High Altitude Observatory and subsequently operated the National Center for Atmospheric Research, 1960 to 1973 at Boulder; instructor at Harvard, Radcliffe, and University of Colorado; Professor of Astro-Geophysics at University of Colorado since 1956; President of the American Association for the Advancement of Science; Trustee of the MITRE Corporation, the Aspen Institute, the International Federation of Institutes for Advanced Study, and the Max Fleischmann Foundation. He is a member of numerous professional societies, and has been involved in extensive scientific collaboration with various institutes in the USSR, Japan, and other countries. He is the author of two books and many scientific and popular articles.

LLOYD SLATER is manager of the Food and Climate Forum, an industry–supported project of the Aspen Institute for Humanistic Studies. After graduating from Cornell University in 1940, he served as a wartime Radio Officer in the U.S. Merchant Marine, then established and managed Honeywell corporation's food division. During the 1950s he was an editor with food and engineering journals in McGraw-Hill publications, and directed an engineering foundation in New York concerned with the development of the automation sciences; in the 1960s, he was Associate Director of Research at Case Institute of Technology, Assistant to the President of Associated Universities Inc. in Washington, and directed the Reston, Virginia, Foundation for Community Development. In 1971, Slater went to Puerto Rico to head the newly established Institute of Social Technology. He returned to journalism in 1974 to edit and publish *Food Engineering International* before assuming his present position in 1978. He is the author of more than 120 articles in scientific and technical journals, mainly on automation, food science, and biomedical engineering.

WILLIAM S. SNEATH is chairman of the board of Union Carbide Corporation. He joined the corporation in 1950 and moved through a number of positions in accounting and finance, becoming vice president and chief financial officer in 1965. He was elected a director in 1969 and president of the corporation in 1971, serving as chief operating officer until 1977 when he was elected chairman of the board and chief executive officer. Mr. Sneath is a director of the Federal Reserve Bank of New York, Metropolitan Life Insurance Company, Rockwell International, and a member of the Business Council, The Business Roundtable, and Chairman of its Task Force on International Trade. He is a Trustee of Williams College, a Board Member of Memorial Sloan-Kettering Cancer Center, and Chairman of the National Fund for Minority Engineering Students. Mr. Sneath is a graduate of Williams College with the degree of B.A. received in 1947. He also has the degree of M.B.A. received in 1950 from the Harvard Graduate School of Business Administration.

MAURICE F. STRONG is a Canadian with extensive experience in both public and private sectors. Formerly President of Power Corporation of Canada, Limited; President of Canadian International Development Agency; Chairman and President of Petro-Canada; he served as Under-Secretary of the United Nations and first Executive Director of United Nations Environment Programme. Currently he is Chairman of International Energy Development Corporation of Geneva, Switzerland; Chairman, AZL Resources, Inc., Phoenix, Arizona; and Chairman of the Executive Bureau, International Union for the Conservation of Nature and Natural Resources, Gland, Switzerland.

MURRAY L. WEIDENBAUM, economist, holds the Edward Mallinckrodt Distinguished Professorship at Washington University in St. Louis, where he is also Director of the Center for the Study of American Business. He spent the academic year 1979-80 as J. E. Lundy Visiting Scholar at the American Enterprise Institute in Washington, D.C. He was chief economist of the Boeing Company in 1958-63 and Assistant Secretary of the Treasury in 1969-71. Since 1964, he has been a member of the Washington University faculty and served as chairman of its economics department in 1966-69. He holds a Ph.D. in economics from Princeton University. His many writings include *The Modern Public Sector* (1969); *The Economics of Peacetime Defense* (1974); *Government-Mandated Price Increases* (1975); *Business, Government, and the Public* (1977); and *The Future of Business Regulation* (1979).

THOMAS W. WILSON, JR., has worked on international problems as journalist, government official, study director, and international civil servant. In the early 1970s he was Director of the Aspen Institute's Program in Environment and Quality of Life while contributing to preparations and proceedings of several world conferences under U.N. auspices. In 1976-77 he served as Principal Officer in the Office of the Secretary General of the United Nations. He has been a consultant to the Presidential Commission on World Hunger during 1979 and 1980. In addition to articles, essays, and studies, he is the author of *Cold War and Common Sense* (1962); *The Great Weapons Heresy* (1971); *Environmental Action: A Global Survey* (1972); *Science Technology and Development: The Politics of Modernization* (1979); and co-author of *Humangrowth: An Essay on Growth, Values and the Quality of Life* (1978).

WILLARD WIRTZ, a partner in the Washington, D.C. firm of Wirtz and Lapointe, is engaged in a variety of consultative activities stemming from his earlier law teaching, legal practice, labor arbitration, and government service. He was Secretary of Labor from 1962 to 1969. His current primary interest in interrelating the learning and the work processes is reflected in his chairmanship of the Board of the National Manpower Institute, a private, nonprofit organization. He is author of *Labor and the Public Interest* (1964) and editor of *The Boundless Resource* (1974). His less formal writings are extensive.

DANIEL YANKELOVICH is President of Yankelovich, Skelly and White, a New York-based social research firm. He is also research professor of psychology at New York University, visiting professor of psychology on the graduate faculty of the New School for Social Research, and a lecturer in psychiatry at Tufts University Medical School. His books include *Changing Values on Campus* (1972), and *The New Morality, a Profile of American Youth in the Seventies* (1974). "The Status of Ressentiment in America" was published in *Social Research*, Winter 1975, and "The New Psychological Contracts at Work" first appeared in *Psychology Today*, May 1978.

ERIC R. ZAUSNER, Senior Vice President and a Director of Booz, Allen & Hamilton Inc., manages the firm's worldwide energy consulting practice. He holds a B.S. in Electrical Engineering from Lehigh University and an M.B.A in Finance from the University of Pennsylvania. Prior to joining Booz, Allen, Mr. Zausner was the Deputy Administrator of the Federal Energy Administration. Mr. Zausner is a member of the Aspen Institute's Energy Committee, the Energy Committee of The Atlantic Council, and has had his articles on energy policy published in *The New York Times, Dunn's Review*, and the *Energy Daily*.